T0189720

Progress in Mathematics

Volume 200

Martin Golubitsky
Ian Stewart

The Symmetry Perspective

From Equilibrium to Chaos in Phase Space and
Physical Space

Springer Basel AG

Authors:

Martin Golubitsky
Department of Mathematics
University of Houston
Houston, TX 77204-3476
USA
e-mail: mg@uh.edu

Ian Stewart
Mathematics Institute
University of Warwick
Coventry CV4 7AL
UK
e-mail: ins@maths.warwick.ac.uk

2000 Mathematics Subject Classification 37-XX; 37G40, 37C80, 37J15, 70H33

A CIP catalogue record for this book is available from the Library of Congress, Washington D.C., USA

Deutsche Bibliothek Cataloging-in-Publication Data

Golubitsky, Martin:
The symmetry perspective : from equilibrium to chaos in phase space and physical space / Martin Golubitsky ; Ian Stewart. – Basel ; Boston ; Berlin : Birkhäuser, 2002
    (Progress in mathematics ; Vol. 200)
    ISBN 978-3-7643-2171-0   ISBN 978-3-0348-8167-8 (eBook)
    DOI 10.1007/978-3-0348-8167-8

The image on the preceding page is adapted from *Fractured Symmetry* in *Symmetry in Chaos* by Field and Golubitsky using the software package prism developed by Mike Field.

© 2002 Springer Basel AG
Originally published by Birkhäuser Verlag in 2002
Softcover reprint of the hardcover 1st edition 2002

Printed on acid-free paper produced of chlorine-free pulp. TCF ∞

ISBN 978-3-7643-2171-0

9 8 7 6 5 4 3 2 1   www.birkhauser-science.com

Ferran Sunyer i Balaguer (1912–1967) was a self-taught Catalan mathematician who, in spite of a serious physical disability, was very active in research in classical mathematical analysis, an area in which he acquired international recognition. His heirs created the Fundació Ferran Sunyer i Balaguer inside the Institut d'Estudis Catalans to honor the memory of Ferran Sunyer i Balaguer and to promote mathematical research.

Each year, the Fundació Ferran Sunyer i Balaguer and the Institut d'Estudis Catalans award an international research prize for a mathematical monograph of expository nature. The prize-winning monographs are published in this series. Details about the prize and the Fundació Ferran Sunyer i Balaguer can be found at

http://www.crm.es/info/ffsb.htm

This book has been awarded the Ferran Sunyer i Balaguer 2001 prize.

The members of the scientific commitee of the 2001 prize were:

Hyman Bass
    University of Michigan

Pilar Bayer
    Universitat de Barcelona

Antonio Córdoba
    Universidad Autónoma de Madrid

Paul Malliavin
    Université de Paris VI

Alan Weinstein
    University of California at Berkeley

Ferran Sunyer i Balaguer Prize winners:

1992   Alexander Lubotzky
        Discrete Groups, Expanding Graphs and
        Invariant Measures, PM 125

1993   Klaus Schmidt
        Dynamical Systems of Algebraic Origin,
        PM 128

1994   The scientific committee decided not to
        award the prize

1995   As of this year, the prizes bear the year
        in which they are awarded, rather than
        the previous year in which they were
        announced

1996   V. Kumar Murty and M. Ram Murty
        Non-vanishing of L-Functions and
        Applications, PM 157

1997   A. Böttcher and Y.I. Karlovich
        Carleson Curves, Muckenhoupt Weights,
        and Toeplitz Operators, PM 154

1998   Juan J. Morales-Ruiz
        Differential Galois Theory and
        Non-integrability of Hamiltonian Systems,
        PM 179

1999   Patrick Dehornoy
        Braids and Self-Distributivity, PM 192

2000   Juan-Pablo Ortega and Tudor Ratiu
        Hamiltonian Singular Reduction,
        (to be published)

Contents

Preface   xi

Chapter 1.   Steady-State Bifurcation   1
- 1.1.   Two Examples   3
- 1.2.   Symmetries of Differential Equations   6
- 1.3.   Liapunov-Schmidt Reduction   16
- 1.4.   The Equivariant Branching Lemma   17
- 1.5.   Application to Speciation   20
- 1.6.   Observational Evidence   25
- 1.7.   Modeling Issues: Imperfect Symmetry   27
- 1.8.   Generalization to Partial Differential Equations   29

Chapter 2.   Linear Stability   33
- 2.1.   Symmetry of the Jacobian   37
- 2.2.   Isotypic Components   37
- 2.3.   General Comments on Stability of Equilibria   38
- 2.4.   Hilbert Bases and Equivariant Mappings   41
- 2.5.   Model-Independent Results for D_3 Steady-State Bifurcation   45
- 2.6.   Invariant Theory for S_N 46
- 2.7.   Cubic Terms in the Speciation Model   48
- 2.8.   Steady-State Bifurcations in Reaction-Diffusion Systems   54

Chapter 3.   Time Periodicity and Spatio-Temporal Symmetry   59
- 3.1.   Animal Gaits and Space-Time Symmetries   59
- 3.2.   Symmetries of Periodic Solutions   61
- 3.3.   A Characterization of Possible Spatio-Temporal Symmetries   65
- 3.4.   Rings of Cells   68
- 3.5.   An Eight-Cell Locomotor CPG Model   72
- 3.6.   Multifrequency Oscillations   78
- 3.7.   A General Definition of a Coupled Cell Network   83

Chapter 4.   Hopf Bifurcation with Symmetry   87
- 4.1.   Linear Analysis   89
- 4.2.   The Equivariant Hopf Theorem   91
- 4.3.   Poincaré-Birkhoff Normal Form   95
- 4.4. $O(2)$ Phase-Amplitude Equations   97
- 4.5.   Traveling Waves and Standing Waves   98
- 4.6.   Spiral Waves and Target Patterns   100

4.7. **O**(2) Hopf Bifurcation in Reaction-Diffusion Equations   105
4.8. Hopf Bifurcation in Coupled Cell Networks   110
4.9. Dynamic Symmetries Associated to Bifurcation   116

Chapter 5. Steady-State Bifurcations in Euclidean
            Equivariant Systems   123
5.1. Translation Symmetry, Rotation Symmetry,
      and Dispersion Curves   126
5.2. Lattices, Dual Lattices, and Fourier Series   128
5.3. Actions on Kernels and Axial Subgroups   132
5.4. Reaction-Diffusion Systems   135
5.5. Pseudoscalar Equations   138
5.6. The Primary Visual Cortex   141
5.7. The Planar Bénard Experiment   148
5.8. Liquid Crystals   149
5.9. Pattern Selection: Stability of Planforms   154

Chapter 6. Bifurcation From Group Orbits   161
6.1. The Couette-Taylor Experiment   165
6.2. Bifurcations From Group Orbits of Equilibria   171
6.3. Relative Periodic Orbits   176
6.4. Hopf Bifurcation from Rotating Waves
      to Quasiperiodic Motion   184
6.5. Modulated Waves in Circular Domains   187
6.6. Spatial Patterns   189
6.7. Meandering of Spiral Waves   192

Chapter 7. Hidden Symmetry and Genericity   201
7.1. The Faraday Experiment   202
7.2. Hidden Symmetry in PDEs   204
7.3. The Faraday Experiment Revisited   209
7.4. Mode Interactions and Higher-Dimensional Domains   209
7.5. Lapwood Convection   211
7.6. Hemispherical Domains   216

Chapter 8. Heteroclinic Cycles   221
8.1. The Guckenheimer-Holmes Example   222
8.2. Heteroclinic Cycles by Group Theory   225
8.3. Pipe Systems and Bursting   227
8.4. Cycling Chaos   237

Chapter 9. Symmetric Chaos   241
9.1. Admissible Subgroups   246
9.2. Invariant Measures and Ergodic Theory   254
9.3. Detectives   257
9.4. Instantaneous and Average Symmetries,
      and Patterns on Average   267
9.5. Synchrony of Chaotic Oscillations and
      Bubbling Bifurcations   268

Chapter 10.   Periodic Solutions of
              Symmetric Hamiltonian Systems   275
   10.1.   The Equivariant Moser-Weinstein Theorem   277
   10.2.   Many-Body Problems   283
   10.3.   Spatio-Temporal Symmetries in Hamiltonian Systems   287
   10.4.   Poincaré-Birkhoff Normal Form   288
   10.5.   Linear Stability   289
   10.6.   Molecular Vibrations   296

Bibliography   301

Index   321

Preface

Pattern formation in physical systems is one of the major research frontiers of mathematics. A tendency towards regular structure, often intricate, sometimes highly complex, seems to be inherent in the natural world and in experimental science. These structures — patterns — are often of considerable importance. The field is enormous, the variety of techniques employed is huge, and the range of applications is extensive.

These are signs of a flourishing field of active research. However, in all branches of science, there is a strong urge to seek general unifying principles and not merely to continue elaborating a growing diversity of examples. This is especially the case in mathematics. A central theme of *The Symmetry Perspective* is that many instances of pattern formation can be understood within a single framework: the viewpoint of *symmetry*.

It has long been recognized that symmetry can be used to *describe* patterns — classical crystallography, with its 230 distinct symmetry classes, was probably the first area of applied science to make systematic use of such ideas. However, description alone is rather a passive use of symmetry. We shall provide evidence that the role of symmetry in pattern-formation can also be used to gain insight into the mechanisms by which patterns form.

Over the past 15 years, we and other authors have been exploring a far more active role for symmetry, in the context of nonlinear dynamical systems. It has become apparent that the symmetries of a system of nonlinear ordinary or partial differential equations can be used, in a systematic and unified way, to analyze, predict, and understand many general mechanisms of pattern-formation. Specifically, the symmetries of a system can be used to work out a 'catalogue' of typical forms of behavior. This catalogue is to a great extent *model-independent*. By this we do not mean that the specific model involved is irrelevant, but that a great deal can be deduced by knowing only the symmetries of that model. In a sense, all models with a given symmetry explore the same range of pattern-forming behaviors, and that range of behaviors can be studied in its own right without reference to many details of the model.

Those details remain crucial, however, because they determine which behaviors the system 'chooses' from its universal pattern-book. From this point of view, it often makes sense to begin with the model-independent generalities, and only later to introduce the specifics. This two-stage approach is not *always* the most sensible, of course. It runs into difficulties when the model-independent analysis becomes too complicated by comparison with a model-specific one. Why tackle a difficult general problem when some simpler method, such as computer simulation, can provide an answer more quickly and more concretely?

On the other hand, excessive focus on specifics is inefficient if the same general phenomenon keeps occurring in many different examples. So the model-independent approach is most useful when it really does provide insight into the observed phenomena, and when a general framework helps with the understanding of the system and its behavior.

Several other issues arise when we wish to compare experimental observations with theoretical predictions, and we emphasize these issues throughout the book. One is the distinction between *phase space* and *physical space*. The theory of nonlinear dynamical systems is largely discussed in terms of trajectories in an abstract phase space — a space of variables that determines the state of the system. However, the theory had its origins in qualitative questions — such as whether a periodic solution exists, or a chaotic one — rather than quantitative questions. As a result, the theoretical work often involves some unspecified change of coordinates in phase space, and the connection between the variables in the theory and the variables that are being observed can be lost. The situation is especially glaring when the system is a PDE. For example, a complex fluid flow-pattern in physical space is usually encoded as a single *point* in phase space. The interpretation of that point (for example, its coordinates might be a series of Fourier coefficients) is highly important in such circumstances.

Moreover, experimentalists typically observe a time series of measurements; most commonly of a single variable, but possibly of several, or of a spatially distributed set of variables (such as a video recording of patterns that change over time). The abstract theory, on the other hand, works with geometric objects such as attractors, homoclinic orbits, or invariant measures. The link between these abstract concepts and feasible observations is often ignored. This is unfortunate, because the link is subtle and indirect.

In this connection, symmetric dynamical systems possess at least one significant advantage. In most (though not all) cases, the symmetry of the system has a well-defined physical interpretation *and* a well-defined interpretation in phase space. For example, if the experimental apparatus has circular symmetry, then a group of rotations acts both in physical space and in phase space. These actions need not be the same — usually they are not — but the elements that act *are* the same. Thus a symmetry property of a solution in phase space usually translates into a symmetry property in physical space, and conversely. For example, the fact that a periodic solution is a rotating wave, or a standing wave, can be deduced from its symmetries in space and time. Model-dependent specifics then tell us what these waves actually look like. Thus symmetries provide an important route between the abstract theory and experimental observations.

The above discussion raises another issue. Knowing that a system of differential equations has a particular symmetry *group* is seldom very informative. The crucial information is how that group *acts* on the appropriate region of phase space. For example, we shall describe six different systems of PDEs in the plane, each being symmetric under the planar Euclidean group. Moreover, all six systems exhibit some common types of solution, for example parallel 'rolls'. Nevertheless, all six systems have very different pattern-forming properties. (This is even the case for isomorphic actions: the actual meaning of the physical variables can be important, not just the abstract action on them.) Again, we place emphasis on such considerations, because a failure to appreciate this issue can be seriously misleading.

Traditionally, symmetry is mostly used to describe and classify static patterns. In the context of dynamical systems, however, symmetry can usefully be employed in connection with dynamic patterns too. We apply symmetry methods to increasingly complex kinds of dynamic behavior as the book progresses: equilibria, period-doubling, time-periodic states, homoclinic and heteroclinic orbits, and chaos. In each case we motivate the type of dynamical behavior being studied by discussing one or more potential applications, drawn from a wide variety of scientific disciplines ranging from theoretical physics to evolutionary biology.

We now summarize the contents of the book.

Chapter 1 focuses on the simplest context for symmetry-based methods: equilibrium states of ODEs. The theory is motivated through a model of speciation in evolution, introducing the concept of a symmetry-breaking bifurcation. We define what we mean by a symmetric dynamical system and discuss the symmetries of solutions — which, in the nonlinear case, need not be the same as the symmetries of the system. Two key results are introduced in this chapter. First, we discuss Liapunov-Schmidt reduction, which — subject to suitable technical hypotheses — reduces the dimension of a bifurcation problem to a size that can often be handled. Indeed, the analogous procedure for PDEs reduces infinite-dimensional problems to finite-dimensional ones. The second key result is the Equivariant Branching Lemma, which provides sufficient conditions for the existence of a bifurcating branch of equilibria with a particular type of symmetry group (said to be 'axial'). We apply the Equivariant Branching Lemma to the speciation model. At this stage we raise and discuss a number of modeling issues, concerning the relevance of symmetry assumptions. We argue that even when the symmetry of the system being modeled is imperfect or approximate, it is often useful to study an idealized, perfectly symmetric model. This contention is supported by numerical experiments in which symmetry is destroyed by introducing small imperfections and/or stochastic effects.

Chapter 2 remains in the equilibrium context, and examines the stability of equilibria, especially those whose existence can be proved by using the Equivariant Branching Lemma. Symmetry has a strong effect on stability because the structure of linear maps that commute with a group action is highly restricted. We also consider the implications of symmetry for nonlinear maps that commute with a group action, which leads into classical invariant theory and the study of equivariant maps. One general principle emerges, which has a strong effect on several applications later in the book: if the symmetry group action has a nontrivial quadratic equivariant map, then generically *all* branches predicted by the Equivariant Branching Lemma are unstable. Taking inspiration from singularity theory, we argue that an effective way to deal with this difficulty is to add higher-order terms that 'stabilize' primary branches, and to treat these higher-order terms as small perturbations that 'unfold' the degenerate case when such higher-order terms are absent.

Chapter 3 employs an application to legged locomotion in animals to motivate the analysis of time-periodic patterns in symmetric systems. We show that time-periodic states have additional 'temporal' symmetry, corresponding to a circle group $\mathbf{S}^1$ of phase shifts (modulo the period). Thus the symmetry group of a time-periodic state involves 'spatio-temporal' symmetries. The central theoretical result is a classification of all possible spatio-temporal symmetry groups for time-periodic states of systems with a given spatial symmetry group. We apply this theorem to deduce, from observations of animal gaits, the architecture of the underlying Central Pattern Generator in the animal's nervous system (subject to a series of explicit assumptions about the symmetries of the Central Pattern Generator, which we argue are plausible, and which lead to testable predictions). We end by discussing multi-rhythms in coupled cell networks, where the presence of spatio-temporal symmetries can lead to surprising frequency-locked states.

One of the standard ways to generate time-periodic states of a system is through Hopf bifurcation, and this is the topic of Chapter 4. Hopf bifurcation occurs when a stable equilibrium loses stability to an oscillatory mode. In the symmetric case, the spatio-temporal symmetries of the bifurcating branches are highly influential. In particular there is an analogue of the Equivariant Branching Lemma for equilibria, which we call the Equivariant Hopf Theorem. We sketch the proof of this theorem by Liapunov-Schmidt reduction from a suitable loop space. The closely related topic of Poincaré-Birkhoff normal form, which sheds light on the stability of bifurcating branches of time-periodic states, is introduced. We illustrate equivariant Hopf bifurcation with a discussion of spiral waves and target patterns in reaction-diffusion systems on a circular domain in the plane. We also apply our results to coupled cell systems. Chapters 3 and 4 both make use of an important idea: certain kinds of dynamics can introduce a new symmetry — in this case the circle groups of phase shifts. Chapter 4 ends with a discussion of other examples of these 'dynamic symmetries', including periodic orbits of maps, and the adding machine dynamics that occurs in a period-doubling cascade at the Feigenbaum point.

In Chapter 5 we study lattice-symmetric equilibrium patterns in Euclidean-invariant PDEs. Although the Euclidean group is non-compact, the assumption of lattice periodicity reduces the problem to one with a compact symmetry group. In fact, we treat six different applications with such symmetry: reaction-diffusion systems, the Navier-Stokes equations, the primary visual cortex, liquid crystals, and Bénard convection with two different kinds of boundary conditions. Each of these systems is symmetric under the Euclidean group in the plane, but their pattern-forming behavior is *different*. The reason for these differences is the main message of this chapter: knowing that a system is invariant under some particular symmetry group is insufficient information for an accurate analysis. At least three further pieces of information are necessary. First: are there additional symmetries (possibly in some reformulation of the problem)? Second: what is the appropriate *representation* of the symmetry group for the problem under study? Third: what is the interpretation of the symmetry in terms of structures that might be observed experimentally? Superficially, all six applications considered here appear to have the same symmetries, so it might be thought that an analysis of the patterns formed in one of them automatically transfers to all the others. In fact, the opposite happens: all six problems have significant differences from the others. We end the chapter with a brief discussion of stability issues.

In a symmetric system of ODEs or PDEs, solutions come in group orbits. In particular, if the group includes continuous symmetries, these solutions may not be isolated. For example, even solutions as apparently straightforward as Taylor vortices in Couette-Taylor flow occur as a continuous group orbit in the usual infinite-cylinder model with periodic boundary conditions, the relevant part of the group being axial translations. Chapter 6 addresses the consequences of this phenomenon. It begins with a model-independent discussion of expected bifurcations in the Couette-Taylor system, which provides a surprisingly detailed and reasonably complete account of the patterns that arise. With this example as motivation, the main mathematical issues are examined. The main new feature here is that when a solution loses stability and other solutions bifurcate, the stability is lost across the entire group orbit. If the group has a continuous subgroup, these orbits form manifolds, and the bifurcation behavior acquires novel features. For example, solutions may 'drift' along a group orbit. One of the most striking applications of these ideas is to meandering spirals in the Belousov-Zhabotinskii experiment. Here the tip of the spiral wave can exhibit an epicyclic motion. There are two types of epicycle, and the transition between them involves a state with unbounded linear drift. We are therefore forced to face up to the non-compact nature of the Euclidean group. Our *results* do not apply to this symmetry group, but our *methods* do — subject to suitable technical modifications.

The main topic of Chapter 7 is the question: which symmetries of a system should be taken into account when applying the methods of this book? At first sight this is a modeling issue: the symmetries of real systems are generally only approximate (as is apparent in both the speciation model and the infinite cylinder model for Couette-Taylor flow), but it is often appropriate to idealize approximate symmetries into exact ones. However, there is a second issue: having chosen a model, it may possess more symmetries than might be anticipated from its construction. In particular, we show that simple PDEs often possess more symmetry than is apparent from the symmetries of the domain, and show how to turn these 'hidden symmetries' to advantage. Applications include the Faraday experiment, in which a shallow dish of liquid is vibrated vertically, and convection in a porous medium.

Equilibrium and time-periodic states can arise by local bifurcation. In Chapter 8 we begin the study of more 'global' dynamics by looking at the effects of symmetry on heteroclinic cycles. Indeed, such cycles are much more common in symmetric systems than they are in asymmetric ones. We discuss the related concept of a 'pipe system' in connection with bursting in neurons. We also discuss 'cycling chaos' in systems of coupled cells, which is related to heteroclinic cycling between chaotic states rather than equilibria.

Chapter 9 opens up the connection between symmetry and chaos. A good motivating example is the Faraday experiment in a square container, which experimentally possesses chaotic states which display no obvious patterns. Where has the symmetry gone? The mathematical answer is that chaotic attractors, being extended objects, have two kinds of symmetry. One, pointwise symmetry, is easily observed: every state looks symmetric but the time-evolution is chaotic. The second, setwise symmetry, is more subtle: now the symmetry of the state cannot be observed in a instantaneous snapshot — an effect often known as spatio-temporal chaos. However, a state corresponding to a symmetric attractor does possess observable symmetries that correspond to those of its attractor. These symmetries become apparent as 'symmetries on average' in a time-series of observations. The main theoretical concerns of this chapter are to characterize the possible symmetry groups of chaotic attractors, and to develop some of the novel types of bifurcation that can occur when chaotic attractors are present, notably the phenomenon of 'bubbling'. This phenomenon occurs, for example, in the much-studied problem of the synchronization of chaotic oscillators. We also discuss how the symmetry of a chaotic attractor can be observed in a practical experiment. One possible answer is the concept of a 'detective'. We discuss the formation of detectives by ergodic sums, and apply these ideas to chaotic solutions of the Brusselator model on a square domain.

Up to this point, generically all systems studied have been dissipative. The final Chapter 10 demonstrates that the methods can also be applied to Hamiltonian systems. The basic principles remain unchanged, but details vary considerably: the main reason is that it is necessary to respect the symplectic structure that is present in Hamiltonian systems. This chapter merely scratches the surface of a deep and beautiful area of mathematics and physics. The main exhibit is an equivariant analogue of the Moser-Weinstein Theorem on the existence of families of periodic solutions near an equilibrium. We discuss applications to many-body problems, the resonant spring pendulum, and molecular vibrations.

We end with an extensive (though by no means comprehensive) bibliography.

How To Approach This Book. Our aim is to present a body of mathematics that we consider to be central to the understanding of symmetry-breaking bifurcations and pattern formation in equivariant dynamics. In a sense *The Symmetry Perspective* is the mathematician's version of our popular book *Fearful Symmetry* [495]. We use the word 'perspective' to emphasize the book's stance: it is a *point of view* on the role of symmetry as an explanation of pattern formation in nonlinear dynamics. This perspective has evolved over the last twenty years, and it leads to a slightly unusual approach to applied science. Instead of starting with very specific models and analysing them as individuals, we begin by contemplating an entire class of models: all those with the appropriate symmetries. We then ask: Which phenomena are typical in such classes? It turns out that many important features of pattern formation and qualitative dynamics are typical, so this question is a sensible one.

Only after establishing the typical properties of symmetry classes of models do we proceed to more detailed analyses of specific models. The advantage of our perspective is that it places the analysis in context, and avoids attributing special features to specific models when in fact those features are universal. Moreover, symmetries of systems (often approximate symmetries, which we idealize in order to carry out an analysis) are often easier to discover and to justify than specific model equations. This is especially the case for biology, and several of our examples are taken from this area of science.

Because our main focus is on a point of view, we have presented the mathematics on several levels: simple examples to build intuition, applications to scientific questions to illustrate the link to the real world, and more technical descriptions where we feel these add to an understanding of the main issues. Thus the choice of what material to include is a personal one, and we often refer to the existing literature for proofs — especially if the source is reasonably accessible or if the area is a standard topic in mathematics. In consequence, *The Symmetry Perspective* is not self-contained as a course text. It is probably more useful for this purpose if it is supplemented by a formal textbook, for example *Singularities and Groups in Bifurcation Theory* volume 2 [237]. We give references to suitable sources for such material as it is needed, but we do not embed it in the kind of logical development that would be required in a traditional postgraduate level course. This book should therefore be considered as a guide to the subject area, not as a comprehensive text.

Acknowledgements. These notes began with a series of lectures, organized by Jerry Marsden at the Fields Institute in the Spring of 1993, then located at the University of Waterloo. Nolan Evans did an excellent job of recording and TeXing the notes from these lectures, including many of the figures. Due to a computer malfunction, the notes were almost lost and would have been lost had it not been for the efforts of Bill Langford and Nolan Evans to resurrect them. Eight years later, in a vastly expanded and changed form, we again used these notes as the basis for a series of lectures — this time at the Summer School on Bifurcations, Symmetry, and Patterns at Coimbra, Portugal in July, 2000 organized by Isabel Labouriau, Jorge

Buescu, Sofia Castro, and Ana Dias, and supported by the Centro de Matematica Aplicada, Universidade do Porto.

The research discussed here has been generously supported during the past decade by the National Science Foundation, the Texas Advanced Research Program, the Office for Naval Research, and the Engineering and Physical Sciences Research Council of the UK, as well as by research supported visits to a number of institutions: the Fields Institute, the Institute for Mathematics and its Applications (University of Minnesota), the University of Houston Department of Mathematics, the Santa Fe Institute, the University of Warwick Mathematics Institute, and the Boston University Center for Biodynamics.

We have had the pleasure to work with many people on the research described in these notes. We wish to thank each of them: Don Aronson, Pete Ashwin, Ernie Barany, Paul Bressloff, Jorge Buescu, Luciano Buono, Ernesto Buzano, Sofia Castro, David Chillingworth, Pascal Chossat, Jack Cohen, Jim Collins, Jack Cowan, John David Crawford, Gerhard Dangelmayr, Michael Dellnitz, Ana Dias, Benoit Dionne, Toby Elmhirst, Mike Field, David Gillis, Gabriela Gomes, Herman Haaf, Andrew Hill, Andreas Hohmann, Ed Ihrig, Mike Impey, Greg King, Edgar Knobloch, Michel Kroon, Martin Krupa, Bill Langford, Victor LeBlanc, Jun Ma, Miriam Manoel, Jerry Marsden, Ian Melbourne, James Montaldi, Matt Nicol, Antonio Palacios, Mark Roberts, David Schaeffer, Mary Silber, Jim Swift, Harry Swinney, Peter Thomas, and André Vanderbauwhede.

Coventry and Houston, December 2000.

Chapter 1

Steady-State Bifurcation

Most research in dynamical systems theory is currently focused on exotic behavior such as chaos. In many applications, however, far simpler kinds of dynamical behavior are of interest — even steady states, in which there is no dynamics, only stasis. By the end of this book we too will be looking at exotic dynamical behavior, with the added ingredient of symmetry; but along the way we will have to rebuild dynamical systems theory from the ground up in the symmetry context. That means starting with steady states — which, for symmetric systems, turn out to be surprisingly interesting and subtle.

One of the guiding principles that underlies *The Symmetry Perspective* is that abstract concepts should be motivated by applications. We therefore introduce each chapter with a problem from some branch of applied science, and we develop the main concepts and techniques of the chapter from that question. The aim here is to make the abstract ideas more accessible, not to provide an accurate historical development of the subject.

In this opening chapter there are three main ideas:

- The notion of symmetry of a system of ordinary differential equations (ODEs).
- The notion of symmetry of a solution of a system of ODEs.
- The phenomenon of symmetry-breaking, in which solutions have less symmetry than equations.

The mathematical ideas involved here come from nonlinear dynamical systems and the theory of group representations. We motivate them through a problem in evolutionary biology — apparently about as far removed from those mathematical areas as it is possible to get. The link between the biology and its mathematical analysis is not yet firmly established, but for the purposes of this book, the scientific status of the models is a secondary issue. What is of primary importance is the mathematical techniques that we introduce in order to understand the models.

In this spirit, we begin this chapter with a question that goes back to Charles Darwin and his epic work *The Origin of Species*. How can new species arise in evolution? Well before Darwin, it was known that animals and plants could be persuaded to change, in small ways, by the artificial application of selective pressure. Breeding techniques could produce bigger or smaller dogs, redder roses, more nutritious cereal crops. However, no amount of selective breeding seemed able to persuade an organism to change species, let alone generate an entirely new species. Darwin's revolutionary insight was to realize that nature provides its own source of selective pressure — competition for survival. He assembled a wealth of evidence for the 'mutability of species', without human intervention and over long periods of time. However, he did not propose any detailed mechanism by which new species could arise, other than the slow accumulation of small changes.

Until recently the conventional wisdom was that it is difficult for gradual changes to cause a split into several species. The argument was that if the nascent species occupy the same habitat, then they will still be able to interbreed. Exchange of genetic material — what Mayr [381, 382] calls *gene-flow* — will cancel out that small divergence as their descendants 'regress to the mean', Galton [205]. So it seems that speciation by gradual changes is a non-starter. Mayr's answer to this dilemma, building on work of Dobzhansky [165], was the mechanism of *allopatric* speciation. Here some (small) 'founder population' becomes isolated from the main group, perhaps by migration or geographical accident. Once separated, the new group evolves independently of the main one — gene-flow between the groups is switched off — until eventually the two groups are no longer able to interbreed even if they are brought back into contact. When and if they come together again, the two groups will remain separate species: gene-flow between them will have ceased, permanently.

The main alternative to this mechanism — rather, class of mechanisms, for there are numerous variations on the original allopatric theory — is *sympatric* speciation. Here the nascent species remain mixed throughout the speciation process, and the unifying effects of gene-flow are nullified by a variety of biological processes. Sympatric speciation is now considered to have been a common — though by no means universal — route to new species, see for example Higashi *et al.* [272], Kondrashov and Kondrashov [320], Dieckmann and Doebeli [158] and references therein, Pennisi [431], Rundle *et al.* [457], and Huey *et al.* [290]. Behind this change of attitude is a growing realization that the old gene-flow arguments involve several tacit assumptions: for example that gradually changing causes must produce gradually changing effects, and that uniform behavior in uniformly changing conditions should remain uniform. These assumptions appear plausible, but they are fallacious. In the terminology of nonlinear dynamics, the first says that jump bifurcation is impossible and the second tells us that symmetries cannot break. In fact, jump bifurcation and symmetry-breaking are generic phenomena in nonlinear dynamical systems.

In order to address the issue of sympatric speciation, we therefore choose to model evolutionary changes in a cluster of organisms using a dynamical system. The observed states of organisms will be assumed to correspond to the stable steady states of this system. Symmetry enters because organisms in the same species are biologically indistinguishable, at least if we view the processes of evolution on geological timescales. Bifurcations occur because the dynamics depends on the environment, which can change. If symmetry is broken, then the original cluster of indistinguishable organisms must split into distinguishable sub-clusters, and this corresponds to speciation. So our model equations for speciation lead directly to the main mathematical topics that we wish to introduce. Not only that: having developed the necessary techniques, we are able to analyze the models effectively and deduce some significant implications for sympatric speciation.

The techniques we develop in this chapter center around the following ideas. The simplest states of a dynamical system are *steady states* (or *equilibria*), where the state $x(t)$ of the system is independent of time: $x(t) = x_0$ for all t. If the system has a symmetry group Γ, then a given equilibrium will be invariant under some subgroup $\Sigma \subseteq \Gamma$. For some equilibria this subgroup is equal to the whole of Γ, but it may be smaller. If so, we say that the equilibrium *breaks symmetry* from Γ to Σ. Symmetry-breaking is an important mechanism for pattern formation, and is therefore a central idea in these notes. In this chapter we describe the beginnings of a general theory of symmetry-breaking bifurcations of equilibria, focusing on two questions.

The first question is: which subgroups Σ can in principle occur? The answer (essentially by definition) is 'isotropy subgroups' — a special class of subgroups defined in terms of the action of Γ. We set up various notions related to isotropy subgroups and illustrate them in the context of a model of speciation proposed by Cohen and Stewart [117] and developed further in Elmhirst [171, 173] and Stewart *et al.* [494]. A similar model was proposed independently by Vincent and Vincent [513]. In this model the appropriate symmetry group is $\Gamma = \mathbf{S}_N$, the symmetric group on N symbols, acting by permutations of coordinates on $\mathbf{R}^N$.

The second question is: which subgroups Σ actually occur in a given problem? The answer usually depends on the details of the dynamical system. However, there is one useful 'model-independent' principle, known as the Equivariant Branching Lemma. This states that (subject to technical genericity conditions) if steady-state bifurcation occurs, then there always exist bifurcating branches of equilibria whose isotropy subgroups belong to a special class, called 'axial'. We state the Equivariant Branching Lemma and give a proof. We also derive its implications for the speciation model.

1.1 Two Examples

We begin by investigating two examples of ODEs with symmetry. The first example is simple enough to analyze by 'bare hands' methods. The second example is more complicated, and for the moment we restrict our analysis to some numerical simulations. Later, we explain the phenomena that these simulations reveal.

Both examples can be interpreted as models of speciation. The variables determine the phenotypes (body-plan and behavioral characteristics) of the organisms concerned. The ODEs describe how those phenotypes change in response to each other and to changes in external parameters (which we interpret as 'environment'). Initially we focus on the mathematics of these examples.

Example 1.1 As a warm-up problem, we illustrate how symmetry principles can constrain the form of a system of ODEs. Suppose, for simplicity, that we are modeling the evolutionary dynamics of three clumps of initially nominally identical organisms, with phenotypes $(x, y, z) \in \mathbf{R}^3$. Then we wish to write down sensible equations for $\mathrm{d}x/\mathrm{d}t, \mathrm{d}y/\mathrm{d}t, \mathrm{d}z/\mathrm{d}t$. For the sake of argument, let us choose

$$\mathrm{d}x/\mathrm{d}t = \lambda x + x^2 - x - y$$

with bifurcation parameter λ. Is this reasonable?

Decompose the vector field into the 'internal dynamics' $\lambda x + x^2 - x$ and a coupling y. In the absence of specific biological knowledge, there is no particular reason to prefer any particular internal dynamic, but the chosen coupling has a disturbing feature. It tells us that the way clump x reacts to clump y is *different* from the way it reacts to clump x. But from x's point of view, the world divides into 'self' (x) and 'non-self' (y and z). There is no reason to distinguish y from z. In other words, the coupling should treat y and z in exactly the same way: it should be invariant under the permutation that swaps y and z. One way to achieve this is to 'average' the coupling over y and z — actually, we just add them, but this is the same up to a constant factor — getting

$$\mathrm{d}x/\mathrm{d}t = \lambda x + x^2 - x - y - z$$

What should the y-equation be? It should have the same internal dynamics as the x-equation, but with x replaced throughout by y, so it should look like

$$dy/dt = \lambda y + y^2 - y + \text{ coupling terms}$$

Moreover, from y's point of view, both x and z are on the same footing, and y should react to z in the same way that x does. (Clump z is non-self for both x and y.) So we should set

$$dy/dt = \lambda y + y^2 - y - x - z$$

Finally, by the same argument, it makes sense to put

$$dz/dt = \lambda z + z^2 - z - x - y$$

Thus we are led to the system

$$
\begin{aligned}
dx/dt &= \lambda x - (x+y+z) + x^2 \\
dy/dt &= \lambda y - (x+y+z) + y^2 \\
dz/dt &= \lambda z - (x+y+z) + z^2
\end{aligned}
\tag{1.1}
$$

The process that leads to this form endows the equations with an evident symmetry, in the following sense: if we permute the variables x, y, z in the same way on both the left- and right-hand sides of the equations, we obtain the same system. Thus the system is symmetric under the group $\mathbf{S}_3$ of all permutations of (x, y, z).

We will see that this symmetry property has a strong effect on the solutions of the equation. However, it is *not* the case that all solutions have the same symmetry as the equations. In fact, a solution is symmetric under $\mathbf{S}_3$ if and only if $x = y = z$. We will show that there exist other solutions.

We often call a steady state *trivial* if it has full symmetry. There is an obvious trivial solution to (1.1), namely $x = y = z = 0$, and this solution is indeed symmetric under $\mathbf{S}_3$. We determine its dynamic stability by finding the eigenvalues of the linearization about $x = y = z = 0$. The linearization has matrix

$$
L = \begin{bmatrix}
\lambda - 1 & -1 & -1 \\
-1 & \lambda - 1 & -1 \\
-1 & -1 & \lambda - 1
\end{bmatrix}
$$

with eigenvectors and eigenvalues as follows:

$$
\begin{aligned}
v_0 &= (1,1,1)^T : \text{eigenvalue } \lambda - 3 \\
v_1 &= (1,-1,0)^T : \text{eigenvalue } \lambda \\
v_0 &= (0,1,-1)^T : \text{eigenvalue } \lambda
\end{aligned}
$$

When $\lambda < 0$ the trivial solution is stable (all eigenvalues negative) but it loses stability when $\lambda > 0$.

We look for equilibria of (1.1) with λ and x near 0. Then

$$\lambda x + x^2 = \lambda y + y^2 = \lambda z + z^2 = x + y + z$$

It is easy to show that near $(\lambda, x, y, z) = 0$ these equations imply that at least two of x, y, z are equal, and that all three are equal only when $x = y = z = 0$.

By symmetry we may assume $x = y$ and $z \neq x$. Then

$$\lambda x + x^2 = \lambda z + z^2 \tag{1.2}$$

$$\lambda x + x^2 = 2x + z \tag{1.3}$$

Equation (1.2) implies that $\lambda(x - z) + (x + z)(x - z) = 0$, so that $x + z = -\lambda$, whence $z = -x - \lambda$. Then (1.3) implies that $\lambda x + x^2 = x - \lambda$. Solving the quadratic equation and retaining only the solution near the origin, we find that a nontrivial solution

$$x = y = \frac{-\lambda + 1 - \sqrt{\lambda^2 - 6\lambda + 1}}{2} \qquad z = \frac{-\lambda - 1 + \sqrt{\lambda^2 - 6\lambda + 1}}{2} \qquad (1.4)$$

exists for all λ near 0. When solutions of a symmetric system of equations possesses less symmetry than the equations themselves, we say that the system has undergone *symmetry-breaking*.

We investigate the geometry of these new solutions. For $\lambda \sim 0$ we have

$$\sqrt{\lambda^2 - 6\lambda + 1} \sim 1 - 3\lambda - 4\lambda^2 + O(\lambda^3)$$

by the binomial theorem, so

$$x = \lambda + 2\lambda^2 + O(\lambda^3)$$
$$y = \lambda + 2\lambda^2 + O(\lambda^3)$$
$$z = -2\lambda - 2\lambda^2 + O(\lambda^3)$$

As λ varies through 0, these equations define a parametrized curve, or *branch*, of solutions. Solutions on this branch exist for all λ near 0. The existence of new branches of solutions near a given branch (here the trivial solution) is called *bifurcation*. A branch such as this one, which exists for all λ near 0 (that is, both $\lambda < 0$ and $\lambda > 0$) is said to be *transcritical*.

One immediate implication of symmetry is that there are two other branches obtained by permuting (x, y, z). We can draw a *bifurcation diagram*, showing how solutions vary with λ. Schematically, the bifurcation diagram looks like Fig. 1.1.

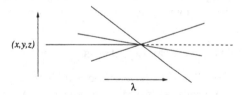

Figure 1.1: Bifurcation diagram illustrating symmetry-breaking.

This example demonstrates a number of general phenomena:
- Solutions of symmetric ODEs need not be symmetric.
- However, the *set* of all solutions is symmetric.
- Branches of symmetry-breaking solutions can bifurcate when a parameter is varied.

Along each bifurcating branch we have

$$x + y + z = 2\lambda^2 + O(\lambda^3)$$

which is quadratic in λ. Therefore the branches are all tangent to the plane $x + y + z = 0$ at the origin. This plane is precisely the kernel of L at $\lambda = x = y = z = 0$. Clearly the kernel is invariant under permutations in $\mathbf{S}_3$. $\diamond$

Example 1.2 We now generalize the above example to a symmetry group $\mathbf{S}_N$, where N can be larger than 3. This time we add a cubic term, and the ODE takes the form:

$$\frac{dx_i}{dt} = \lambda x_i - (x_1 + \cdots + x_N) + x_i^2 - x_i^3 \qquad (1 \le i \le N)$$

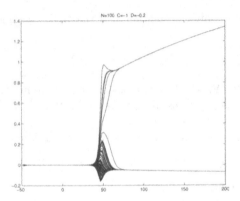

Figure 1.2: Simulation of speciation for $N = 100$ clumps of organisms. Note division of population into two species of 16 and 84 clumps each.

Figure 1.2 shows a numerical simulation of a solution of these equations for $N = 100$. The picture is obtained by increasing λ by a tiny amount at each step of a numerical integration, while using the values of x_i obtained on the previous step as initial values for the next step. This technique is called *ramping the bifurcation parameter*.

Initially the system is at the origin. There is a relatively rapid jump bifurcation, after which the values of the x_i settle down into two clusters. One cluster contains 16 of the 100 variables x_i, and the common value is positive; the other cluster contains 84 of the variables and the common value is negative.

Again we see symmetry-breaking. In fact the symmetry group of the bifurcating branch is $\mathbf{S}_{16} \times \mathbf{S}_{84} \subseteq \mathbf{S}_{100}$, as we discuss later in this chapter. $\diamond$

1.2 Symmetries of Differential Equations

The first step in understanding the above phenomena is to formalize the meaning of 'symmetry' of a dynamical system. We also formalize what we mean by the symmetry of a solution.

The symmetries of a system of ODEs (or indeed PDEs) are specified in terms of a group of transformations of the variables that in some sense preserves the structure of the equation — in particular, its solutions. We therefore require some background ideas about transformation groups (or group *representations*). Good references on Lie groups and their representations are Bröcker and tom Dieck [77] and Miller [399].

Typical Groups. In these notes we mainly concentrate on compact Lie groups. Lie groups divide into two classes: continuous and finite. The main finite groups that we shall consider are:

- $\mathbf{D}_m$, the *dihedral group* of order $2m$ (rotations and reflections in the plane that preserve a regular m-gon).
- $\mathbf{Z}_m$ is the *cyclic group* or order m (rotations only).
- $\mathbf{S}_m$, the *symmetric group* consisting of all permutations on m symbols; order $m!$

Note that $\mathbf{S}_3 \cong \mathbf{D}_3$ since the symmetries of an equilateral triangle permute its three vertices in all six possible ways.

The compact continuous groups are:

- S^1, the *circle group*.
- $SO(2)$, the *special orthogonal group* in $R2$, consisting of all rotations

$$R_\theta = \begin{bmatrix} \cos\theta & -\sin\theta \\ \sin\theta & \cos\theta \end{bmatrix}$$

  in the plane, which is isomorphic to S^1.
- $O(n)$, the *orthogonal group* in R^n, consisting of all $n \times n$ matrices A such that $AA^T = I$.
- $SO(n)$, the *special orthogonal group*, where in addition $\det A = 1$.
- $T^m = \underbrace{S^1 \times \cdots \times S^1}_{m}$, the *m-torus*.

We also consider two non-compact continuous Lie groups:

- $E(2)$, the *Euclidean group* in the plane (all rigid motions, that is, generated by rotations, reflections, and translations).
- $SE(2)$, the *special Euclidean group*, which is the subgroup of orientation-preserving rigid motions (generated by rotations and translations).

Group Representations or Actions. A recurrent theme of these notes is that one of the most important features of a symmetric bifurcation problem is not just the abstract structure of its symmetry group, but how that group *acts*. That is, we must consider the group in a fairly concrete form as a group of transformations of some space. To begin with, this space is usually a vector space, but later we also consider affine spaces. More generally, the space may be a manifold.

Definition 1.3 Let Γ be a Lie group and V a vector space. An *action* of Γ on V is a smooth homomorphism $\rho : \Gamma \to GL(V)$. To simplify notation we write $\gamma v = \rho(\gamma)(v)$. Then the condition that ρ be a homomorphism implies the following identities:

(1) $(\gamma\delta)v = \gamma(\delta v)$
(2) $1v = v$ where 1 is the identity element of Γ.   ◇

The group $\rho(\Gamma) = \hat{\Gamma} \subseteq GL(V)$ is a subgroup, and a group of matrices. If ρ is one-to-one then $\hat{\Gamma} \cong \Gamma$.

Example 1.4 We return to Example 1.2. Here the symmetry group is S_N, and it acts by permuting the coordinates x_j. Specifically, let $x = (x_1, \ldots, x_N)$. Then the S_N-action is

$$\sigma x = (x_{\sigma^{-1}(1)}, \ldots, x_{\sigma^{-1}(N)}) \tag{1.5}$$

The inverse σ^{-1} occurs here because that ensures property (1) of a group action. To verify this point let

$$(y_1, \ldots, y_N) = \sigma x$$

where $y_j = x_{\sigma^{-1}(j)}$. Then

$$\begin{aligned} \gamma(\sigma x) &= \gamma y \\ &= (y_{\gamma^{-1}(1)}, \ldots, y_{\gamma^{-1}(N)}) \\ &= (x_{\sigma^{-1}\gamma^{-1}(1)}, \ldots, x_{\sigma^{-1}\gamma^{-1}(N)}) \\ &= (x_{(\gamma\sigma)^{-1}(1)}, \ldots, x_{(\gamma\sigma)^{-1}(N)}) \\ &= (\gamma\sigma)x \end{aligned}$$

◇

It is easier (less abstract) to think of $\hat{\Gamma}$ as a group of matrices on V and to think of $\hat{\Gamma}$ as acting on V by matrix multiplication. If Γ is compact then $\hat{\Gamma}$ is compact. Without loss of generality, we may assume that a compact Lie group acts by orthogonal transformations, and this is a useful simplification:

Theorem 1.5 *Every compact Lie group Γ acting on $\mathbf{R}^n$ may be identified with a subgroup of $\mathbf{O}(n)$.*

Sketch of Proof. The idea is to average an inner product on $\mathbf{R}^n$ over Γ. (There exists a canonical invariant measure, *Haar measure*, and the average is defined in terms of that measure.) With respect to the averaged inner product, it is easy to see that $\Gamma \subseteq \mathbf{O}(n)$. For details see Golubitsky *et al.* [237] pages 30-31.   □

Equivariant Mappings. What does it mean for a system of ODEs to be symmetric under a group action? Suppose we have a parametrized system of ODEs

$$\dot{x} = f(x, \lambda) \tag{1.6}$$

where $f : \mathbf{R}^n \times \mathbf{R}^r \to \mathbf{R}^n$ and throughout we assume $f \in C^\infty$ unless otherwise stated. Here $\lambda \in \mathbf{R}^r$ is an r-dimensional parameter. Later, the role of λ becomes crucial, but for the moment we suppress λ in our notation.

Definition 1.6 The group element $\gamma \in \mathbf{O}(n)$ is a *symmetry* of (1.6) if for every solution $x(t)$ of (1.6), $\gamma x(t)$ is also a solution.   ◇

We derive a useful condition for γ to be a symmetry. Let $y(t) = \gamma x(t)$. Then

$$\dot{y}(t) = f(y(t)) = f(\gamma x(t))$$

Moreover,

$$\dot{y}(t) = \gamma \dot{x}(t) = \gamma f(x(t))$$

So

$$f(\gamma x(t)) = \gamma f(x(t))$$

for all solutions $x(t)$ of (1.6). Since solutions exist for arbitrary initial conditions, this is equivalent to

$$f(\gamma x) = \gamma f(x)$$

for all $x \in \mathbf{R}^n$. If this holds, we say that f commutes with γ or f is γ-*equivariant*. This leads to the following definition:

Definition 1.7 Let Γ act on $\mathbf{R}^n$ and let $f : \mathbf{R}^n \times \mathbf{R}^k \to \mathbf{R}^n$. Then f is Γ-*equivariant* if $f(\gamma x, \alpha) = \gamma f(x, \alpha)$ for all $\gamma \in \Gamma, x \in \mathbf{R}^n$.   ◇

It is easy to verify that according to this definition, the equation in Example 1.1 is $\mathbf{S}_3$-equivariant, and the equation in Example 1.2 is $\mathbf{S}_N$-equivariant.

Isotropy Subgroups. Our next task is to formalize the notion of the symmetry of a solution to an equivariant ODE. For the moment we consider only equilibria. Equilibria of (1.6) are the solutions of

$$f(x) = 0$$

Suppose that x is such an equilibrium. Then we can express the symmetries of this *solution* of the system (as opposed to the symmetries of the *system*) as follows: a symmetry σ of x is an element of Γ that leaves x invariant. The set of all such σ is a subgroup of Γ, known as the 'isotropy subgroup' of x. In fact, we can define the isotropy subgroup for any x, whether or not it is an equilibrium:

Definition 1.8 Let $v \in \mathbf{R}^n$. The *isotropy subgroup* of v is

$$\Sigma_v = \{\gamma \in \Gamma : \gamma v = v\}$$

$\diamond$

Example 1.9 In Example 1.1 we found four equilibrium solutions: $(0,0,0)$, (u,u,v), (u,v,u), (v,u,u) where

$$u = \frac{-\lambda + 1 - \sqrt{\lambda^2 - 6\lambda + 1}}{2} \qquad v = \frac{-\lambda - 1 + \sqrt{\lambda^2 - 6\lambda + 1}}{2}$$

The isotropy subgroup of $(0,0,0)$ is the whole of $\mathbf{S}_3$. The isotropy subgroup of (u,u,v) consists of the identity and the 2-cycle (12); that of (u,v,u) consists of the identity and the 2-cycle (13); and that of (v,u,u) consists of the identity and the 2-cycle (23). Thus we can distinguish these solutions by their isotropy subgroups.

The final three solutions are images of each other under the action of $\mathbf{S}_3$. Their isotropy subgroups are conjugate in $\mathbf{S}_3$: for example, $(23)^{-1}(12)(23) = (13)$. This phenomenon generalizes: see (1.7) below. $\diamond$

Example 1.10 Let $\Gamma = \mathbf{O}(2)$ act on $\mathbf{R}^2 \equiv \mathbf{C}$ so that rotations θ and the standard reflection κ act by

$$\theta z = e^{i\theta} z$$
$$\kappa z = \bar{z}$$

If $z = 0$ is the origin, then

$$\Sigma_0 = \mathbf{O}(2)$$

whereas if z is real and nonzero, say $z = 1$, then

$$\Sigma_1 = \{0, \kappa\}$$

which is cyclic subgroup of order 2. More generally, if

$$z = re^{i\theta}$$

with $r \neq 0$, then a simple calculation shows that

$$\Sigma_z = \{0, \kappa(-2\theta)\}$$

which is again cyclic of order 2, generated by a reflection $\kappa(-2\theta)$ (which is one of the conjugates of κ). $\diamond$

Example 1.11 Equation (1.2) is equivariant under the permutation action of $\mathbf{S}_N$. Numerically, we have found a solution x corresponding to two values a, b where $a > 0, b < 0$. If $A = \{j : x_j = a\}$ and $B = \{j : x_j = b\}$ then $|A| = 16, |B| = 84$. Clearly the isotropy subgroup of this solution is $\mathbf{S}_{16} \times \mathbf{S}_{84}$ where $\mathbf{S}_{16}$ acts by permuting the indices in A and $\mathbf{S}_{84}$ acts by permuting the indices in B. $\diamond$

In the biological interpretation of (1.2) the variables x_j represent values of some phenotypic observable (for example, beak length in a bird). The content of the previous example is that (in an idealized case) different species can be distinguished by symmetries — not of individual organisms, but of the entire model 'ecosystem'.

Example 1.12 Consider two related species of (imaginary) birds: *longbeaks* and *shortbeaks*. They can be distinguished by the phenotypic variable 'beak length'. Imagine a tree containing eight longbeaks, and a second tree containing five longbeaks and three shortbeaks. Assume (the idealization) that all longbeaks look identical, and all shortbeaks look identical (but differ from longbeaks). Specifically, assume that in longbeaks the beak length is always 2cm, but in shortbeaks it is always 1cm.

Potential symmetries of these systems are permutations of the birds If the permutation swaps a longbeak with a longbeak, or a shortbeak with a shortbeak, we cannot tell the difference. However, if a permutation swaps a shortbeak and a longbeak, we do notice the difference.

Formally, we can set up these transformations by attaching labels to the birds and permuting the labels. We use the labels $\{1, 2, 3, 4, 5, 6, 7, 8\}$, so all permutations belong to $\mathbf{S}_8$. In the first tree (all longbeaks) everything looks the same after any permutation of the labels, so the symmetry group is $\mathbf{S}_8$. In the second tree, assume for the sake of argument that we use labels 1-5 for the longbeaks and 6-8 for the shortbeaks. Then everything looks the same if we permute $\{1, 2, 3, 4, 5\}$ in any way and also permute $\{6, 7, 8\}$ in any way. That is, symmetries must preserve the 'block structure' that separates the shortbeaks from the longbeaks.

The symmetry group in the second case is therefore the subgroup

$$\mathbf{S}_5 \times \mathbf{S}_3 \subseteq \mathbf{S}_8$$

and the symmetry of the first situation is broken in the second.

In the real world, organisms that belong to the same species are not precisely identical, but they resemble each other much more closely than they do organisms from some different species — so the model is reasonable 'up to equivalence', where organisms are equivalent if they belong to the same species.

This set-up can be formalized in terms of the permutation action of $\mathbf{S}_N$ on a space of 'phenotypic vectors'. Given a set of N organisms, we can define their phenotypes by a vector $(x_1, \ldots, x_N)$, where for simplicity we look at a single phenotypic variable, so each $x_j \in \mathbf{R}$. (To include more phenotypic variables, let the x_j be vectors in some $\mathbf{R}^k$. The discussion generalizes to this case, but at the expense of some technicalities.) The entry x_j is the value of this phenotypic variable for the jth organism. The symmetric group $\mathbf{S}_N$ acts on $\mathbf{R}^N$ by permutations.

We compute the isotropy subgroup Σ_x for arbitrary x. If $\sigma.x = x$ then the only entries x_i, x_j of x that can be permuted by σ are those with equal values, $x_i = x_j$. We therefore partition $\{1, \ldots, N\}$ into disjoint blocks $B_1, \ldots, B_k$ with the property that $x_i = x_j$ if and only if i, j belong to the same block. (A slick way to do this is to define an equivalence relation $\sim_x$ by $i \sim_x j \Leftrightarrow x_i = x_j$ and let the blocks be the equivalence classes.) Letting $b_\ell = |B_\ell|$, we find that

$$\Sigma_x = \mathbf{S}_{b_1} \times \cdots \times \mathbf{S}_{b_k}$$

where $\mathbf{S}_{b_\ell}$ is the symmetric group on block B_ℓ. $\diamond$

The key conclusion of the above example is that we can detect the presence of more than one species by looking at symmetries. A collection of organisms belonging to a single species has more symmetry than an equal-sized collection that contains two (or more) species.

More generally, the isotropy subgroup of a solution of an equivariant system of ODEs provides useful information about the form of that solution. We shortly see that it also leads to a technique for *finding* solutions with given symmetries. First we make a few more remarks about isotropy subgroups.

The Isotropy Lattice. If $x \in \mathbf{R}^n$ is an equilibrium of (1.6) and $\gamma \in \Gamma$, then so is γx, because $f(\gamma x) = \gamma f(x) = \gamma 0 = 0$. The *group orbit* of x is

$$\Gamma x = \{\gamma x : \gamma \in \Gamma\}$$

It is easy to see that x and γx have conjugate isotropy subgroups; indeed

$$\Sigma_{\gamma x} = \gamma \Sigma_x \gamma^{-1}$$

For many purposes we consider conjugate equilibria to be different manifestations of 'the same' equilibrium, and we classify solutions in terms of conjugacy classes of isotropy subgroups — by which we mean the set of all conjugates of a given subgroup. An important relation between subgroups is that of containment — whether or not one subgroup is contained in another. Carrying this relation over to conjugacy classes of subgroups, we are led to define the following abstract structure:

Definition 1.13 Let $H = \{H_i\}$ and $K = \{K_j\}$ be two conjugacy classes of (isotropy) subgroups of Γ. Define a partial ordering $\preceq$ on the set of such conjugacy classes by

$$H \preceq K \Leftrightarrow H_i \subseteq K_j$$

for some representatives H_i, K_j. The *isotropy lattice* of Γ in its action on $\mathbf{R}^n$ is the set of all conjugacy classes of isotropy subgroups, partially ordered by $\preceq$.   ◇

Strictly speaking, this set need not be a lattice in the formal algebraic sense: it is a partially ordered set (which is finite for all compact Γ). However, the term 'lattice' is standard. The isotropy lattice classifies the possible ways for equilibria to break symmetry, and arranges them in a hierarchy with the property that smaller isotropy subgroups correspond to breaking more symmetries.

Example 1.14 In Example 1.12, up to conjugacy we may assume that the block corresponds to consecutive intervals of labels; that is:

$$\begin{aligned}
B_1 &= \{1, \ldots, b_1\} \\
B_2 &= \{b_1 + 1, \ldots, b_1 + b_2\} \\
& \ldots \\
B_k &= \{b_1 + \cdots + b_{k-1} + 1, \ldots, b_N\}
\end{aligned}$$

Moreover, we can assume that $b_1 \geq b_2 \geq \cdots \geq b_k$ so that the block sizes are in descending order. Therefore conjugacy classes of isotropy subgroups of $\mathbf{S}_N$ are in one-to-one correspondence with partitions of N into nonzero natural numbers arranged in descending order.

We can transfer the partial order $\preceq$ to these partitions through the one-to-one correspondence. We then find that if P_1, P_2 are two partitions then $P_1 \preceq P_2$ if and only if P_1 is a *refinement* of P_2, where we obtain a refinement of a partition P by breaking some of its parts into smaller parts and rearranging all parts into descending order. For example if $N = 8$ and

$$P_1 = \{5, 3\}, \quad P_2 = \{4, 2, 1, 1\}$$

then $P_2 \preceq P_1$ since $5 = 4+1$ and $3 = 2+1$. It follows that the isotropy lattice of $\mathbf{S}_5$, say, contains seven conjugacy classes, corresponding to the partitions

$$\{5\}, \ \{4, 1\}, \ \{3, 2\}, \ \{2, 2, 1\}, \ \{3, 1, 1\}, \ \{2, 1, 1, 1\}, \ \{1, 1, 1, 1, 1\}$$

and the lattice ordering is as in Figure 1.3.

◇

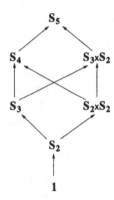

Figure 1.3: Isotropy lattice of S_5.

Fixed-Point Subspaces. We now explain how to use the structure of the isotropy lattice to carry out a systematic search for solutions with a given symmetry. We also show that for certain special but common types of isotropy subgroup, we can guarantee in advance that generically solution branches with such symmetries exist.

Definition 1.15 Let $\Sigma \subseteq \Gamma$ be a subgroup. Then the *fixed-point subspace* of Σ is

$$\text{Fix}(\Sigma) = \{v \in \mathbf{R}^n : \sigma v = v, \quad \forall \sigma \in \Sigma\}$$

$\diamond$

Example 1.16 In Example 1.1 the fixed-point subspace of S_3 is $\{(0,0,0)\}$. The fixed-point subspace of $\langle(12)\rangle$ is $\{(u,u,v) : u, v \in \mathbf{R}\}$. $\diamond$

This definition leads to the following theorem, which although simple to prove, is central to the development of the theory of equivariant dynamics.

Theorem 1.17 Let $f : \mathbf{R}^n \to \mathbf{R}^n$ be Γ-equivariant and let $\Sigma \subseteq \Gamma$ be a subgroup. Then

$$f(\text{Fix}(\Sigma)) \subseteq \text{Fix}(\Sigma)$$

Proof. Let $v \in \mathbf{R}^n$ and $\sigma \in \Sigma$. Then $\sigma v = v$ implies that

$$\sigma f(v) = f(\sigma v) = f(v)$$

so that $f(v) \in \text{Fix}(\Sigma)$. $\square$

Note that

$$\text{Fix}(\gamma \Sigma \gamma^{-1}) = \gamma \text{Fix}(\Sigma) \qquad (1.7)$$

A trivial but important consequence of Theorem 1.17 is:

Proposition 1.18 Let $x(t)$ be a solution trajectory of an equivariant ODE. Then

$$\Sigma_{x(t)} = \Sigma_{x(0)}$$

for all $t \in \mathbf{R}$. That is, isotropy subgroups remain constant along trajectories.

Proof. Since $\text{Fix}(\Sigma_{x(0)})$ is flow-invariant by Theorem 1.17, we have $x(t) \in \text{Fix}(\Sigma_{x(0)})$ for all t. Therefore $\Sigma_{x(0)} \subseteq \Sigma_{x(t)}$. Since $x(0)$ lies on the trajectory through $x(t)$, the same argument yields $\Sigma_{x(t)} \subseteq \Sigma_{x(0)}$. Therefore, $\Sigma_{x(0)} = \Sigma_{x(t)}$. $\square$

Theorem 1.17 tells us that we can find an equilibrium solution with isotropy subgroup Σ by restricting the original vector field to the subspace $\text{Fix}(\Sigma)$. Unless $\text{Fix}(\Sigma)$ is the whole space $\mathbf{R}^n$, the restricted problem is posed on a space of lower dimension, and therefore 'ought to be' simpler. The bigger Σ is, in terms of the partial ordering on the isotropy lattice, the smaller $\dim \text{Fix}(\Sigma)$ is. So we can search for solutions by starting with the largest isotropy subgroups and systematically working down the lattice.

Example 1.19 Continuing our analysis of $\mathbf{S}_N$ acting on $\mathbf{R}^N$, suppose that Σ is an isotropy subgroup corresponding to the simplest nontrivial partition $P = \{p, q\}$ where $p + q = N$ and $p \geq N/2$. It is then easy to see that $\text{Fix}(\Sigma)$ consists of all vectors

$$\underbrace{(u, \ldots, u,}_{p} \underbrace{v, \ldots, v)}_{q}$$

for real numbers u, v. (Here u and v may be equal, or not.) Therefore $\dim \text{Fix}(\Sigma) = 2$.

Similarly, if Σ corresponds to a partition of N into k blocks, then $\dim \text{Fix}(\Sigma) = k$. The reason is that a vector $x \in \text{Fix}(\Sigma)$ if and only if all of its entries for a given block of the partition corresponding to Σ are equal, so $\text{Fix}(\Sigma)$ has a basis consisting of vectors v^ℓ for which $v_j^\ell = 1$ if $j \in B_\ell$, and $v_j^\ell = 0$ otherwise. $\diamond$

Invariant Subspaces. While analyzing the bifurcations of Example 1.1 we observed that the kernel of the linearization L at the origin is invariant under the action of $\mathbf{S}_3$. Similarly in Example 1.2 it can be seen that the kernel of the linearization at the origin is invariant under the action of $\mathbf{S}_N$.

We now generalize this observation to any equivariant system of ODEs. Invariant subspaces are important because they are the spaces that support bifurcations, so a systematic theory of invariant subspaces will be essential. All of the ideas discussed in this subsection are standard in the representation theory of compact Lie groups: as it happens, they are also tailor-made for the bifurcation analysis of equivariant ODEs.

Definition 1.20 $V \subseteq \mathbf{R}^n$ is a Γ-*invariant subspace* if $\gamma V = V, \forall \gamma \in \Gamma$. $\diamond$

Some examples of invariant subspaces are:
(a) $V = \{0\}$ and $V = \mathbf{R}^n$ are always Γ-invariant.
(b) $\text{Fix}(\Gamma)$ is Γ-invariant. Indeed, Γ fixes every vector in $\text{Fix}(\Gamma)$.
(c) Let $\mathbf{SO}(2)$ act on $\mathbf{R}^4 = \mathbf{R}^2 \times \mathbf{R}^2$ by

$$\theta(v, w) = (R_\theta(v), R_\theta(w)) \qquad \theta \in [0, 2\pi)$$

Fix $\psi \in \mathbf{SO}(2)$ and let $V_\psi = \{(v, R_\psi(v)) : v \in \mathbf{R}^2\}$. The subspace V_ψ is $\mathbf{SO}(2)$-invariant. To see this calculate

$$\theta(v, R_\psi(v)) = (R_\theta(v), R_\theta R_\psi(v))$$
$$= (R_\theta(v), R_\psi R_\theta(v)) \in V_\psi$$

In this example there is a continuum of two-dimensional invariant subspaces.
(d) Let $\mathbf{S}_N$ act on $\mathbf{R}^N$ by permuting coordinates as above. The subspaces

$$\begin{aligned} V_0 &= \mathbf{R}(1, 1, \ldots, 1) \\ V_1 &= \{(x_1, \ldots, x_N) : x_1 + \cdots + x_N = 0\} \end{aligned} \tag{1.8}$$

are $\mathbf{S}_N$-invariant. To see this, note that the definitions of these spaces remain unchanged if we permute coordinates. $\diamond$

Irreducible Subspaces. When a (compact) group acts on some space via a system of transformations, we can often decompose the space into subspaces, so that the group acts separately on each subspace. The building-blocks for such a decomposition are said to be 'irreducible'. More precisely:

Definition 1.21 The subspace $W \subseteq \mathbf{R}^n$ is Γ-*irreducible* if the only Γ-invariant subspaces of W are W and $\{0\}$. $\diamond$

Note that if Γ acts irreducibly on $\mathbf{R}^n$, then $\mathrm{Fix}(\Gamma) = \{0\}$ or $\mathrm{Fix}(\Gamma) = \mathbf{R}^n$. If Γ acts trivially, then $\mathrm{Fix}(\Gamma) = \mathbf{R}^n$. If Γ acts nontrivially, then $\mathrm{Fix}(\Gamma) = \{0\}$. In any case, if $\mathrm{Fix}(\Gamma) = \{0\}$, then $f(0) = 0$ by Theorem 1.17. Γ-irreducible subspaces are important for the following reason.

Theorem 1.22 *Let* $\Gamma \subseteq \mathbf{O}(n)$ *be a compact Lie group. Then there exist* Γ-*irreducible subspaces* $V_1, \ldots, V_s$ *such that*

$$R^n = V_1 \oplus \cdots \oplus V_s$$

Proof. There exists a Γ-invariant inner product $(\cdot, \cdot)$. If Γ acts irreducibly on $\mathbf{R}^n$, then we are done. Suppose not, then there exists a Γ-invariant subspace $W \neq \{0\}$ or $\mathbf{R}^n$. Form $W^\perp = \{v \in \mathbf{R}^n : (v, w) = 0, \forall w \in W\}$. The subspace $W^\perp$ is Γ-invariant because for $v \in W^\perp, w \in W$ and $\gamma \in \Gamma$ we have

$$(\gamma v, w) = (v, \gamma^{-1}w)$$
$$= (v, w') \qquad (w' \in W)$$
$$= 0$$

Thus, $\gamma v \in W^\perp$.

So $\mathbf{R}^n = W \oplus W^\perp$. By finite dimensionality this method of decomposition will eventually stop. Note that this decomposition need not be unique. $\square$

Example 1.23 Let $\mathbf{S}_N$ act on R^N by permuting coordinates. Define V_0, V_1 as in (1.8). Then each of V_0, V_1 is $\mathbf{S}_N$-irreducible, and

$$\mathbf{R}^N = V_0 \oplus V_1$$

$\diamond$

Let $\mathbf{S}_N$ act on V_1. Then the isotropy subgroups of $\mathbf{S}_N$ are the same as for the action of $\mathbf{S}_N$ on $\mathbf{R}^N$, but the dimension of $\mathrm{Fix}(\Sigma)$ is now $k - 1$ when Σ corresponds to a partition into k blocks, because the relation $x_1 + \cdots + x_N = 0$ reduces all relevant dimensions by 1.

Commuting Linear Maps. The first step towards understanding a symmetric bifurcation problem is to linearize the system, so it is important to understand the class of linear maps that arises. It turns out that these maps commute with the set of transformations by which the group acts.

Definition 1.24 Let Γ act on V and let $A : V \to V$ be a linear map. A *commutes with* Γ if $A\gamma = \gamma A, \forall \gamma \in \Gamma$. Equivalently, A is Γ-*equivariant*. $\diamond$

Remark 1.25 (1) $\ker A$ is a Γ-invariant subspace, since

$$Av = 0 \Rightarrow A(\gamma v) = \gamma(Av) = 0, \text{ so } \gamma v \in \ker A$$

If Γ acts irreducibly, then $\ker A = \{0\}$ or $\ker A = \mathbf{R}^n$. Observe that $\ker A = \{0\}$ implies that A is invertible, while $\ker A = \mathbf{R}^n$ implies that $A = 0$.

(2) If A commutes with Γ and A^{-1} exists, then A^{-1} also commutes with Γ:

$$A^{-1}\gamma = (\gamma^{-1}A)^{-1} = (A\gamma^{-1})^{-1} = \gamma A^{-1}$$

$\diamond$

Group representation theory reveals a fundamental classification of irreducible subspaces, according to the structure of the set of commuting linear maps. There are three basic types, and the type has a major effect on bifurcation behavior. More precisely, let D be the set of commuting linear maps when the group acts on an irreducible subspace. It can be shown (easily) that $D \subseteq GL(\mathbf{R}^n)$ is a linear subspace and is also a skew field. This statement is known as Schur's Lemma, and it implies that D is either $\mathbf{R}, \mathbf{C}$ or the quaternions (which, respectively, are 1-, 2- or 4-dimensional), [237].

Definition 1.26 Γ acts *absolutely irreducibly* on $\mathbf{R}^n$ if $D = \mathbf{R}$; that is, the only commuting linear maps are $\{cI : c \in \mathbf{R}\}$. $\diamond$

Absolute irreducibility has an important implication for equivariant bifurcation problems. Consider (1.6) where f is Γ-equivariant, and now take account of the parameter λ. Apply the chain rule:

$$(df)_{\gamma x, \lambda}\gamma = \gamma(df)_{x,\lambda}$$

and

$$(df)_{0,\lambda}\gamma = \gamma(df)_{0,\lambda}$$

Thus $(df)_{0,\lambda}$ commutes with Γ. Therefore, if Γ acts absolutely irreducibly on $\mathbf{R}^n$, then

$$(df)_{0,\lambda} = c(\lambda)I$$

Compare this with Lemma 1.31(ii).

Genericity Results. A *steady-state* bifurcation occurs at an equilibrium where the linearized equation has a zero eigenvalue and no other eigenvalues on the imaginary axis. Absolutely irreducible representations play a key role in steady-state bifurcations, because they determine the 'generic' situation. To see why, suppose that there is a branch of group-invariant equilibria $y(\lambda)$ to the ODE

$$\dot{y} = F(y, \lambda)$$

and that at some parameter value $\lambda = \lambda_0$ there is a steady-state bifurcation; that is,

$$A_0 = (dF)_{y(\lambda_0), \lambda_0}$$

is singular. Then the following theorem is valid, see [237], p. 82.

Theorem 1.27 *If the above conditions hold, then generically the following are true.*
(a) 0 is the only eigenvalue of A_0 *on the imaginary axis.*
(b) The generalized eigenspace corresponding to 0 *is* $\ker A_0$.
(c) Γ acts absolutely irreducibly on $\ker A_0$. $\square$

We now show that finding equilibria in equivariant bifurcation problems can be reduced to finding the zeros of mappings on $\ker A_0$.

1.3 Liapunov-Schmidt Reduction

In Example 1.1 we observed that the bifurcating branches from the origin are tangent to the kernel of the linearization L. This is a general property of equivariant ODEs, and it can be combined with the Implicit Function Theorem to provide an important technique for analyzing bifurcations. We discuss that technique in this subsection. It is known as Liapunov-Schmidt Reduction. The search for bifurcating branches of equilibria reduces to solving a system

$$f(x, \lambda) = 0$$

where $f : \ker A_0 \times \mathbf{R} \to \ker A_0$.

An alternative approach is known as Center Manifold Reduction: see Guckenheimer and Holmes [255]. This procedure has the advantage that it applies to vector fields, not just zeros of mappings, so it leads to a reduced equation

$$\frac{dx}{dt} = g(x, \lambda)$$

where $g : \ker A_0 \times \mathbf{R} \to \ker A_0$. However, Center Manifold Reduction suffers from some technical disadvantages, namely possible lack of uniqueness and lack of smoothness. Nevertheless, the two reduction techniques have a great deal in common. In particular, both of these reductions can be performed in such a way that the symmetries of the original system are preserved, that is, f and g are also Γ-equivariant. Center Manifold Reduction guarantees that the asymptotic dynamics is preserved, while Liapunov-Schmidt reduction guarantees only that equilibria (and sometimes their stability) are preserved by reduction. We now discuss Liapunov-Schmidt reduction.

Consider a system of ordinary differential equations:

$$\dot{y} = F(y, \alpha) \quad (y \in \mathbf{R}^N, \alpha \in \mathbf{R}^k)$$

where $F(0, 0) = 0$ and $(dF)_{0,0}$ is singular. Let

$$\mathcal{K} = \ker(dF)_{0,0} \neq \{0\} \qquad \mathcal{R} = \operatorname{range}(dF)_{0,0}$$

We consider two splittings of $\mathbf{R}^N$:

$$\mathbf{R}^N = \mathcal{K} \oplus \hat{\mathcal{K}}$$
$$\mathbf{R}^N = \mathcal{R} \oplus \hat{\mathcal{R}}$$

we call $\hat{\mathcal{K}}$, respectively $\hat{\mathcal{R}}$, the *cokernel* and *corange* of $(dF)_{0,0}$. Then $F(x, w, \alpha) = 0$ ($x \in \mathcal{K}, w \in \hat{\mathcal{K}}$) if and only if

(1) $EF(x, w, \alpha) = 0$
(2) $(I - E)F(x, w, a) = 0$

where $E : \mathbf{R}^N \to \mathcal{R}$ is the projection with kernel $\ker E = \hat{\mathcal{R}}$. Now differentiate (1) with respect to w:

$$\frac{d}{dw} EF(x, w, \alpha)|_0 = E(dF)_{0,0}|\hat{\mathcal{K}} = (dF)_{0,0}|\hat{\mathcal{K}}$$

Since $(dF)_{0,0}|\hat{\mathcal{K}}$ is nonsingular we can apply the Implicit Function Theorem to show that there exists a unique $w : \mathcal{K} \times \mathbf{R}^k \to \hat{\mathcal{K}}$ with $w(0, 0) = 0$ such that

$$EF(x, w(x, \alpha), \alpha) \equiv 0. \tag{1.9}$$

Therefore, solving $F = 0$ is equivalent to solving

$$f(x, \alpha) = (I - E)F(x, w(x, \alpha), \alpha) = 0 \tag{1.10}$$

where $f : \mathcal{K} \times \mathbf{R}^k \to \hat{\mathcal{R}}$, $f(0,0) = 0$ and $(\mathrm{d}f)_{0,0} = 0$. Note that $\dim \mathcal{K} = \dim \hat{\mathcal{R}} = n$, so we have $f : \mathbf{R}^n \times \mathbf{R}^k \to \mathbf{R}^n$.

Using the Implicit Function Theorem, we have proved that the solutions to $F(y, \alpha) = 0, y \in \mathbf{R}^N$ near the origin are in 1:1 correspondence with the solutions of $f(x, \alpha), x \in \mathbf{R}^n$ near the origin.

We next show that the reduction of F to f can be carried out in a manner that preserves symmetry.

Theorem 1.28 *Suppose that $\Gamma \subseteq \mathbf{O}(N)$ is a compact Lie group and that $F : \mathbf{R}^N \times \mathbf{R}^k \to \mathbf{R}^N$ is Γ-equivariant. Then the reduced bifurcation equation $f : \mathbf{R}^n \times \mathbf{R}^k \to \mathbf{R}^n$ defined in (1.10), where $\mathbf{R}^n = \ker(\mathrm{d}F)_{(0,0)}$, can be chosen to be Γ-equivariant.*

Proof. The matrix $(\mathrm{d}F)_{0,0}$ commutes with Γ and so $\mathcal{K}$ and $\mathcal{R}$ are both Γ-invariant subspaces. We can choose $\hat{\mathcal{K}}$ and $\hat{\mathcal{R}}$ to be Γ-invariant, for example, let $\hat{\mathcal{K}} = \mathcal{K}^\perp$ and $\hat{\mathcal{R}} = \mathcal{R}^\perp$.

We claim that $w(\gamma x, \alpha) = \gamma w(x, \alpha)$ for all $\gamma \in \Gamma$ where w is the implicit function defined in (1.9). Let $w_\gamma(x, \alpha) = \gamma^{-1} w(\gamma w, \alpha)$, then $w_\gamma : \mathcal{K} \to \hat{\mathcal{K}}$. Compute

$$EF(x, w_\gamma(x, \alpha), \alpha) = EF(x, \gamma^{-1} w(\gamma x, \alpha), \alpha))$$
$$= E\gamma^{-1} F(\gamma x, w(\gamma x, \alpha), \alpha)$$
$$= \gamma^{-1} EF(\gamma x, w(\gamma x, \alpha), \alpha) = 0$$

We can interchange γ^{-1} and the projection E as we have chosen $\hat{\mathcal{K}}$ and $\hat{\mathcal{R}}$ to be Γ-invariant. Therefore, by uniqueness of the implicit function, we have $w_\gamma = w$ or $\gamma w(x, \alpha) = w(\gamma x, \alpha)$. It follows directly from (1.10) that f is Γ-equivariant. $\qquad \square$

Note that now the action of Γ is on $\mathcal{K}$, so a partial answer to our question is that only the action of the symmetries on $\ker(\mathrm{d}F)_{0,0}$ matters. The action of Γ on $\ker(\mathrm{d}F)_{0,0}$ could, for instance, turn out to be trivial, in which case all of the symmetries have disappeared from the problem. Nonetheless, the interpretation of solutions remains nontrivial: in this case, *every* solution that we find has full symmetry.

The Liapunov-Schmidt procedure can also be performed on a system of PDEs, subject to certain technical assumptions of a functional-analytic nature.

1.4 The Equivariant Branching Lemma

In Example 1.2 we observed numerically that symmetry-breaking solutions to a problem with $\mathbf{S}_{100}$ symmetry could have $\mathbf{S}_{16} \times \mathbf{S}_{84}$ symmetry. We now explain this curious observation: the key point is that $\mathbf{S}_{16} \times \mathbf{S}_{84}$ is a maximal isotropy subgroup, hence has a minimal-dimensional fixed-point space. More strongly still, in the appropriate Liapunov-Schmidt reduction it has a one-dimensional fixed-point space. The main result of this section is the Equivariant Branching Lemma, which asserts the generic existence of bifurcating branches of steady states for a particular class of isotropy subgroups:

Definition 1.29 An isotropy subgroup Σ is *axial* if $\dim \mathrm{Fix}(\Sigma) = 1$. $\qquad \diamond$

The name arises by analogy with a rotation in 3-dimensional space, which fixes a 1-dimensional 'axis'. We return to the example of $\mathbf{S}_N$:

Example 1.30 The axial subgroups of $\mathbf{S}_N$ acting on

$$V_1 = \{x \in \mathbf{R}^N : x_1 + \cdots + x_N = 0\} \subseteq \mathbf{R}^N$$

are, up to conjugacy, those of the form $\mathbf{S}_p \times \mathbf{S}_q$ where $p + q = N$ and $1 \leq p \leq [N/2]$. the reason is that these subgroups correspond to a partition of N into 2 blocks, and by Example 1.23 the corresponding fixed-point space has dimension $2 - 1 = 1$. $\Diamond$

The simplest general existence theorem for symmetry-breaking branches of equilibria in a symmetric dynamical system was proved originally by Vanderbauwhede [508] and Cicogna [112] and is:

Lemma 1.31 (Equivariant Branching Lemma) *Let $\Gamma \subseteq \mathbf{O}(n)$ be a compact Lie group.*
(a) Assume Γ acts absolutely irreducibly on $\mathbf{R}^n$.
(b) Let $f : \mathbf{R}^n \times \mathbf{R} \to \mathbf{R}^n$ be Γ-equivariant. This implies:

$$f(0, \lambda) \equiv 0$$
$$(\mathrm{d}f)_{0,\lambda} = c(\lambda)I$$

Assume $c(0) = 0$ (this is the condition for a bifurcation to occur).
(c) Assume $c'(0) \neq 0$ (eigenvalue crossing condition).
(d) Assume $\Sigma \subseteq \Gamma$ is an axial subgroup.
Then there exists a unique branch of solutions to $f(x, \lambda) = 0$ emanating from $(0, 0)$ where the symmetry of the solutions is Σ.

The result is useful because axial subgroups often exist and are relatively easy to determine. It provides a general class of subgroups that occur as the symmetry group of a symmetry-breaking branch of steady states.

Proof. By (d) we have $\dim \mathrm{Fix}(\Sigma) = 1$, and let $v \in \mathrm{Fix}(\Sigma)$ be nonzero. Then $\mathrm{Fix}(\Sigma) = \mathbf{R}\{v\}$. So

$$f(tv, \lambda) = h(t, \lambda)v \qquad h : \mathbf{R} \times \mathbf{R} \to \mathbf{R}$$

and

$$f(0, \lambda) = h(0, \lambda) = 0$$

By Taylor's Theorem we have

$$h(t, \lambda) = tk(t, \lambda)$$

where $k(0, 0) = c(0) = 0$ by (b) and $k_\lambda(0, 0) = c' \neq 0$ by (c). By the Implicit Function Theorem there exists a unique $\lambda(t)$ such that $\lambda(0) = 0$ and $k(t, \lambda(t)) \equiv 0$. So $f(tv, \lambda(t)) = 0$.

Finally, the solutions obtained lie in $\mathrm{Fix}(\Sigma)$ and hence have symmetries including Σ, so $\Sigma_v \supseteq \Sigma$. Since Σ is an isotropy subgroup, $\Sigma_v = \Sigma$. $\square$

Applications. We now describe a few straightforward examples of the Equivariant Branching Lemma. We return to the speciation model in §1.5.

Example 1.32 Let $\Gamma = \mathbf{Z}_2$ with action $x \mapsto -x$. For $\Sigma = 1$ we have $\mathrm{Fix}(\Sigma) = \mathbf{R}$ and $\dim \mathrm{Fix}(\Sigma) = 1$. Therefore by the Equivariant Branching Lemma, there exists a branch of solutions with trivial symmetry. Note that

$$f(-x, \lambda) = -f(x, \lambda)$$

implies that

$$f(0, \lambda) = -f(0, \lambda)$$

so $f(0, \lambda) = 0$. By Taylor's Theorem
$$f(x, \lambda) = a(x, \lambda)x$$
and $a(-x, \lambda) = a(x, \lambda)$. Therefore one can write $a(x, \lambda) = b(x^2, \lambda)$ (see [237]). Generically $b_{x^2} \neq 0 \neq b_\lambda$ so after rescaling
$$f(x, \lambda) \sim (x^2 \pm \lambda)x + \cdots$$
It can be proved that the higher order terms do not affect the bifurcation. Thus we arrive at the normal form
$$f(x, \lambda) = x^3 \pm \lambda x$$
of the *pitchfork* bifurcation.   ◇

Example 1.33 Let $\Gamma = \mathbf{O}(2)$ act on $\mathbf{R}^2 \cong \mathbf{C}$. The action on $\mathbf{C}$ is generated by:
$$\theta z \equiv e^{i\theta} z \qquad \theta \in [0, 2\pi)$$
and
$$\kappa z = \bar{z}$$
For $\Sigma = \mathbf{Z}_2(\kappa) = \{1, \kappa\}$ we have $\mathrm{Fix}(\Sigma) = \mathbf{R}$. Thus, at an $\mathbf{O}(2)$ steady-state bifurcation, generically, there exist equilibria with a reflectional symmetry.   ◇

Example 1.34 Let $\Gamma = \mathbf{D}_m$ act on $\mathbf{C} = \mathbf{R}^2$. The action on $\mathbf{C}$ is generated by:
$$\theta z \equiv e^{i\theta} z, \quad \theta = 2\pi/m$$
and
$$\kappa z = \bar{z}$$
For $\Sigma = \mathbf{Z}_2(\kappa)$ we have $\mathrm{Fix}(\Sigma) = \mathbf{R}$, and there exist solutions with reflectional symmetry.   ◇

Next consider $\mathbf{D}_m$ symmetry for $m = 4$ and 5. (Note that $m = 4$ is typical of even m and $m = 5$ is typical of odd m.)

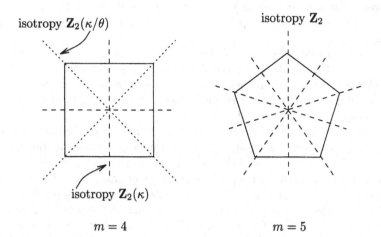

isotropy $\mathbf{Z}_2(\kappa/\theta)$ isotropy $\mathbf{Z}_2$

isotropy $\mathbf{Z}_2(\kappa)$

$m = 4$ $m = 5$

Figure 1.4: Pictures of isotropy subgroups for $\mathbf{D}_m$ acting on $\mathbf{R}^2$.

Group related solutions have conjugate isotropy subgroups. If $f(v, \lambda) = 0$, then $f(\gamma v, \lambda) = \gamma f(v, \lambda) = 0$. Suppose $\sigma v = v$ then $(\gamma \sigma \gamma^{-1})\gamma v = \gamma \sigma v = \gamma v$. So
$$\Sigma_{\gamma v} = \gamma \Sigma_v \gamma^{-1} \tag{1.11}$$

Also if $T = \gamma\Sigma\gamma^{-1}$ then $\text{Fix}(T) = \gamma(\text{Fix}(\Sigma))$. So when $m = 5$ there is a single conjugacy class of isotropy subgroups. However when $m = 4$ there are two conjugacy classes. The bifurcation diagrams are given in Figure 1.5.

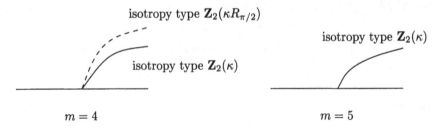

isotropy type $\mathbf{Z}_2(\kappa R_{\pi/2})$

isotropy type $\mathbf{Z}_2(\kappa)$

isotropy type $\mathbf{Z}_2(\kappa)$

$m = 4$ $m = 5$

Figure 1.5: Bifurcation diagrams for $\mathbf{D}_m$ acting on $\mathbf{R}^2$.

To verify this statement note that when $m = 4$ there are two types of lines of symmetry: those connecting vertices and those connecting midpoints of opposite sides. This is the geometric realization of the two conjugacy classes. When $m = 5$ there is only one type of line of symmetry.

1.5 Application to Speciation

We now present an extended example: a system of $\mathbf{S}_N$-equivariant ODEs that models sympatric speciation, introduced by Cohen *et al.* [117]. We begin with some general background to put the model in context.

A classic example of speciation is 'Darwin's finches' in The Galápagos Islands, where over a period of about 5 million years a single species of finch has diversified into 14 species (one is actually in the Cocos Islands). Figure 1.6 shows the evolutionary history (phylogeny) of these finches, as deduced from their phenotypes by standard methods ('cladistics'). Evolutionary changes in Darwin's finches can still be observed today, over periods of just a few years: see Ridley [446].

In the biological literature, speciation is usually discussed primarily in terms of *genotype* — genetics.

Here, however, we consider only the *phenotype* — the organism's form and behavior. We do this on the grounds that the dynamics of evolution is driven by natural selection, which acts on phenotypes: the role of genes is to make it possible for the phenotype to change. For recent support for this view, see Pennisi [431], Rundle *et al.* [457], and Huey *et al.* [290].

Until very recently, most theories of speciation have explained the evolution of new species in terms of geographical or environmental discontinuities or nonuniformities. For example, the mechanism known as *allopatry* involves an initial species being split into two geographically isolated groups — say by one group moving to new territory, later isolated from the original territory by floods or other geographical changes. Once separated, the two groups can evolve independently, for they cannot interbreed. After several millions of years of gradual evolution, the two groups can evolve such different characteristics (in response to the different environments in the two localities) that they become genuinely distinct species. If later geographical changes once more allow the two groups to mix, they will remain separate species because they still do not interbreed — but this time because of the evolved differences, not because of geographical barriers. See Mayr [381, 382] for details.

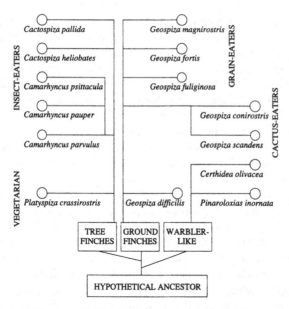

Figure 1.6: Phylogeny of Darwin's finches.

Ultimately, such theories are based on the belief that discontinuous or non-uniform effects must have discontinuous or non-uniform causes. If the organisms of two nascent species are not isolated, they will be able to interbreed, and 'gene flow' will — it used to be thought — maintain them as a single species. Therefore gene-flow must be disrupted in some way, and the obvious possibility is geographical isolation. However, it is now recognized that discontinuous or non-uniform effects can have continuous or uniform causes. Indeed, these are the phenomena addressed in bifurcation theory and symmetry-breaking. Towards the end of the 1990s evolutionary biologists increasingly began to consider mechanisms for *sympatric* speciation. This term refers to situations in which the groups of organisms remain intermingled throughout the process of speciation. In sympatric speciation, gene-flow is disrupted by more subtle mechanisms than geographical isolation: the most significant such mechanism is natural selection, which eliminates 'hybrid' offspring arising from matings between members of the two different speciating groups.

Cohen and Stewart [117] developed a model in which sympatric speciation is explicitly represented as a form of spontaneous symmetry-breaking in a system with all-to-all coupling ($\mathbf{S}_N$ symmetry). Elmhirst [171] studied the stability of primary branches in this model. Cohen *et al.* [494] made numerical studies of relatively concrete models of this general type. Elmhirst [178, 172] took the analysis further, and linked it to a biologically specific model of speciation in a population of birds.

The equations of these models illustrate a number of issues related to symmetry assumptions in the biological sciences (where *identical* organisms or systems are rare, so symmetries are approximate in some sense). We introduce the equations, describe the main results of simulations, and discuss these modeling issues. The story continues in the next chapter, where the stability of bifurcating equilibria is discussed.

Placeholders for Organism Dynamics. Example 1.12 shows that in some sense a single species is invariant under all permutations of its organisms, whereas a mixture of species is invariant only under the smaller group of permutations that preserve each species. The appropriate symmetry group is the symmetric group $\mathbf{S}_N$ of all permutations on N symbols. Since individual organisms can die or breed, their numbers can change, so it is unsatisfactory to model the system with a fixed number of immortal organisms. For this reason, Cohen and Stewart [117] interpret the symbols as representing *PODs* — Placeholders for Organism Dynamics — which are coarse-grained clusters (in phenotypic space) of related organisms. PODs act as carriers for phenotypes. The model assumes that the relevant phenotypes can be described by continuous characters, such as beak length for birds, and may therefore be inappropriate for characters that are determined by a single gene or a small gene-complex.

The model demonstrates that sympatric speciation can occur in a panmictic population (meaning that all organisms can potentially interbreed) and in an environment that is uniform at any instant but can change as time passes. ('Uniform' here does not rule out environmental diversity: all it means is that at any given instant of time each organism experiences the *same* diversity of environment.) A speciation event (bifurcation) is triggered if environmental changes render the uniform state (a single species) unstable, so that the symmetry of the uniform state breaks. Such an instability occurs if the organisms can survive more effectively by adopting different strategies, rather than by all adopting the same strategy (subject to genetic feasibility).

To model such a symmetry-breaking bifurcation we consider a system of N PODs. The state of POD j is described by a vector x_j belonging to *phenotypic space* $\mathbf{R}^r$, where $1 \leq j \leq N$. A point in phenotypic space represents an entire phenotype, not just one character. Each entry x_j^i of the vector $x_j = (x_j^1, \ldots, x_j^r)$ represents a phenotypic character. For example a population of birds might be coarse-grained into $N = 100$ PODs, described by phenotypic vectors $x_1, \ldots, x_{100}$. A given vector, say x_{17}, lists r characters of POD 17 — for instance, weight, wingspan, size of beak, size of foot, and daily food intake (in which case $r = 5$).

Throughout the following discussion, we assume that the x_i are normalized to represent the *deviation* from the initial phenotype, so all $x_i = 0$ prior to bifurcation. For simplicity we focus on just one character, $r = 1$ (in fact, the general case reduces to $r = 1$ by applying Liapunov-Schmidt reduction.

Preliminary Simulations. We set up the general model and describe some promising numerical simulations.

Let $a = (a_1, \ldots, a_s)$ be a vector of s parameters, representing 'environmental' influences (such as climate, food resources, other organisms, and so on). Assume that on the appropriate time scale the changes in phenotype can be described by a dynamical system parametrized by the environmental variables. In general, this will be of the form

$$\frac{\mathrm{d}x_j}{\mathrm{d}t} = f_j(x_1, \ldots, x_N; a_1, \ldots, a_s) \tag{1.12}$$

for suitable functions $f_j : \mathbf{R}^N \times \mathbf{R}^s \to \mathbf{R}^N$, determined by biological considerations. The key observation is that whatever these functions may be, the system should have $\mathbf{S}_N$-symmetry. Intuitively, this just means that the dynamical equations should treat all PODs in the same way. Thus we assume that $F = (f_1, \ldots, f_N)$ is $\mathbf{S}_N$-equivariant.

The $\mathbf{S}_N$-equivariant mappings will be classified in Section 2.6, and we briefly summarize the relevant information here. (It is in any case easy to check that all polynomials listed below are equivariant, which is all we need for the moment.)

For $k = 1, \ldots, N$ let

$$\pi_k = \sum_{i=1}^{N} x_i^k$$

and for $k = 1, \ldots, N - 1$ let

$$E_k = (x_1^k, x_2^k, \ldots, x_N^k)^T$$

Then the π_k generate the $\mathbf{S}_N$-invariants and the E_k generate the $\mathbf{S}_N$-equivariants. Thus there are:

- One constant equivariant E_0.
- Two linear equivariants $\pi_1 E_0, E_1$.
- Four quadratic equivariants $\pi_1^2 E_0, \pi_2 E_0, \pi_1 E_1, E_2$.
- Six cubic equivariants $\pi_1^3 E_0, \pi_1 \pi_2 E_0, \pi_3 E_0, \pi_1^2 E_1, \pi_1 E_2, E_3$.

There are further equivariants of all higher degrees. The general cubic-degree truncation of F has 13 parameters:

$$
\begin{aligned}
F \quad = \quad & a_0 E_0 + \\
& b_1 \pi_1 E_0 + b_2 E_1 + \\
& c_1 \pi_1^2 E_0 + c_2 \pi_2 E_0 + c_3 \pi_1 E_1 + c_4 E_2 + \\
& d_1 \pi_1^3 E_0 + d_2 \pi_1 \pi_2 E_0 + d_3 \pi_3 E_0 + d_4 \pi_1^2 E_1 + d_5 \pi_1 E_2 + d_6 E_3
\end{aligned}
\qquad (1.13)
$$

Numerical integrations show that simple choices of the parameters produce stable bifurcation to the expected primary branches. Indeed, Figure 1.2 is a case in point. Figure 1.7 illustrates two other simulations for $N = 25$, with different parameter values (which we will not specify at this stage). In the left-hand figure the bifurcation does not break symmetry: all PODs bifurcate to a new $\mathbf{S}_{25}$-invariant state. In the right-hand figure the symmetry breaks to $\mathbf{S}_9 \times \mathbf{S}_{16}$. Notice the rapidity of the initial change in the symmetry-breaking case.

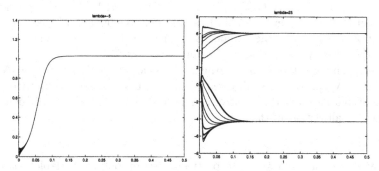

Figure 1.7: Bifurcation to two species in model with $N = 25$ PODs. Time series of all PODs are superimposed. (Left) one species; (right) two species (one with 9 PODs and one with 16 PODs).

The results of many different simulations of this kind reveal several noteworthy general features.

- The initial bifurcation is a jump transition (which in evolutionary parlance is called 'punctuated equilibrium'). This feature is visible in the rapid broadening of the relevant bifurcation diagrams.

- Even though the phenotypes cluster in a jump transition, the mean value of all phenotypes remains approximately constant. Thus the speciating clumps 'pull apart' in *opposite* directions in phenotypic space.
- An eventual divergence into *two* species seems universal in this model, though for some parameter values there can be a transitory separation into three or more discernible clumps at intermediate λ-values.
- The numbers (p, q) of PODs in each eventual clump depends on the parameters, and to a lesser extent on initial conditions. There is no global restriction on the sizes of the clumps: p can be anything between 1 and $[N/2]$.
- There can be multistability: with fixed parameter values, final clumps of different sizes can occur for different initial conditions.

In this chapter and the next we investigate these features analytically using the techniques of equivariant bifurcation theory, and explain why they occur and in what circumstances.

Symmetry-Based Analysis. Even without using the detailed form of an $\mathbf{S}_N$-equivariant mapping, we can make progress by exploiting the generalities of symmetry.

Bifurcation of (1.13) occurs when the linearization $dF|_0$ becomes singular. Generically the kernel of the linearization is one of the two irreducible components

$$\begin{aligned} V_0 &= \mathbf{R}(1, 1, \ldots, 1) \\ V_1 &= \{(x_1, \ldots, x_N) : x_1 + \cdots + x_N = 0\} \end{aligned}$$

of $\mathbf{R}^N$. Both of these are absolutely irreducible. If the kernel is V_0, then symmetry does not break: we just get a new branch of $\mathbf{S}_N$-invariant equilibria and the population remains a single species, though the species as a whole experiences a qualitative change in phenotype. The case where the kernel is V_1 leads to symmetry-breaking, hence bifurcation to more than one species. We therefore restrict our analysis to this case.

We can find symmetry-breaking equilibria supported by V_1 by applying the Equivariant Branching Lemma. As in Example 1.12, the number of species present in a solution x is given by the number of distinct entries in x, so we can find the number of species by computing the isotropy subgroup Σ_x.

Up to conjugacy, the axial subgroups of $\mathbf{S}_N$ are $\mathbf{S}_p \times \mathbf{S}_q$ where $q = N - p$ and $1 \le p \le [N/2]$. By the Equivariant Branching Lemma, generically there exist branches of solutions with these isotropy subgroups. The fixed-point space of $\mathbf{S}_p \times \mathbf{S}_q$ is spanned by all vectors of the form

$$(u, \ldots, u; v, \ldots, v) \tag{1.14}$$

where there are p u's and q v's. In particular the phenotypic variables for these solutions take exactly *two* distinct values u and v. The axial (hence primary) solution branches therefore correspond to a split of the population of N identical PODs into two distinct species consisting of p and q PODs respectively. One species has the phenotype u and the other species has the phenotype v. Since these are the primary branches, the model predicts that *dimorphism* — bifurcation to a two-species state — will be common.

Analytic equations for these branches can be derived by substituting this form for x in (1.13): we postpone this analysis to Chapter 2 where we carry it out for the Liapunov-Schmidt reduced problem.

We can also make an interesting universal *quantitative* prediction: on the above branches the mean value of the phenotypic variables changes smoothly during the bifurcation. We noted this behavior in Example 1.1; we now show that it is a general consequence of the form of V_1 and simple properties of Liapunov-Schmidt reduction. Solutions of the form (1.14) that lie in V_1 must have entries that sum to zero, so

$$pu + qv = 0 \qquad (1.15)$$

Therefore, the mean value of the phenotypic variable after bifurcation is $pu + qv)/N = 0$. Before bifurcation it is also 0, because we are looking for bifurcations from the trivial equilibrium and have normalized the phenotypic variables to be zero for the original single species. So here the mean remains constant throughout the bifurcation. However, we are working with the Liapunov-Schmidt reduced problem, which involves a nonlinear change of variables. Therefore, the mean varies *smoothly* in the original phenotypic variables, and is thus approximately constant, as illustrated schematically in Figure 1.8.

Note that we speak here of two means: the mean over all N PODs (which is the average of coordinates in $\mathbf{R}^N$) and the weighted mean of the phenotypes (which is the weighted average (1.15) over the coordinates u, v in the fixed-point subspace). The two means are equal but their interpretation is slightly different.

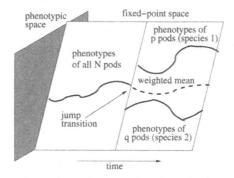

Figure 1.8: Generic $\mathbf{S}_N$-equivariant steady-state bifurcation: jump transition from one species to two, with smoothly varying mean.

1.6 Observational Evidence

Is there any evidence in favor of the symmetry-breaking model of sympatric speciation? Some studies reported in the existing literature are consistent with its predictions, especially the 'constant mean' prediction. These include various biologically motivated models, such as Higashi *et al.* [272], Kondrashov and Kondrashov [320], Dieckmann and Doebeli [158], and Tregenza and Butlin [504]. The predictions of the model are also consistent with some field observations, for instance Beauchamp and Ulyett [57], Huey *et al.* [291, 290], Huey and Pianka [292], Bantock and Bayley [38], Bantock *et al.* [39], Fenchel [180, 181] (but see Barnes [48, 49, 50] for caveats), and Rundle *et al.* [457].

However, in general it is very difficult, if not impossible, to observe the speciation process in action — except perhaps 'in miniature' in such organisms as Darwin's finches, African lake cichlids, sticklebacks, and fruit flies. What we can observe fairly easily is something that appears to be the end result of a speciation transition, with two closely related species or subspecies coexisting in a given environment (which may be a 'hybrid zone' where the two species's normal habitats are adjacent). This is the 'allopatric' context. Having observed the phenotypes that occur in the allopatric context, we can compare them with the corresponding phenotypes for the two species when only one of them exists (ideally in the same environment as for the allopatric situation). This is the 'sympatric' context.

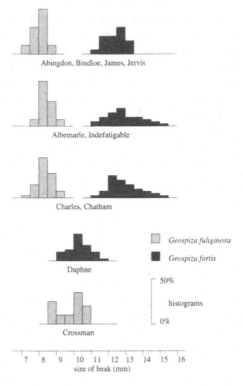

Figure 1.9: Beak sizes in allopatric and sympatric populations of *Geospiza* in the Galápagos Islands. Redrawn from Lack [336].

We describe one case from the literature: Darwin's finches. The prediction of smoothly changing mean is consistent with observations of these finches. As just explained, we observe a surrogate for actual evolution: differences in the phenotype of a given species in allopatric and sympatric populations. We expect to see approximately the same mean in either situation. This is the case for the two species *Geospiza fortis* and *G. fuliginosa*, which occur in both sympatric and allopatric populations. *G. fortis* is allopatric on the island known as Daphne, and *G. fuliginosa* is allopatric on Crossman. The two species are sympatric on a number of islands which Lack placed in three groups for data analysis: Abingdon, Bindloe, James, Jervis; Albemarle, Indefatigable; and Charles, Chatham. Fig. 1.9, adapted from Lack [336], shows the differences in beak size between these species on the cited groups of islands.

The mean beak sizes of both *G. fortis* and *G. fuliginosa* are approximately 10mm in allopatric populations. In all three (groups of) sympatric populations, the mean for *G. fortis* is about 12mm, while that for *G. fuliginosa* is about 8mm. These figures are consistent with the 'constant mean' prediction.

1.7 Modeling Issues: Imperfect Symmetry

A common criticism of symmetry methods, especially in biology, is that real systems are seldom perfectly symmetric. Part of the answer is that the behavior of a dynamical system that is 'nearly symmetric' is much more like that of an idealized symmetric system than it is like the typical behavior of a completely asymmetric one. To illustrate this point we consider two different ways to make the symmetry of the speciation model imperfect, and argue that they lead to much the same conclusions as the ideal symmetric model — even when both sources of imperfection are present together. We also sketch theoretical reasons for this kind of robustness.

We begin with a general point about modeling. The predictive and explanatory power of a mathematical model depends only weakly on the validity of its *assumptions*. What matters are its *conclusions*, and it is those that must compare favorably with reality. For example, many astronomical studies model a star as a point mass, which in literal terms is nonsense — yet for most purposes of celestial mechanics this model works extremely well. It is, however, useless for astrophysics. That said, a good way to investigate such issues is to make the assumptions of the model more realistic, and see to what extent the conclusions change. If they survive, then this strengthens confidence in the original model. If the original model is simpler to understand than the more realistic one, or simpler to analyze or compute with, then the use of the simplified model is still justified.

We consider two ways — one deterministic, one stochastic — to introduce variability among organisms in the same species:

- Change the equations so that they are no longer perfectly symmetric in the N variables.
- Add stochastic terms to introduce a random element.

We will argue that the main phenomena associated with (1.13) survive such modifications, so the quest for greater biological realism does not alter the main conclusions derived from the less realistic, but far more tractable, idealized equations.

We can break the $\mathbf{S}_N$ symmetry of (1.13) by making the coefficients vary slightly with the index i, and by replacing terms like $(x_1 + \cdots + x_N)$ and $(x_1^2 + \cdots + x_N^2)$ by $(r_1 x_1 + \cdots + r_N x_N)$ and $(s_1 x_1^2 + \cdots + s_N x_N^2)$ where the r_j and s_j differ slightly from 1. Here 'slightly' is governed by a new parameter, which is typically 0.1 or thereabouts, indicating a 10% variation of the parameter values. In simulations, these variations are defined at the start of each run using a random number generator. Figure 1.10 shows typical bifurcation diagrams for this case. The main change is that after bifurcation the traces do not converge as strongly, but they still clump. All three universal predictions of the model remain valid.

An alternative way to introduce variability within species is to convert the model into a stochastic one by adding random noise. In simulations we discretize and iterate:

$$x_i(t + \varepsilon) = x_i(t) + \varepsilon F(x) \sigma r_i(t)$$

Here ε is small, σ determines the size of the noise, and r_i is a random variable distributed uniformly between -0.5 and 0.5.

Figure 1.11 shows the effect of noise on a typical bifurcation diagram. The trajectories become irregular, and certain details change — here we see a secondary bifurcation in which one of the PODs suddenly changes from one species to the other — but the split into two species remains robust. The prediction of approximately constant mean remains valid, and so does that of jump bifurcation. For further discussion of secondary bifurcations, see Chapter 2.

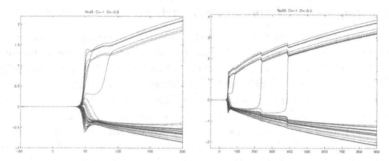

Figure 1.10: Effect of approximate symmetry on the speciation model. Throughout $N = 25, C \sim -1, D \sim -0.2$ with 10% variation in C and D. (Left) λ in the range $[-50, 200]$; (right) λ in the range $[-50, 900]$.

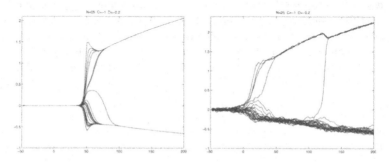

Figure 1.11: Effect of random perturbations on the speciation model. (Left) no noise; (right) noise included. The noise has triggered a secondary bifurcation but the split into two clumps still persists.

Small Noise and Small Symmetry Breaking. We briefly discuss why a small amount of noise, or small symmetry breaking, should not affect the conclusions of the idealized model. We give references to the literature where the ideas introduced here are more fully developed.

First, there are general theoretical reasons for expecting the effects of small amounts of noise on the model to be small. For the addition of random terms, the theory of stochastic nonlinear differential equations

$$dx = F(x, a)dt + \sigma dw \tag{1.16}$$

where dw indicates white noise can be invoked. The statistical behavior is governed by the *Fokker-Planck equation*

$$\frac{\partial u}{\partial t} = \sigma \Delta u - \nabla \cdot (uF)$$

Here the function $u = u(t)$ represents a time-varying smooth probability density, Δ is the Laplacian, the term $\sigma \Delta u$ represents random diffusion, and $\Delta \cdot (uF)$ represents the deterministic flow ($\sigma = 0$). Intuitively, solutions of (1.16) are like solutions of the deterministic equation subjected to random kicks at all instants of time, so for low noise ($\sigma << 1$) they look like slightly irregular versions of deterministic solutions. Formalizing these ideas requires substantial technical effort: see Itô [300], McKean [383], Gihman and Skorohod [207]. It is known that as t tends to infinity, any solution of the Fokker-Planck equation converges to a unique steady state U.

This implies that if F is Γ-equivariant for some group Γ and the noise is Γ-invariant then U must be Γ-invariant. In particular if for specific parameter values a the attractors of the deterministic system are hyperbolic equilibria and $\sigma \ll 1$, then U has peaks near all stable equilibria and is small everywhere else, see Kifer [308], Zeeman [533]. This result reflects the very long term behavior: because Brownian motion is long-range, the stochastic system can climb out of any potential well given enough time. However, that time scale is very long indeed, and what we see in simulations is more limited. Numerical evidence supports the conjecture that all attractors of (1.13) that bifurcate from the origin are equilibria, and these are generically hyperbolic: we assume this conjecture is true for the following discussion. In simulations, where the noise is taken to be short range (arguably more realistic for evolutionary biology), almost all trajectories eventually become trapped inside the basin of attraction of some equilibrium once the attraction towards that equilibrium exceeds the discretized noise level. This explains why the bifurcation behavior of the stochastic system resembles that of the deterministic one when $\sigma \ll 1$.

Second, we can also explain why the bifurcation behavior of the deterministic modification resembles that of the ideal symmetric one when the amount g of imposed symmetry-breaking is small. The theoretical reason is normal hyperbolicity, which is a generic property in this context, Arrowsmith and Place [11]. This property guarantees that equilibria move continuously if the system is perturbed (including symmetry-breaking perturbations).

1.8 Generalization to Partial Differential Equations

In the final section of this chapter we discuss a far-reaching generalization of the theory that has been developed. The generalities of equivariant bifurcation theory can also be extended to PDEs. The main technical difference is that solutions no longer live in a finite-dimensional vector space $\mathbf{R}^n$, but in some function space or Banach space. For certain classes of PDEs ('elliptic') these technical considerations cause few complications, but for other kinds of PDE the functional-analytic difficulties create serious new problems and lead to new bifurcation phenomena. Later we will study pattern-formation in certain PDEs, mostly reaction-diffusion equations, where the theory developed so far is quite effective, so we set up some of the basic ideas now. This section can be omitted or postponed.

As motivating example we consider a reaction-diffusion equation posed on a circular domain $\Omega = \mathbf{R}/2\pi\mathbf{Z}$:

$$\frac{\partial u}{\partial t} = \Delta u + g(u, \lambda)$$

Here Δ is the Laplacian, and $u = u(\theta, t)$ is a sufficiently smooth function $u : \Omega \to \mathbf{R}$, which we can lift to a 2π-periodic function $\mathbf{R} \to \mathbf{R}$, which we call $\hat{u}$. Consider only steady states, in which u is independent of t, so that

$$\Delta u + g(u, \lambda) = 0 \qquad\qquad (1.17)$$

We can write (1.17) in operator form:

$$\mathcal{P}(u, \lambda) = 0 \qquad\qquad (1.18)$$

where $\mathcal{P} = \Delta + g$. The technical question is: to what space should u belong? Let $\mathcal{C}^k_{2\pi}$ denote the space of k-times continuously differentiable functions $w : \Omega \to \mathbf{R}$, or equivalently k-times continuously differentiable 2π-periodic functions $\hat{w} : \mathbf{R} \to \mathbf{R}$. Equip $\mathcal{C}_{2\pi}^k$ with the C^k norm

$$||u||_k = \sup_{\theta \in \Omega} |u(\theta)| + \sup_{\theta \in \Omega} |u'(\theta)| + \ldots + \sup_{\theta \in \Omega} |u^{(k)}(\theta)| + \ldots$$

Then $\mathcal{C}_{2\pi}^k$ is a Banach space for any finite k. We can consider $\mathcal{P}$ as a parametrized nonlinear operator

$$\begin{aligned} \mathcal{P} : \mathcal{C}_{2\pi}^2 &\to \mathcal{C}_{2\pi}^0 \\ (u, \lambda) &\mapsto \Delta u + g(u, \lambda) \end{aligned}$$

Then the solutions of (1.17) are the zeros of $\mathcal{P}$.

Already we see a technical difference from ODEs: the domain and range of $\mathcal{P}$ are different Banach spaces. Unfortunately the space of C^∞ functions on Ω is not a Banach space for any sensible norm, so we cannot work in $\mathcal{C}_{2\pi}^\infty$. Thus the use of two different spaces is essential here (and typical of the PDE theory).

Symmetry enters the picture because the operator $\mathcal{P}$ is equivariant under an action of the symmetry group of Ω, which is $\mathbf{O}(2)$. Specifically, let $\gamma \in \mathbf{O}(2)$ and define

$$(\gamma u)(\theta) = u(\gamma^{-1}\theta) \tag{1.19}$$

Here

$$\begin{aligned} \gamma^{-1}\theta &= \theta - \gamma \ \text{ if } \gamma \in \mathbf{SO}(2) \\ \kappa\theta &= -\theta \end{aligned}$$

Now $\mathbf{O}(2)$ acts on $\mathcal{C}_{2\pi}^2$ and $\mathcal{C}_{2\pi}^0$, and

$$\mathcal{P}(\gamma u, \lambda) = \Delta(\gamma u) + g(\gamma u, \lambda)$$

so that if $\gamma \in \mathbf{SO}(2)$ we compute

$$\begin{aligned} [\mathcal{P}(\gamma u, \lambda)](\theta) &= [\Delta(\gamma u)](\theta) + g(\gamma u(\theta), \lambda) \\ &= \frac{\partial^2}{\partial\theta^2} u(\theta - \gamma) + g(u(\theta - \gamma), \lambda) \\ &= \frac{\partial^2}{\partial(\theta - \gamma)^2} u(\theta - \gamma) + g(u(\theta - \gamma), \lambda) \\ &= [\gamma \mathcal{P}(u, \lambda)](\theta) \end{aligned}$$

A similar calculation works for κ since

$$\frac{\partial^2}{\partial(-\theta)^2} = \frac{\partial^2}{\partial(\theta)^2}$$

Hence for all $\gamma \in \mathbf{O}(2)$

$$\mathcal{P}(\gamma u, \lambda) = \gamma \mathcal{P}(u, \lambda)$$

and $\mathcal{P}$ is $\mathbf{O}(2)$-equivariant.

The key point here, of course, is that Δ is *invariant* under $\mathbf{O}(2)$. Observe how this translates into equivariance of $\mathcal{P}$.

The concepts of isotropy subgroup and fixed-point space carry over immediately to a Banach space setting, and have the same basic properties. The representation theory of *compact* Lie groups on Banach spaces is closely analogous to the finite-dimensional theory (but with extra topological considerations). An especially useful feature of the infinite-dimensional setting is that the Implicit Function Theorem remains valid for Banach spaces. Therefore we can apply Liapunov-Schmidt reduction, just as before.

A necessary condition for steady-state bifurcation at (x_0, λ_0) is that the derivative (often known as the Fréchet derivative) $\mathrm{d}_x \mathcal{P}|_{(x_0,\lambda_0)}$ should be nonsingular as a linear operator. If so, then $\mathcal{K} = \ker \mathrm{d}\mathcal{P} \neq 0$. Let $\mathcal{R} = \text{range } \mathrm{d}\mathcal{P}$ and as usual consider two splittings

$$\begin{aligned} \mathcal{C}^2_{2\pi} &= \mathcal{K} \oplus \hat{\mathcal{K}} \\ \mathcal{C}^0_{2\pi} &= \mathcal{R} \oplus \hat{\mathcal{R}} \end{aligned} \tag{1.20}$$

Then we obtain a Liapunov-Schmidt reduced bifurcation equation

$$p : \mathcal{K} \times \mathbf{R} \to \hat{\mathcal{R}}$$

In the case of ODEs, the spaces $\mathcal{K}$ and $\hat{\mathcal{R}}$ are finite-dimensional and have the same dimension. In infinite dimensions these properties need not hold. However, they do hold in one important case:

Definition 1.35 Let $\mathcal{B}, \mathcal{B}'$ be Banach spaces. A linear operator $Q : \mathcal{B} \to \mathcal{B}'$ is *Fredholm of index zero* if

1) $\ker Q$ is finite-dimensional.
2) range Q is closed.
3) $\dim \ker Q = \text{codim range } Q$.

$\diamond$

When $\mathrm{d}\mathcal{P}$ is Fredholm (as it is in the reaction-diffusion example under consideration) then $\mathcal{K}$ and $\hat{\mathcal{R}}$ are finite-dimensional and have the same dimension. Indeed we can identify $\mathcal{K}$ and $\hat{\mathcal{R}}$. Moreover, if $\mathcal{P}$ is Γ-equivariant for a Lie group Γ, not necessarily compact, then we can choose splittings (1.20) to be Γ-invariant, in which case the reduced mapping p is Γ-equivariant on some $\mathbf{R}^n$. Thus in the Fredholm context steady-state equivariant bifurcation of PDEs *reduces* to steady-state equivariant bifurcation of ODEs on $\mathbf{R}^n$.

Other results from the finite-dimensional theory can now be generalized, in particular Theorem 1.27 (in generic steady-state bifurcation the kernel is absolutely irreducible) and the Equivariant Branching Lemma (Lemma 1.31). We pursue these ideas further in Section 2.8.

Chapter 2

Linear Stability

Having proved the Equivariant Branching Lemma, we consider how to compute more detailed information at bifurcations. This information includes the linear stability of equilibria and the existence and nonexistence of other equilibria.

Stability is important in most areas of application, because unstable states are usually not physically realizable. This does not mean that unstable states are physically irrelevant: for instance in Chapter 8 we show that unstable equilibria may sometimes be linked together to form heteroclinic cycles, and that these cycles may be stable in a quasi-global sense. But if we are looking for physically realizable equilibria of a system, it is the stable ones that we expect to observe in experiments.

The previous chapter centered on a general existence theorem for symmetry-breaking equilibria, the Equivariant Branching Lemma. We motivated this result in terms of a model of speciation. We now use the same model to motivate the next step in analyzing such problems, which is to determine the stability of equilibria in an equivariant ODE, with particular emphasis on 'primary' branches of equilibria derived using the Equivariant Branching Lemma. Two main issues arise:

- Symmetry can be exploited systematically to simplify computations of stabilities.
- Sometimes, symmetry constraints force *all* bifurcating branches found using the Equivariant Branching Lemma to be unstable near the bifurcation point.

The first issue concerns the analysis of a given model. The second issue is more subtle: it affects the choice of model. Models that lack stable states are inadequate, and must be modified to incorporate additional features. The standard procedure in such cases is to consider the absence of stable branches as a 'degeneracy condition', and to 'unfold' the model by including new higher-order terms.

We show in Example 2.3 that all primary branches of equilibria derived for S_3 symmetry in Example 1.1 are unstable near bifurcation. The underlying reason for this is traced to the existence of a non-trivial quadratic equivariant, see Theorem 2.14. Example 2.15 applies this general result to any S_N-equivariant bifurcation problem on the standard irreducible representation $\mathbf{R}^{N-1}$, showing that again all primary branches of equilibria are unstable near bifurcation. In Section 2.7 we confirm this abstract result by direct computation, and obtain an improved model by passing to the general *cubic* order S_N-equivariant ODE. Here the initially unstable branches given by the Equivariant Branching Lemma become stabilized by the higher-order terms, through two mechanisms. One is that a branch may 'turn round' at a fold point (saddle-node bifurcation), a phenomenon that we illustrate in Example 2.2. The second is that the presence of secondary branches can lead to exchanges of stability, which also — in appropriate circumstances — can stabilize primary branches. See Example 2.16.

The key mathematical problem here is to understand the constraints that symmetry imposes on the Jacobian df at an equilibrium with given isotropy, and in particular the implications of those constraints for the eigenvectors and eigenvalues of the Jacobian. This is a natural problem to investigate, because the stability of an equilibrium is determined by the eigenvalues of the Jacobian at that equilibrium. If all eigenvalues have negative real part, then the equilibrium is stable; if some eigenvalue has positive real part then it is unstable. (If an eigenvalue lies on the imaginary axis, then the situation is more complicated. We generally take this as a sign of bifurcation and ignore stability issues at such points.)

If f has symmetry, then the computation of these eigenvalues can often be simplified by working out how symmetry constrains the form of the Jacobian. Roughly speaking, the symmetry provides good coordinates that make the computation simpler. These constraints show up in three ways:

- The Jacobian has natural invariant subspaces given by the isotypic decomposition for the isotropy subgroup of the equilibrium.
- The Jacobian has zero eigenvalues forced by symmetry.
- In certain circumstances, all solutions obtained from the Equivariant Branching Lemma are generically unstable.

Examples of 'Bare Hands' Stability Calculations. In simple cases, we can compute stabilities directly:

Example 2.1 Consider the pitchfork bifurcation, $\Gamma = \mathbf{Z}_2$, for which $n = 1$. The non-identity element of Γ maps $x \in \mathbf{R}$ to $-x$. Equivariance implies that

$$f(-x, \lambda) = -f(x, \lambda)$$

so that

$$f(x, \lambda) = a(x^2, \lambda)x$$

as in Example 1.32. Assume that a bifurcation occurs at the origin (that is, $a(0) = 0$), the eigenvalues cross the imaginary axis with nonzero speed, and the trivial solution is stable for $\lambda < 0$ (that is, $a_\lambda(0) > 0$). Then the system can be written in the form

$$\frac{\mathrm{d}x}{\mathrm{d}t} = (a_z x^2 + a_\lambda \lambda)x + \text{h.o.t.}$$

where h.o.t. denotes higher order terms and $z = x^2$. Ignoring these and solving $\mathrm{d}x/\mathrm{d}t = 0$ we obtain the bifurcation equation

$$\lambda = -\frac{a_z}{a_\lambda}x^2 + \cdots$$

If we retain the higher order terms we get the same equation to quadratic order, and this truncation determines the topology of the bifurcation diagram.

Figure 2.1: Bifurcation diagrams of pitchfork bifurcations. (Left) supercritical. (Right) subcritical. Solid line shows stable states, dotted line shows unstable states.

The two possible bifurcation diagrams are shown in Figure 2.1. If $a_z < 0$ the bifurcation is *supercritical* and the nontrivial solution is stable when $\lambda > 0$; if $a_z > 0$ the bifurcation is *subcritical* and the nontrivial solution is unstable when $\lambda < 0$. This demonstrates *exchange of stability* at the pitchfork. $\diamond$

In the next example, we omit the calculations of stabilities, which are routine, and summarize the results. This example shows that when an unstable branch 'turns around' at a saddle-node bifurcation, an eigenvalue changes sign. In some circumstances, including this example, the result is to stabilize the branch.

Example 2.2 Consider the ODE

$$\frac{\mathrm{d}x}{\mathrm{d}t} = \lambda x + 2x^2 - x^3$$

The dynamics of this system is illustrated in Fig. 2.2. The solid and dotted lines show how the equilibria $\mathrm{d}x/\mathrm{d}t = 0$ vary with λ: a solid line indicates a stable equilibrium, a dotted line an unstable one. The arrows show the direction in which x changes when the system is not in equilibrium. Bifurcations occur at $\lambda = 1, 0$.

Point A, where $\lambda = 0$, is a point of *transcritical* bifurcation. For $\lambda < 0$ the equilibrium $x = 0$ is stable, but it becomes unstable for $\lambda > 0$. Meanwhile a previously unstable branch — a segment of the parabola $\lambda = x^2 - 2x$ between points B and A — becomes stable for $\lambda > 0$. There is an 'exchange of stability' between the two branches.

Point B, where $\lambda = -1$, is a *fold* or *saddle-node* point, see Arrowsmith and Place [11]. For $\lambda < -1$ there are no equilibria near B (only the distant one at $x = 0$). As λ increases through -1, two new branches — one stable, the other unstable — separate from B.

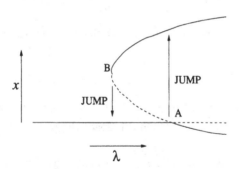

Figure 2.2: Bifurcation diagram illustrating jumps and hysteresis.

Associated with this combination of bifurcations are two effects: jump bifurcation and hysteresis. We briefly describe these effects, which can be important in applications.

Suppose the system starts out with $\lambda < -1$, and λ slowly increases. We assume *quasi-static* variation, that is, x settles down to the closest available equilibrium much faster than λ is changing. It therefore remains at (strictly: very close to) $x = 0$ when λ becomes greater than -1, even though a new stable equilibrium has appeared. However, as λ passes through 0, the value of x suddenly changes to the only available stable equilibrium, on the top branch that emanates from B. This is a *jump bifurcation*, and we see that when λ makes a very small change though -1, the value of x changes substantially. Indeed, in an idealization where the timescale for λ is infinitely slow compared to that for x, we see that a continuous change in λ causes a discontinuous change in x.

Suppose that after the jump has occurred, we decrease λ back below 0. Because the top arc of the parabola is stable for $\lambda > -1$, the system does not reverse its previous change and jump back to $x = 0$ when λ passes through 0. Instead, it remains on the top branch until λ gets below -1. Only then does it jump back to $x = 0$. We therefore observe 'irreversible' changes in behavior as the parameter varies — the phenomenon of *hysteresis.*   ◇

Example 2.3 We return to the system (1.1) of S_3-equivariant ODEs

$$\begin{aligned}
dx/dt &= \lambda x - (x + y + z) + x^2 \\
dy/dt &= \lambda y - (x + y + z) + y^2 \\
dz/dt &= \lambda z - (x + y + z) + z^2
\end{aligned}$$

with a nontrivial solution branch

$$x = y = \frac{-\lambda + 1 - \sqrt{\lambda^2 - 6\lambda + 1}}{2} \qquad z = \frac{-\lambda - 1 + \sqrt{\lambda^2 - 6\lambda + 1}}{2}$$

together with its two conjugates. We now compute the stabilities of solutions along this branch.

The Jacobian is

$$J = \begin{bmatrix} \lambda - 1 + 2x & -1 & -1 \\ -1 & \lambda - 1 + 2y & -1 \\ -1 & -1 & \lambda - 1 + 2z \end{bmatrix}$$

One eigenvector is $(1, -1, 0)^T$ with eigenvalue

$$\varepsilon_1 = \lambda + 2x$$

(we here use the fact that $x = y$). We compute the other two eigenvalues by defining

$$u = (1, 1, 0)^T \qquad v = (0, 0, 1)^T$$

We have

$$Ju = (\lambda - 2 + 2x)u - 2v \qquad Jv = -u + (\lambda - 1 + 2z)v$$

Thus $\mathbf{R}\{u, v\}$ is invariant under J. On this space the matrix of J is

$$K = \begin{bmatrix} \lambda - 2 + 2x & -2 \\ -1 & -\lambda - 1 - 2x \end{bmatrix}$$

where we have replaced z by $-x - \lambda$. The trace of K is

$$T = -3$$

and the determinant is

$$D = -(\lambda + 2x - 2)(\lambda + 2x + 1) - 2$$

Therefore the eigenvalues of K are

$$\varepsilon_2 = \frac{-3 + \sqrt{9 - 4D}}{2} \qquad \varepsilon_3 = \frac{-3 - \sqrt{9 - 4D}}{2}$$

When λ is small, $\varepsilon_2 \sim 0$ and $\varepsilon_3 \sim -3$. Computing to first order in λ, we find

$$\varepsilon_1 = 3\lambda + O(\lambda^2) \qquad \varepsilon_2 = -\lambda + O(\lambda^2) \qquad \varepsilon_3 = -3 + \lambda + O(\lambda^2)$$

Near the origin, ε_1 and ε_2 have opposite signs, so the solution branch is unstable near the origin. By symmetry, the same applies to the two conjugate branches. So locally there are no nontrivial stable branches other than the origin, and the origin is stable only when $\lambda < 0$.   ◇

This example shows that although the equations of Example 1.1 have branches of equilibria corresponding to two species, those branches are unstable. As it stands, this is a highly unsatisfactory conclusion: we want the speciation process to lead to stable branches. To make matters worse, it turns out that the same difficulty arises for S_N-symmetric models: generically, all primary branches of equilibria are unstable near the origin.

2.1 Symmetry of the Jacobian

In order to explain how we can prove this, and to show that this apparent obstacle to defining an acceptable model for speciation is actually a blessing in disguise, we need to develop better theoretical techniques. When the group action is more complicated, 'bare hands' computation of eigenvalues is less successful. Computer algebra systems may be able to perform the calculations unintelligently, but give little insight into the results. Explicit use of symmetry constraints on the Jacobian is more satisfactory.

Suppose that we have a Γ-equivariant system $f(x, \lambda)$ with equilibrium (x_0, λ_0) such that $\Sigma_{x_0} \subseteq \Gamma$. To analyze stability we need to compute the eigenvalues of $(df)_{x_0, \lambda_0}$ at a solution (x_0, λ_0) of the equation $f(x, \lambda) = 0$.

Proposition 2.4 *With the above notation, $(df)_{x_0, \lambda_0}$ commutes with Σ_{x_0}.*

Proof. Suppose $\sigma \in \Sigma_{x_0}$ is a symmetry of x_0. Differentiating

$$f(\sigma x, \lambda) = \sigma f(x, \lambda) \qquad \text{yields} \qquad (df)_{\sigma x_0, \lambda_0} \sigma = \sigma (df)_{x_0, \lambda_0}$$

Since $\sigma x_0 = x_0$ it follows that $(df)_{x_0, \lambda_0}$ commutes with Σ_{x_0}. □

2.2 Isotypic Components

In order to deduce the restrictions placed on $(df)_{x_0, \lambda_0}$ by this commutativity, we must set up some basic ideas from representation theory.

Definition 2.5 Suppose that Γ acts on V and also on W. We say that V is Γ-*isomorphic* to W, denoted $V \cong W$, if there exist a linear isomorphism $A : V \to W$ such that $A(\gamma x) = \gamma A(x)$ for all $x \in V$. ◇

By abuse of notation we can write this condition as $A\gamma = \gamma A$ and say that A 'commutes' with γ. Note, however, that the γ here refers to two different maps: one on V, one on W.

Let Γ act on $\mathbf{R}^n$. We can decompose $\mathbf{R}^n$ into a direct sum of Γ-irreducible subspaces, but in general this decomposition is not unique. However, if we use components that combine together all of the Γ-irreducible subspaces that lie in a fixed isomorphism class, we obtain a decomposition that *is* unique. It also has pleasant invariance properties: see Lemma 2.10.

Definition 2.6 Choose a Γ-irreducible representation $V \subseteq \mathbf{R}^n$. Let $\hat{V}$ be the sum of all Γ-irreducible subspaces of $\mathbf{R}^n$ that are isomorphic to V. Then $\hat{V}$ is the *isotypic component* of $\mathbf{R}^n$ corresponding to V. ◇

By definition, isotypic components are unique. They have a (usually non-unique) decomposition as a direct sum of isomorphic irreducibles:

Lemma 2.7 *Let V be an irreducible subspace. Then there exist irreducible subspaces $V_1, \ldots, V_t$, all isomorphic to V, such that $\hat{V} = V_1 \oplus \cdots \oplus V_t$.*

Proof. Let $V_1 = V \subseteq \hat{V}$. If every $W \cong V$ is contained in V_1, we are done. Otherwise choose $V_2 \cong V$ so that $V_2 \not\subset V_1$. By irreducibility of V_2 we have $V_2 \cap V_1 = 0$ so the sum $V_1 \oplus V_2$ is direct. Continue inductively, so that at stage p we have a direct sum $V_1 \oplus \cdots \oplus V_p$ with all $V_j \cong V (j = 1, \ldots, p)$. If every $W \cong V$ is contained in $V_1 \oplus \cdots \oplus V_p$, we are done. Otherwise choose $V_{p+1} \cong V$ so that $V_{p+1} \not\subset V_1 \oplus \cdots \oplus V_p$. By irreducibility of V_{p+1} we have $V_{p+1} \cap (V_1 \oplus \cdots \oplus V_p) = 0$, so the sum $V_1 \oplus V_{p+1}$ is direct. Since the dimension increases at each stage, the inductive process stops, so $\hat{V} = V_1 \oplus \cdots \oplus V_t$ for some t. $\qquad\square$

Within a given isotypic component, all irreducible subspaces are isomorphic:

Lemma 2.8 *Let U, V be irreducible subspaces of $\mathbf{R}^n$ and suppose that $U \subseteq \hat{V}$. Then $U \cong V$.*

Proof. Lemma 2.7 implies that $\hat{V} = V_1 \oplus \cdots \oplus V_t$, so $U \subseteq V_1 \oplus \cdots \oplus V_t$. Choose a subset $\{j_1, \ldots, j_p\} \subseteq \{1, \ldots, t\}$ that is minimal subject to $U \subseteq V_{j_1} \oplus \cdots \oplus V_{j_p}$. Then $U \not\subset V_{j_1} \oplus \cdots \oplus V_{j_{p-1}}$. By irreducibility, $U \cap (V_{j_1} \oplus \cdots \oplus V_{j_{p-1}}) = 0$, so U is mapped one-to-one by the projection $\pi_p : V_{j_1} \oplus \cdots \oplus V_{j_p} \to V_{j_p}$. By irreducibility of U and V_{j_p}, this map is an isomorphism. Therefore $U \cong V_{j_p} \cong V$. $\qquad\square$

We can now show that $\mathbf{R}^n$ is the direct sum of its distinct isotypic components.

Lemma 2.9 *Choose irreducible subspaces $V_j \subseteq \mathbf{R}^n, j = 1, \ldots, s$, so that every irreducible subspace is isomorphic to exactly one of the V_j. Then*

$$\mathbf{R}^n = \hat{V}_1 \oplus \cdots \oplus \hat{V}_s$$

Proof. Clearly $\mathbf{R}^n$ is the sum of the $\hat{V}_j$. It remains to prove that the sum is direct. This is the case provided $(\hat{V}_1 \oplus \cdots \oplus \hat{V}_j) \cap \hat{V}_{j+1} = 0$ for $j = 1, \ldots, s-1$. Suppose the intersection is nonzero, and let U be an irreducible contained in the intersection. By Lemma 2.8 $U \cong V_{j+1}$. By an argument similar to that used in Lemma 2.8, $U \cong V_p$ for some $p \leq j$. This contradicts the choice of the V_j. $\qquad\square$

Lemma 2.10 *If $A : \mathbf{R}^n \to \mathbf{R}^n$ is a linear map that commutes with Γ, then $A(\hat{V}) \subseteq \hat{V}$.*

Proof. $V_\alpha \subseteq \hat{V}$ so $\ker(A|V_\alpha)$ is Γ-invariant. Irreducibility implies that

$$\ker(A|V_\alpha) = \begin{cases} 0 & \Rightarrow A(V_\alpha) \cong V_\alpha \cong V \\ V_\alpha & \Rightarrow A \equiv 0 \end{cases}$$

In either case $A(V_\alpha) \subseteq \hat{V}$, so $A(\hat{V}) \subseteq \hat{V}$. $\qquad\square$

Definition 2.11 The decomposition $\mathbf{R}^n = \hat{V}_1 \oplus \cdots \oplus \hat{V}_s$ is the *isotypic decomposition* of $\mathbf{R}^n$ with respect to the Γ-action. The subspaces $\hat{V}_j$ are *isotypic components*. $\diamond$

2.3 General Comments on Stability of Equilibria

We can now state and prove three general theorems about the stability of equilibria. The first reduces the computation of eigenvalues of the Jacobian to the computation of its restriction to isotypic components with respect to the isotropy subgroup:

Theorem 2.12 *The isotypic decomposition of $\mathbf{R}^n$ with respect to Σ_{x_0} can be used to block-diagonalize $(\mathrm{d}f)_{x_0, \lambda_0}$.*

Proof. This follows since $(\mathrm{d}f)_{x_0, \lambda_0}$ commutes with Σ_{x_0}. Then Lemma 2.10 implies that each isotypic component is invariant under $(\mathrm{d}f)_{x_0, \lambda_0}$. $\qquad\square$

The second theorem shows that certain eigenvalues must be zero:

Theorem 2.13 *There are* $\dim \Gamma - \dim \Sigma_{x_0}$ *zero eigenvalues of* $(\mathrm{d}f)_{x_0,\lambda_0}$ *and the corresponding eigenvectors may be computed explicitly.*

Proof. Suppose $f(x_0, \lambda_0) = 0$; then $f(\gamma x_0, \lambda_0) = 0$. Let $\gamma \in \Gamma$ depend on a parameter t, that is, $\gamma = \gamma_t$ with $\gamma_0 = 1$. Then $\dot\gamma_0 x_0$ is a tangent vector to the group orbit Γx_0 through x_0 and every such tangent vector has this form. To see why, calculate

$$0 = \left. \frac{\mathrm{d}}{\mathrm{d}t} f(\gamma_t x_0, \lambda_0) \right|_{t=0} = (\mathrm{d}f)_{x_0,\lambda_0}(\dot\gamma_0 x_0)$$

to see that $(\mathrm{d}f)_{x_0,\lambda_0}$ vanishes on the tangent space to Γx_0. Since $\gamma x_0 = \delta x_0$ if and only if $\delta^{-1}\gamma \in \Sigma_{x_0}$, there are $\dim \Gamma x_0 = \dim \Gamma - \dim \Sigma_{x_0}$ linearly independent vectors $\dot\gamma x_0$.   □

A theorem of Ihrig and Golubitsky [293] provides the rationale behind the computations of Example 1.1. It shows that under certain circumstances the bifurcating solutions obtained by the Equivariant Branching Lemma are (generically) unstable. In the statement of the theorm we omit certain technical hypotheses, but these are 'usually' true. For a complete statement see [237] p. 90.

Theorem 2.14 *Let* Γ *act absolutely irreducibly on* $\mathbf{R}^n$ *where* $n \geq 2$. *Suppose there exists a nonzero quadratic* Γ-*equivariant polynomial mapping* $Q : \mathbf{R}^n \to \mathbf{R}^n$. *Then generically all solutions found by the Equivariant Branching Lemma are unstable.*

Sketch of Proof. Consider $\mathrm{d}Q$, a matrix whose coefficients are linear maps. Define $h : \mathbf{R}^n \to \mathbf{R}$ by $h(x) = \mathrm{tr}(\mathrm{d}Q)_x$. Then h is a linear function and

$$\begin{aligned}
h(\gamma x) &= \mathrm{tr}(\mathrm{d}Q)_{\gamma x} \\
&= \mathrm{tr}\,\gamma(\mathrm{d}Q)_x \gamma^{-1} \\
&= \mathrm{tr}(\mathrm{d}Q)_x \\
&= h(x)
\end{aligned}$$

Thus h is Γ-invariant and $\ker h$ is Γ-invariant. Further, $\ker h$ is at least $(n-1)$ dimensional. Thus $h = 0$ by Γ-irreducibility. This implies that the trace (that is, the sum of the eigenvalues) of $\mathrm{d}Q$ is zero. Thus if some eigenvalue of $\mathrm{d}Q$ has nonzero real part, then there must be eigenvalues with real parts of both signs.   □

Example 2.15 Consider $\mathbf{S}_N$ acting on $V_1 = \{x : x_1 + \cdots + x_N = 0\} \subseteq \mathbf{R}^N$, as in our speciation example, Section 1.5. The mapping

$$\begin{bmatrix} Nx_1^2 - (x_1^2 + \cdots + x_N^2) \\ Nx_2^2 - (x_1^2 + \cdots + x_N^2) \\ Nx_3^2 - (x_1^2 + \cdots + x_N^2) \\ \vdots \\ Nx_n^2 - (x_1^2 + \cdots + x_N^2) \end{bmatrix}$$

is a quadratic equivariant. Therefore by Theorem 2.14, generically the 'two species' solutions obtained are unstable near bifurcation.   ◇

This situation is reasonably common in equivariant bifurcation theory. Roughly speaking, it occurs because a quadratic truncation of the equations fails to capture the global geometry of bifurcating branches. These can regain stability through either of the mechanisms mentioned earlier: a branch can turn round and change the sign of one eigenvalue, or the branch can meet a secondary branch, which also changes the sign of an eigenvalue.

Example 2.16 Both phenomena occur in the case $N = 3$, and we briefly describe the scenario. See Golubitsky *et al.* [237] Chapter XV Section 4 for details. For $\mathbf{S}_3 \cong \mathbf{D}_3$ we can identify V_1 with $\mathbf{C}$. The relevant normal form is

$$F(z) = (|z|^2 - \lambda)z + (|z|^2 + \mu \operatorname{Re}(z^3) - \alpha)\bar{z}^2$$

where α, μ are parameters and λ is the bifurcation parameter. Depending on the signs of α, μ, there can exist secondary branches with trivial symmetry (that is, three-species branches). The geometry of the bifurcation diagrams is shown in Figure 2.3. When $\alpha < 0, \mu > 0$ the three-species branch can be stable, but it exists only for a finite interval of values of λ.

Notice that the presence of cubic terms in the normal form here causes the primary branch to turn round at a fold point, and (in this case) it regains stability there. It can also *lose* stability by encountering a secondary branch, or regain stability by encountering a secondary branch, depending on delicate features of the equations.   ◇

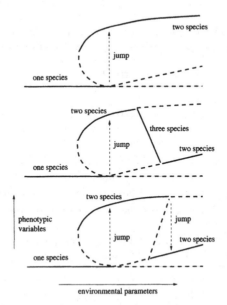

Figure 2.3: Existence of 3-species secondary bifurcations in the case $N = 3$. (Top) $\alpha > 0$. (Middle) $\alpha < 0, \mu > 0$. (Bottom) $\alpha < 0, \mu < 0$.

We show later that similar (but slightly more complicated) features occur in $\mathbf{S}_N$-equivariant bifurcation problems for $N > 3$, with interesting implications for our model of sympatric speciation.

2.4 Hilbert Bases and Equivariant Mappings

In order to explore the consequences of including higher order terms than quadratic, we must first develop systematic methods for classifying equivariant mappings.

We have seen that the existence and stability of equilibria in an equivariant bifurcation problem depend on various features of the equivariant maps. We now show that in principle it is possible to determine *all* Γ-equivariant maps algorithmically. In fact the determination of the smooth Γ-equivariant maps reduces to that of the polynomial Γ-equivariant maps, and this is a purely algebraic question with an algorithmic solution. There is a simple theoretical characterization of such maps; however, explicit computations can be difficult. The ideas, hence some of the terminology, go back to early work by David Hilbert in invariant theory.

Definition 2.17 A polynomial $h : \mathbf{R}^n \to \mathbf{R}$ is Γ-*invariant* if $h(\gamma x) = h(x)$ for all $x \in \mathbf{R}^n$. $\diamond$

We let $\mathcal{P}(\Gamma)$ denote the space of all Γ-invariant polynomials. It forms a commutative ring with the usual operations of addition and multiplication.

Definition 2.18 The Γ-invariant polynomials $u_1(x), \ldots, u_k(x)$ form a *Hilbert basis* for $\mathcal{P}(\Gamma)$ if for every $h \in \mathcal{P}(\Gamma)$ there exists a polynomial $p : \mathbf{R}^k \to \mathbf{R}$ such that

$$h(x) = p(u_1(x), \ldots, u_k(x))$$

$\diamond$

Theorem 2.19 (Hilbert-Weyl Theorem) *Let Γ be a compact Lie group acting on a vector space. Then there exists a finite Hilbert basis for the Γ-invariant functions.*

Proof. The result is standard: see for example [237] Chapter XII Section 6(a). $\square$

Example 2.20 (a) If $\Gamma = \mathbf{Z}_2$ acting on $\mathbf{R}$ then $u(x) = x^2$ is a Hilbert basis.

(b) If $\Gamma = \mathbf{S}_N$ acting on $(x_1, \ldots, x_N)$ then it is well known that a Hilbert basis is given by the *elementary symmetric polynomials* $\sigma_1(x), \ldots, \sigma_N(x)$, where

$$
\begin{aligned}
\sigma_1(x) &= x_1 + \cdots + x_N \\
\sigma_2(x) &= x_1 x_2 + \cdots + x_{N-1} x_N \\
&\vdots \\
\sigma_N(x) &= x_1 \cdots x_N
\end{aligned}
\tag{2.1}
$$

For a proof see any text on Galois Theory, for example Stewart [486] Exercises 2.13 and 2.14. The σ_k can also be defined by the identity

$$(t - x_1)(t - x_2) \cdots (t - x_N) = t^N - \sigma_1 t^{N-1} + \sigma_2 t^{N-2} - \cdots + (-1)^N \sigma_N \tag{2.2}$$

for any indeterminate t. We discuss $\mathbf{S}_N$-equivariants in Section 2.6. $\diamond$

Hilbert bases are almost never unique. The smooth invariant functions can be characterized in terms of a Hilbert basis of polynomial invariant functions:

Theorem 2.21 (Schwartz's Theorem) *Let $u_1(x), \ldots, u_k(x)$ be a Hilbert basis for $\mathcal{P}(\Gamma)$. Then for each Γ-invariant C^∞ function $f : \mathbf{R}^n \to \mathbf{R}$ there exists a C^∞ map $p : \mathbf{R}^k \to \mathbf{R}$ such that*

$$f(x) = p(u_1(x), \ldots, u_k(x)).$$

Proof. See Schwartz [467], Mather [377], or Bierstone [62]. An incomplete sketch is given in [237] Chapter XII Section 6(b). $\square$

Corollary 2.22 (Poénaru's Theorem) *Let $u_1(x), \ldots, u_k(x)$ be a Hilbert basis and let $u(x) = (u_1(x), \ldots, u_k(x))$. There exist Γ-equivariant polynomial mappings $X_1, \ldots, X_l : \mathbf{R}^n \to \mathbf{R}^n$ such that for each C^∞ Γ-equivariant map $f : \mathbf{R}^n \to \mathbf{R}^n$, there exist C^∞ maps $p_1, \ldots, p_l : \mathbf{R}^k \to \mathbf{R}$ such that*

$$f(x) = p_1(u(x))X_1(x) + \cdots + p_l(u(x))X_l(x).$$

Proof. See Poénaru [436] and [237] Chapter XII Section 6(c). □

Remark 2.23 (1) We may assume that the u_j and X_j are homogeneous polynomials.

(2) In general it is difficult to compute a Hilbert basis. However, the *dimensions* of the spaces of homogeneous invariants and equivariant polynomials of given degree can be computed by integration over the group, see Molien [403], Sattinger [462], Sattinger and Weaver [463], and Worfolk [527]. ◇

The Dihedral Group $\mathbf{D}_m$ Acting on $\mathbf{R}^2$. Consider $\mathbf{D}_m$ acting on $\mathbf{R}^2 \cong \mathbf{C}$ with the action

$$\begin{aligned} \theta z &= e^{i\theta} z \quad (\theta = \tfrac{2\pi}{m}) \\ \kappa z &= \bar{z} \end{aligned} \tag{2.3}$$

In this subsection we prove the following theorem.

Theorem 2.24 *Let $f : \mathbf{C} \to \mathbf{C}$ be $\mathbf{D}_m$-equivariant. Then there exist $p, q : \mathbf{R}^2 \to \mathbf{R}$ such that*

$$f(z) = p(u, v)z + q(u, v)\bar{z}^{m-1}$$

where $u = z\bar{z}$ and $v = z^m + \bar{z}^m$.

The Use of Theorem 2.24 in Bifurcation Calculations. We can apply Theorem 2.24 to classify equilibria in a general $\mathbf{D}_m$-equivariant bifurcation problem. We use the explicit form of $\mathbf{D}_m$-equivariants in two ways. First, we show that generically the only solutions expected at bifurcation are those supplied by axial subgroups; second, we compute the stability of solutions on these branches.

To verify the form of all solutions, let f be $\mathbf{D}_m$-equivariant. Equilibria are solutions of $f(x) = 0$. There are three cases:

(a) $z = 0$, the trivial equilibrium.
(b) z is parallel to $\bar{z}^{m-1}$. This implies that $\text{Im}(z^m) = 0$; thus we may take $z = x \neq 0$ and we have

$$f(x) = p(x^2, 2x^m)x + q(x^2, 2x^m)x^{m-1} = 0$$

But $x \neq 0$, so we may write

$$p(x^2, 2x^m) + q(x^2, 2x^m)x^{m-2} = 0$$

(c) z is not parallel to $\bar{z}^{m-1}$. In this case we must have $p = q = 0$. Recall that $p(0) = 0$ is a necessary condition for bifurcation and generically $q(0) \neq 0$. Thus generically there are no solutions of this type.

In particular, generically there are no solutions other than those predicted by the Equivariant Branching Lemma.

Next, we compute the stability of branches of solutions with $\Sigma = \mathbf{Z}_2(\kappa)$ symmetry. In real coordinates we can write

$$\kappa = \begin{bmatrix} 1 & 0 \\ 0 & -1 \end{bmatrix}$$

The isotypic decomposition is

$$V_1 = \mathbf{R}(1,0) \qquad V_2 = \mathbf{R}(0,1)$$

and df is diagonal.

We now apply a useful general trick. An easy computation proves that in complex coordinates

$$(df)(w) = f_z w + f_{\bar{z}} \bar{w} \tag{2.4}$$

The eigenvalues of df are $(df)(1) = \lambda_1$ and $(df)(i) = \lambda_2 i$. So

$$\lambda_1 = f_z + f_{\bar{z}}$$
$$\lambda_2 = (f_z i + f_{\bar{z}}(-i))(-i) = f_z - f_{\bar{z}}$$

Using the form previously calculated for f we have

$$f_z = p + p_z z + q_z \bar{z}^{m-1}$$
$$f_{\bar{z}} = p_{\bar{z}} z + q_{\bar{z}} \bar{z}^{m-1} + (m-1)q\bar{z}^{m-2}$$

We are computing along solutions of the type $p + q\bar{z}^{m-2} = 0$, so

$$f_z = p_z z + q_z \bar{z}^{m-1} - q\bar{z}^{m-2}$$

The eigenvalues are therefore

$$f_z + f_{\bar{z}} = (p_z + p_{\bar{z}})z + (q_z + q_{\bar{z}})\bar{z}^{m-1} + (m-2)q\bar{z}^{m-2}$$
$$f_z - f_{\bar{z}} = (p_z - p_{\bar{z}})z + (q_z - q_{\bar{z}})\bar{z}^{m-1} - mq\bar{z}^{m-2}$$

Now

$$p_z = p_u \bar{z} + p_v m z^{m-1}$$
$$p_{\bar{z}} = p_u z + p_v m \bar{z}^{m-1}$$

so

$$p_z + p_{\bar{z}} = p_u(z + \bar{z}) + p_v m(z^{m-1} + \bar{z}^{m-1})$$
$$p_z - p_{\bar{z}} = p_u(z - \bar{z}) + p_v m(z^{m-1} - \bar{z}^{m-1})$$

and similarly for $q_z + q_{\bar{z}}$ and $q_z - q_{\bar{z}}$. Note, however, that $z = x$, so $p_z - P_{\bar{z}} = q_z - q_{\bar{z}} = 0$. Therefore

$$f_z + f_{\bar{z}} = 2x^2 p_u + 2m p_u x^{m-1} + 2q_u x^m + 2m q_u x^{2m-2} + (m-2)qx^{m-2}$$
$$f_z - f_{\bar{z}} = -mqx^{m-2}$$

The genericity assumption $q(0) \neq 0$ implies that

$$\mathrm{sgn}(f_z - f_{\bar{z}}) = -\mathrm{sgn}(q(0))$$

Assuming, in addition, that $p_u(0) \neq 0$ when $m \geq 5$, or $p_u(0) + q(0) \neq 0$ when $m = 4$, we get

$$\mathrm{sgn}(f_z + f_{\bar{z}}) = \begin{cases} \mathrm{sgn}(p_u(0)) & m \geq 5 \\ \mathrm{sgn}(p_u(0) + q(0)) & m = 4 \\ \mathrm{sgn}(q(0)) & m = 3 \end{cases}$$

When $m = 3$ we have $\mathrm{sgn}(f_z + f_{\bar{z}}) = -\mathrm{sgn}(f_z - f_{\bar{z}})$. Thus the bifurcating solutions are unstable, as expected.

We begin the verification of Theorem 2.24 by proving:

Lemma 2.25 *The polynomials $u(z) = z\bar{z}$ and $v(z) = z^m + \bar{z}^m$ are a Hilbert basis for the action (2.3) of $\mathbf{D}_m$ on $\mathbf{C}$.*

Proof. Consider an arbitrary $\mathbf{D}_m$-invariant polynomial $h : \mathbf{C} \to \mathbf{R}$. It will have the following properties:

(a) $h(z) = h(\bar{z})$
(b) $h(e^{i\theta} z) = h(z)$
(c) $h(z) = \overline{h(z)}$ (the requirement that $h(z)$ is real)

Every polynomial h is of the form

$$h(z) = \sum_{\alpha,\beta} a_{\alpha\beta} z^{\alpha} \bar{z}^{\beta}$$

Invariance therefore yields the conditions

$$h(z) = \overline{h(z)} \Rightarrow \bar{a}_{\alpha\beta} = a_{\beta\alpha}$$

$$h(z) = h(\bar{z}) \Rightarrow a_{\alpha\beta} = a_{\beta\alpha}$$

Together (a) and (c) imply $a_{\alpha\beta} \in \mathbf{R}$. We choose $u_1(z) = z\bar{z} = |z|^2$ to be our first Hilbert basis element. Assuming $\beta \le \alpha$, the function h can now be written

$$h(z) = \sum_{\alpha,\beta} a_{\alpha\beta} (z\bar{z})^{\alpha} (z^{\gamma} + \bar{z}^{\gamma}) \qquad \gamma = \alpha - \beta$$

Therefore, $u_1(z) = z\bar{z}$, $u_2(z) = z + \bar{z}$, $u_3(z) = z^2 + \bar{z}^2$, ... is a set of generators for $\mathcal{P}(\mathbf{D}_m)$. Now use θ-invariance:

$$h(e^{i\theta} z) = \sum_{\alpha,\beta} a_{\alpha\beta} (z\bar{z})^{\alpha} (e^{i\gamma\theta} z^{\gamma} + e^{-i\gamma\theta} \bar{z}^{\gamma})$$

Therefore either $e^{i\gamma\theta} = 1$ or $a_{\alpha\beta} = 0$. Hence $\gamma \equiv 0 \pmod{m}$. Now the generators reduce to $u_1(z) = z\bar{z}$, $u_2(z) = z^m + \bar{z}^m$, $u_3(z) = z^{2m} + \bar{z}^{2m}$, ... In order to find a minimal basis, consider the identity

$$z^{m(j+1)} + \bar{z}^{m(j+1)} = (z^{mj} + \bar{z}^{mj})(z^m + \bar{z}^m) - z^{mj} \bar{z}^m - \bar{z}^{mj} z^m$$

$$= (z^{mj} + \bar{z}^{mj})(z^m + \bar{z}^m) - (z\bar{z})^m (z^{m(j-1)} + \bar{z}^{m(j-1)})$$

$$= v(z)(z^{mj} + \bar{z}^{mj}) - u(z)^m (z^{m(j-1)} + \bar{z}^{m(j-1)})$$

By induction $z^{mj} + \bar{z}^{mj} = h_j(u(z), v(z))$ for some polynomial $h_j : \mathbf{R}^2 \to \mathbf{R}$. $\qquad \square$

Proof of Theorem 2.24 Consider a general $\mathbf{D}_m$-equivariant mapping $f : \mathbf{C} \to \mathbf{C}$ written as

$$f(z) = \sum_{\alpha,\beta} b_{\alpha\beta} z^{\alpha} \bar{z}^{\beta}$$

Applying the symmetries,

$$f(z) = \overline{f(z)} \Rightarrow b_{\alpha\beta} = \bar{b}_{\alpha\beta} \Rightarrow b_{\alpha\beta} \in \mathbf{R}$$

$$e^{-i\theta} f(e^{i\theta} z) = f(z)$$

Thus $b_{\alpha\beta} = 0$ unless $\alpha = \beta + 1 \pmod{m}$, so the equivariants are $z, \bar{z}^{m-1}, \dots$. To find a minimal set of generators, consider

$$f(z) = \sum_{\alpha,\gamma} b_{\alpha\gamma} (z\bar{z})^{\alpha} \bar{z}^{\gamma} + c_{\alpha\gamma} (z\bar{z})^{\alpha} z^{\gamma}$$

where for the first term $\gamma + 1 \equiv 0 \pmod{m}$, where $\gamma = mk - 1$; and for the second term $\gamma - 1 \equiv 0 \pmod{m}$ where $\gamma = mk + 1$. Now

$$z^{mk+1} = (z^{mk} + \bar{z}^{mk})z - \bar{z}^{mk}z$$

$$= \underbrace{(z^{mk} + \bar{z}^{mk})z}_{\text{invariant}} - \underbrace{z\bar{z}}_{\text{invariant}} \, \bar{z}^{mk-1}$$

and

$$z^{mk-1} = \bar{z}^{m-1}\underbrace{(z^{(k-1)m} - \bar{z}^{(k-1)m})}_{\text{invariant}} - \underbrace{(z\bar{z})}_{\text{invariant}}^{m-1} z^{(k-2)m+1}$$

By induction, the equivariant generators are $X_1(z) = z$ and $X_2(z) = \bar{z}^{m-1}$. $\square$

2.5 Model-Independent Results for $\mathbf{D}_3$ Steady-State Bifurcation

For a given group Γ acting on a given space, there is a fixed 'catalog' of possible generic behavior. *Any* model that is consistent with this Γ-action can be expected to display generic behavior from that catalog. Exactly which behavior turns up for which parameter values depends on the model, but the range of phenomena is *model-independent*. Our philosophy is to determine the range of model-independent behavior once and for all, for a given group action, rather than re-deriving it many times for different models. Any extra model-specific results can then be found by other methods: the model-independent approach provides a general framework into which special models must fit.

For example, suppose that we have a system of equations with $\mathbf{D}_3$ symmetry, and we wish to discover the types of steady-state bifurcations that should be expected. Without some kind of genericity assumption, almost any type of bifurcation could happen. However, if we consider only generic behavior it is often possible to say something more specific. For example, in the case of steady-state bifurcation in systems with $\mathbf{D}_3$ we can make the following general observations. Let $\mathcal{K}$ be the kernel of the linearization at a $\mathbf{D}_3$-invariant equilibrium.

1) Generically, in the world of $\mathbf{D}_3$ symmetry, $\mathbf{D}_3$ acts absolutely irreducibly on $\mathcal{K}$.

2) There are three irreducible representations of $\mathbf{D}_3$, one 2-dimensional and two 1-dimensional.

These points are proved using the following theorem, whose proof can be deduced from Bröcker and tom Dieck [77] Chapter 2 Section 4.

Theorem 2.26 *Let Γ be a finite group of order m. Suppose there exist distinct absolutely irreducible representations of Γ on vector spaces $V_1, \ldots, V_s$ such that*

$$m = (\dim V_1)^2 + \cdots + (\dim V_s)^2$$

Then, up to isomorphism, the V_j enumerate all irreducible representations of Γ.

Since $6 = 1 + 1 + 2^2$ it follows that $\mathbf{D}_3$ has only these three different irreducible representations.

$\mathbf{D}_3$ *Acts Trivially on* $\mathbf{R}$. In this case the generic behavior is a limit point bifurcation and all solutions have $\mathbf{D}_3$ symmetry. See Figure 2.4.

$\mathbf{D}_3$ *Acts Nontrivially on* $\mathbf{R}$. In this case $\mathbf{Z}_3$ acts trivially, κ acts nontrivially. Since $\mathbf{D}_3/\mathbf{Z}_3 \cong \mathbf{Z}_2$, the nontrivial action of $\mathcal{K}$ is $\mathbf{Z}_2$, which implies a pitchfork bifurcation. The nontrivial solutions have $\mathbf{Z}_3$ symmetry (Figure 2.5).

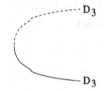

Figure 2.4: Limit point bifurcation when $\mathbf{D}_3$ symmetry is not broken.

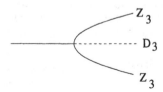

Figure 2.5: Pitchfork bifurcation when $\mathbf{D}_3$ symmetry breaks to $\mathbf{Z}_2$.

$\mathbf{D}_3$ *Acts Irreducibly on* $\mathbf{R}^2 = \mathbf{C}$. This is the case covered in the previous section. The bifurcating solutions have a reflectional symmetry (Figure 2.6). The bifurcation diagram shows a *transcritical* bifurcation.

Figure 2.6: Transcritical bifurcation when $\mathbf{D}_3$ symmetry breaks to $\mathbf{1}$.

A general discussion of model-independent uses of symmetry is given in [228, 237].

2.6 Invariant Theory for $\mathbf{S}_N$

The speciation model (1.13) is an example of a system with *all-to-all coupling*. In such systems, N identical subsystems are coupled together so that each influences all of the others, and in the same way. This is a common assumption in applications. We abstract this type of coupling as equivariance under the symmetric group $\mathbf{S}_N$, acting by permutation of N variables. We now derive the algebraic form of the invariants and equivariants for this action. The invariants are described in (2.1) and (2.6); the equivariants are stated in Proposition 2.27 below.

Let $\mathbf{S}_N$ act on $V = \mathbf{R}^N$ by permutation of coordinates:

$$\rho(x_1, \ldots, x_N) = (x_{\rho^{-1}(1)}, \ldots, x_{\rho^{-1}(N)}) \tag{2.5}$$

The inverses are included so that (2.5) defines a group action. The elementary symmetric functions (2.1) freely generate the polynomial $\mathbf{S}_N$-invariants over $\mathbf{R}$. Equivalently the sums of kth powers

$$\pi_k = \sum_{i=1}^{N} x_i^k \tag{2.6}$$

where $k = 1, \ldots, N$ are polynomial generators. Note that the invariant polynomial $p(x) = 1$ is included by virtue of the phrase 'generated over $\mathbf{R}$'.

Consider the $\mathbf{S}_N$-equivariants. The group $\mathbf{S}_N$ is generated by the N-cycle $\alpha = (123\ldots N)$ and the subgroup $\mathbf{S}_{N-1}$ consisting of all permutations of $\{2, \ldots, N\}$. Therefore a mapping is equivariant under $\mathbf{S}_N$ if and only if it is equivariant under $\mathbf{S}_{N-1}$ and α^{-1}. Suppose that $F : \mathbf{R}^N \to \mathbf{R}^N$ is an $\mathbf{S}_N$-equivariant polynomial mapping with components $F = (F_1, \ldots, F_N)$, and write F_1 in the form

$$F_1(x) = \sum_d G_d(x_2, \ldots, x_N) x_1^d$$

Equivariance under $\mathbf{S}_{N-1}$ implies that

$$F_1(x) = \sum_d G_d(x_{\rho(2)}, \ldots, x_{\rho(N)}) x_1^d$$

for all $\rho \in \mathbf{S}_{N-1}$, since $\mathbf{S}_{N-1}$ fixes the symbol 1. Therefore each G_d is *invariant* under $\mathbf{S}_{N-1}$. In addition, α^{-1}-equivariance implies that

$$
\begin{aligned}
F_i(x) &= \sum_d G_d(x_{i+1}, \ldots, x_{i-1}) x_i^d \\
&= \sum_d G_d(x_1, \ldots, x_{i-1}, x_{i+1}, \ldots, x_N) x_i^d
\end{aligned}
$$

For any polynomial $P(x_1, \ldots, x_N)$ we define

$$\lceil P(x_1, \ldots, x_N) \rceil = \begin{bmatrix} P(x_1, x_2, \ldots, x_{N-1}, x_N) \\ P(x_2, x_3, \ldots, x_N, x_1) \\ \vdots \\ P(x_N, x_1, \ldots, x_{N-2}, x_{N-1}) \end{bmatrix}$$

If P is $\mathbf{S}_{N-1}$-invariant, then $\lceil P \rceil$ is $\mathbf{S}_N$-equivariant, and conversely.

We now define $\mathbf{S}_N$-equivariants E_k, for $k = 0, 1, 2, \ldots$ by

$$E_k = \lceil x_1^k \rceil = [x_1^k, x_2^k, \ldots, x_N^k]^T \tag{2.7}$$

Proposition 2.27 *The $\mathbf{S}_N$-equivariant polynomial mappings are generated over the $\mathbf{S}_N$-invariant polynomial functions by $E_0, \ldots, E_{N-1}$.*

Proof. Let p be any $\mathbf{S}_{N-1}$-invariant polynomial in $(x_2, \ldots, x_N)$, and denote its degree by k. We claim that there exists an $\mathbf{S}_N$-invariant polynomial $\tilde{p}$ in $(x_1, \ldots, x_N)$, whose degree is the same as that of p, and $\mathbf{S}_N$-invariant polynomials $q_1, \ldots, q_k$ in $(x_2, \ldots, x_N)$, such that

$$p = \tilde{p} + x_1 q_1 + \cdots + x_1^k q_k \tag{2.8}$$

To prove the claim, write p in the form

$$p(x) = G(\hat{\pi}_1, \ldots, \hat{\pi}_{N-1})$$

where $\hat{\pi}_l = x_2^l + \cdots x_N^l$. (This is possible since the $\hat{\pi}_l$ for $l = 1, \ldots, N-1$ generate the $\mathbf{S}_{N-1}$-invariants.) Define

$$\tilde{p} = G(\pi_1, \ldots, \pi_{N-1})$$

Observe that by construction

$$\tilde{p}(0, x_2, \ldots, x_N) = p(x_2, \ldots, x_N)$$

Therefore, considered as a polynomial in x_1 with coefficients that are polynomials in $x_2, \ldots, x_N$, the difference $p - \tilde{p}$ is divisible by x_1 (using the remainder theorem, see for example Fraleigh [200] page 276). Expanding in powers of x_1, the claim follows, except that at the moment the q_j are only known to be $\mathbf{S}_{N-1}$-invariant. However, since p and $\tilde{p}$ have the same degree, all the q_j have smaller degree than p. Inductively, it follows that we can write p in the form (2.8) where now the q_j are $\mathbf{S}_N$-invariant polynomials in $(x_1, \ldots, x_N)$. This proves the claim.

Suppose that $E(x)$ is equivariant, and write it in the form $E(x) = \lceil Q(x) \rceil$. Expanding Q as a polynomial in x_1 over the remaining variables, we can write

$$Q(x) = Q_0 + x_1 Q_1 + \cdots + x_1^k Q_k$$

for some k, where the Q_j are $\mathbf{S}_{N-1}$-invariants. By the claim above, we can rewrite Q in the form

$$Q(x) = R_0 + x_1 R_1 + \cdots + x_1^l R_l$$

where the R_j are $\mathbf{S}_N$-invariants. Therefore

$$
\begin{aligned}
E(x) &= \lceil Q(x) \rceil \\
&= \lceil R_0 + x_1 R_1 + \cdots + x_1^l R_l \rceil \\
&= R_0 \lceil 1 \rceil + R_1 \lceil x_1 \rceil + \cdots + R_l \lceil x_1^l \rceil \\
&= R_0 E_0 + R_1 E_1 + \cdots + R_l E_l
\end{aligned}
$$

so that the $\mathbf{S}_N$-equivariants are generated by the E_j over the $\mathbf{S}_N$-invariants.

Finally, we must show that only $E_0, \ldots, E_{N-1}$ are necessary as generators. This is a consequence of identity (2.2). Setting $t = x_1$ implies that

$$x_1^N = \sigma_1 x_1^{N-1} - \sigma_2 x_1^{N-2} + \cdots + (-1)^{N+1} \sigma_N$$

Repeatedly multiplying by x_1 and applying induction, it follows that for all $n \geq N$

$$x_1^n = S_1 x_1^{N-1} + S_2 x_1^{N-2} + \cdots + S_N$$

where the S_j are $\mathbf{S}_N$-invariants. Applying the operator $\lceil \ \rceil$ we see that E_n is a linear combination of $E_0, \ldots, E_{N-1}$ over the $\mathbf{S}_N$-invariants. $\qquad\square$

2.7 Cubic Terms in the Speciation Model

We now return to the speciation model and discuss stability issues. As previewed in Chapter 1, the general cubic-degree truncation of F has 13 parameters:

$$
\begin{aligned}
F = \ & a_0 E_0 + \\
& b_1 \pi_1 E_0 + b_2 E_1 + \\
& c_1 \pi_1^2 E_0 + c_2 \pi_2 E_0 + c_3 \pi_1 E_1 + c_4 E_2 + \\
& d_1 \pi_1^3 E_0 + d_2 \pi_1 \pi_2 E_0 + d_3 \pi_3 E_0 + d_4 \pi_1^2 E_1 + d_5 \pi_1 E_2 + d_6 E_3
\end{aligned}
\tag{2.9}
$$

indicating the likely complexity of any detailed analysis.

Linear Analysis. We begin by carrying out a linear analysis. We normalize the phenotypic variables to represent the deviation from the initial phenotype (single species), which removes the constant term E_0. Thus we assume that prior to bifurcation the phenotype is $x = (0, \ldots, 0)$.

There are two linear equivariants because $\mathbf{R}^N$ is a direct sum of two absolutely irreducible subspaces. The map E_1 is the identity, and $\sigma_1 E_0 = M$ where

$$
M = \begin{bmatrix}
1 & 1 & \cdots & 1 \\
1 & 1 & \cdots & 1 \\
\vdots & \vdots & \ddots & \vdots \\
1 & 1 & \cdots & 1
\end{bmatrix}
$$

The general linear equivariant is therefore $\alpha I + \beta M$ where $\alpha, \beta \in \mathbf{R}$ are our previous b_2, b_1 respectively. The eigenvalues and eigenvectors are:

$$
\begin{aligned}
u_0 &= (1,1,\ldots,1)^T \quad \text{eigenvalue } \alpha + N\beta \\
u_0 &= (1,-1,0,\ldots,0)^T \quad \text{eigenvalue } \alpha \\
u_0 &= (0,\ldots,0,1,-1)^T \quad \text{eigenvalue } \alpha
\end{aligned}
$$

There are two potential local bifurcation points: one where $\alpha + N\beta = 0$ and one where $\alpha = 0$. These are distinct when $\beta \neq 0$, a nondegeneracy condition that prevents the occurrence of a mode interaction point and is generically valid. If $\alpha + N\beta = 0$ then the kernel is V_0, whereas if $\alpha = 0$ then the kernel is V_1. We therefore choose α to be the bifurcation parameter, and arrange for the first bifurcation to have kernel V_1, with the trivial solution stable for $\alpha < 0$. This is the case provided $\beta < 0$. By scaling x we may assume $\beta = -1$, leading to a vector field of the form

$$
F(x) = \left[\begin{array}{c} \lambda x_1 - (x_1 + \cdots + x_N) \\ \lambda x_1 - (x_1 + \cdots + x_N) \\ \cdots \\ \lambda x_N - (x_1 + \cdots + x_N) \end{array} \right] + \text{h.o.t.}
$$

Liapunov-Schmidt Reduction. As already indicated, the general cubic truncation of an $\mathbf{S}_N$-equivariant ODE involves 13 parameters, which is too complicated for detailed analysis. Here we describe work of Elmhirst [171] on the dynamics of the Liapunov-Schmidt reduction of such a cubic truncation to the space V_1, see also Cohen and Stewart [117]. After scaling, the Liapunov-Schmidt reduction has only two parameters, plus the bifurcation parameter, so a comprehensive analysis is feasible. In fact we will interpret the equations as the cubic truncation of a center manifold reduction, which yields the same general form but preserves the dynamics).

We can obtain the general form of the Liapunov-Schmidt reduction onto V_1 by restriction and projection from the general $\mathbf{S}_N$-equivariant mapping on $\mathbf{R}^N$. To cubic order, we do this by imposing the relation $\pi_1 = 0$ on (2.9) and then projecting the results onto V_1. It is easy to check that only the following terms survive:

$$
G = b_2 E_1 + c_4 \left(E_2 - \frac{1}{N} \pi_2 E_0 \right) + d_5 \left(\pi_2 E_1 - \frac{1}{N} \pi_1 \pi_2 E_0 \right) + d_6 \left(E_3 - \frac{1}{N} \pi_3 E_0 \right)
$$

Cohen and Stewart [117] and Elmhirst [171] used different methods to obtain the equivalent form

$$
\begin{aligned}
dx_i/dt &= \lambda x_i + \varepsilon(Nx_i^2 - \Sigma_2) + C(Nx_i^3 - \Sigma_3) + \\
&\quad D(Nx_i(x_1^2 + \cdots + x_{i-1}^2 + x_{i+1}^2 + \cdots + x_N^2) - \Sigma_{12})
\end{aligned} \tag{2.10}
$$

where $i = 1, \ldots, N$, $\varepsilon = \pm 1, 0$ (after scaling x_i, λ by positive factors, which preserves orientation of the bifurcation diagram and stability), and

$$
\begin{aligned}
\Sigma_2 &= x_1^2 + \cdots + x_N^2 \\
\Sigma_3 &= x_1^3 + \cdots + x_N^3 \\
\Sigma_{12} &= \sum_{i \neq j} x_i^2 x_j
\end{aligned}
$$

For consistency with the calculations of Elmhirst [171] we will work with (2.10) from now on. We also restrict attention to the case $\varepsilon = +1$. The case $\varepsilon = -1$ can be recovered by transforming x into $-x$, which turns the bifurcation diagrams upside-down, and we consider the case $\varepsilon = 0$ to be non-generic.

Axial Subgroups. We study the primary branches of equilibria of (2.10), on the spaces $W_p = \text{Fix}(\mathbf{S}_p \times \mathbf{S}_q)$ where $p + q = N$ and $1 \leq p \leq N/2$. We also study the stabilities of these branches. We coordinatize W_p by $\alpha \in \mathbf{R}$, corresponding to the point

$$x_\alpha = \alpha(\underbrace{q, \ldots, q}_{p}; \underbrace{-p, \ldots, -p}_{q}) \qquad (2.11)$$

Define

$$
\begin{aligned}
c &= C(N^2 - 3Np + 3p^2) \\
d &= D(N^2 - N(N+3)p + (N+3)p^2)
\end{aligned}
$$

We find the branching equation on W_p by substituting (2.11) into the right-hand side of (2.10) and selecting the 1st and $(p+1)$st components. Using $pu + qv = 0$ we obtain the branching equation

$$\lambda = -\alpha N(N - 2p) - \alpha^2 N(c - d)$$

which is a parabola passing through the origin. Let α_0 be the α-coordinate of the other point at which the parabola crosses the α-axis, and let (λ_c, α_c) be the coordinates of the vertex of the parabola. Then

$$
\begin{aligned}
\alpha_0 &= \frac{N - 2p}{d - c} \\[2mm]
\alpha_c &= \frac{N - 2p}{2(d - c)} \\[2mm]
\lambda_0 &= \frac{N(N - 2p)^2}{4(c - d)}
\end{aligned}
$$

Stability. We next compute the eigenvalues of dG along the branch in W_p. Let

$$G_{ij} = \frac{\partial G_i}{\partial x_j}\Big|_{x_\alpha}$$

By direct computation:

$$G_{ii} = \lambda + 2\alpha(N-1)q + \alpha^2[3C(N-1)q^2 + D(N(N-1)pq + (3-N)q^2]$$

if $1 \leq i \leq p$, whereas

$$G_{ii} = \lambda + 2\alpha(N-1)p + \alpha^2[3C(N-1)p^2 + D(N(N-1)pq + (3-N)p^2]$$

if $p + 1 \leq i \leq N$. Also, for $i \neq j$ we obtain

$$
G_{ij} = \begin{cases}
-2\alpha q - \alpha^2[3Cq^2 - D((2N+3)q^2 - Npq)] & \text{if } 1 \leq i, j \leq p \\
-2\alpha q - \alpha^2[3Cq^2 - D(3q^2 - 3Npq)] & \text{if } 1 \leq j \leq p < i \leq N \\
2\alpha p - \alpha^2[3Cq^2 - D(3p^2 - 3Npq)] & \text{if } 1 \leq i \leq p < j \leq N \\
2\alpha p - \alpha^2[3Cp^2 - D((2N+3)p^2 - Npq)] & \text{if } p + 1 \leq i, j \leq N
\end{cases}
$$

In order to compute the eigenvalues along the primary branches we appeal to Theorem 2.12: we can use the isotypic components of the action of the isotropy subgroup to block-diagonalize the Jacobian. We therefore compute the isotypic components. The isotypic decomposition of V_1 for the action of $\Sigma = \mathbf{S}_p \times \mathbf{S}_q$ is

$$X = W_0 \oplus W_1 \oplus W_2 \qquad (2.12)$$

where

$$W_0 = \{(\underbrace{qu,\ldots,qu}_{p}; \underbrace{-pu,\ldots,-pu}_{q}) : u \in \mathbf{R}\}$$

$$W_1 = \{(\underbrace{x_1,\ldots,x_p}_{p}; \underbrace{0,\ldots,0}_{q}) : x_1 + \cdots + x_p = 0\}$$

$$W_2 = \{(\underbrace{0,\ldots,0}_{p}; \underbrace{x_{p+1},\ldots,x_N}_{q}) : x_{p+1} + \cdots + x_N = 0\}$$

Note that when $p = 1$ the component W_1 should be omitted. Note also that $W_0 = \text{Fix}(\Sigma)$. The action of Σ is absolutely irreducible on each isotypic component, and trivial on W_0.

Thus there are (at most) three distinct eigenvalues, one for each W_j. We may therefore compute these eigenvalues by applying $dG|_{x_\alpha}$ to appropriate eigenvectors, from which we deduce that

$$\mu_0 = (G_{11} + \cdots + G_{1p}) - \tfrac{p}{q}(G_{1,p+1} + \cdots + G_{1N})$$
$$\mu_1 = G_{11} - G_{12}$$
$$\mu_2 = G_{p+1,p+1} - G_{p+1,p+2}$$

Putting all this information together we find that the eigenvalues of the linearization along the primary branch in W_p are:

$$\mu_0 = \alpha N(N - 2p) + 2\alpha^2 N(c - d)$$
$$\mu_1 = \alpha N^2 + \alpha^2 N^2 (2N - 3p)(C - D) \qquad (2.13)$$
$$\mu_2 = -\alpha N^2 + \alpha^2 N^2 (3p - N)(C - D)$$

with multiplicities $1, p-1, q-1$ respectively. Note than when $N = 3$ we have $p = 1$ so μ_1 does not occur. The computations to support these statements may be found in Elmhirst [171].

Observe that near the origin, where the linear term in α dominates, μ_1 and μ_2 have opposite signs. Therefore the bifurcating branches are always unstable near the origin, as predicted by Theorem 2.14. This is why we must work with (at least) the cubic truncation.

We summarize a few of the main consequences of these computations. Ignoring degenerate cases when $C = D$ or $c = d$, each eigenvalue changes sign twice along the branch: once at the origin and once somewhere else. Let the nonzero value of α at which this sign change occurs be β_j for μ_j. Then

$$\beta_0 = \frac{N - 2p}{2(d - c)} = \alpha_c$$

$$\beta_1 = \frac{1}{(2N - 3p)(D - C)}$$

$$\beta_2 = \frac{1}{(3p - N)(C - D)}$$

Thus β_0 changes sign at the vertex of the parabola (the fold point).

Stability at Infinity. A detailed analysis of the bifurcations and stabilities in this system is complicated, see Elmhirst [171]. Here, we illustrate the results with one computation, the condition that the branch should be stable for sufficiently large α ('stable at infinity'). This ensures that the speciation event persists for all sufficiently large λ — it is 'permanent'. It is easy to derive a necessary and sufficient condition for stability at infinity, as follows:

Proposition 2.28 *Necessary and sufficient conditions for a primary branch of type* (p, q) *to be stable at infinity are:*

(a) *If* $D > 0$ *then* $C < D(1 - \frac{Np(N-p)}{N^2 - 3Np + 3p^2})$ *and* $p > N/3$

(b) *If* $D < 0$ *then* $C < D$ *and* $p > N/3$

Proof. The sign of μ_j near infinity is dominated by the coefficient of α^2, and we want all $\mu_j < 0$. Therefore we require

$$c - d \ < \ 0$$
$$(2N - 3p)(C - D) \ < \ 0$$
$$(3p - N)(C - D) \ < \ 0$$

where we are assuming $C \neq D$ and $p \neq N/3$. If $C > D$ then $p > 2N/3$, contrary to $p \leq N/2$, so we need

$$C < D \tag{2.14}$$

This being so, we deduce that $p > N/3$, and these conditions ensure that $\mu_1 < 0$ and $\mu_2 < 0$. It remains to consider μ_0. This is negative provided $c < d$, or equivalently

$$C < D \left(1 - \frac{Np(N - p)}{N^2 - 3Np + 3p^2} \right) \tag{2.15}$$

Both $Np(N - p)$ and $N^2 - 3Np + 3p^2$ are positive for all p, and it is easy to prove that whenever $1 \leq p \leq N/2$ we have $Np(N - p) > N^2 - 3Np + 3p^2$, so $(1 - \frac{Np(N-p)}{N^2 - 3Np + 3p^2}) < 0$. Thus if $D > 0$ inequality (2.15) implies (2.14), but if $D < 0$ then inequality (2.14) implies (2.15). The conditions stated then follow. $\qquad \square$

The most interesting condition here is that in either case $p > N/3$, which goes a long way towards verifying a conjecture of Cohen and Stewart [117], to the effect that stability forces p and q to be fairly large in comparison with N. Here, the bifurcating species must contain more than one third of the total number of PODs, and (since $p + q = N$) less than two thirds of the total number of PODs. So in this model, on the extra assumption of stability at infinity, the 'founder populations' are *large*. This is very different from the common assumption of a small founder population in the allopatric mechanism.

Of course, stability at infinity is an artificial condition, because it allows λ to increase without limit. Stable speciation can occur for bounded λ when $p \leq N/3$; indeed it can occur for $p = 1$. Nonetheless, there is a general tendency for speciation to 'prefer' large values of p in this model.

Simulations of the Cubic Truncation. Next we describe the results of some numerical simulations of (2.9). In these simulations we begin with a random initial condition near 0. At each time-step we 'ramp' the bifurcation parameter. That is, we increment λ by some fixed small amount and use the previous value of x as the initial condition for a single time-step in a numerical algorithm for solving the ODE (2.10). Here we have used the Euler method because of its simplicity, although a Runge-Kutta algorithm would be more usual. We add a small amount of random noise to each component of x. Finally, we found that if (2.10) is integrated without further precautions numerical solutions can diverge from V_1 and blow up. We therefore project x back onto V_1 at each integration step by subtracting the mean of the $x(j)$ from each component.

Figure 2.7(left) shows a typical speciation event, occurring for the parameter values shown in the caption. Here the number of PODs is $N = 25$, and the initial single-species state splits into a state with $p = 6, q = 19$. The bifurcation is rapid enough to be described as a jump. The jump does not set in at the bifurcation point $\lambda = 0$ because, as is well known, ramping can lead to the phenomenon of 'tunneling through the bifurcation'. Prior to bifurcation, $x = 0$ is a stable equilibrium so solutions converge rapidly towards it. Immediately after bifurcation $x = 0$ is only weakly unstable, so it takes some time before x starts to diverge from 0. We have not taken steps to eliminate this effect because we feel that ramping the bifurcation parameter is very much in the spirit of real evolutionary dynamics, where each generation forms the 'initial conditions' for the next, and the environment slowly changes.

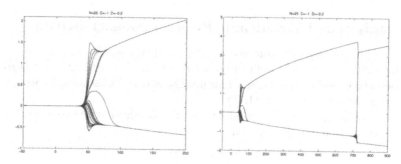

Figure 2.7: Numerical simulation of (2.10) with parameters $N = 25, C = -1, D = -0.2$. Each component of x is plotted on the same vertical axis. (Left) $\lambda \in [-50, 200]$; (right) $\lambda \in [-50, 900]$ and we observe a secondary bifurcation.

Since $6 < 25/3$ we know from general theory that when λ becomes sufficiently large, the solution depicted must become unstable. Nevertheless, we see that it can remain stable for a broad range of λ values. If we plot the bifurcation diagram for larger λ, we find that where the branch becomes unstable there is a secondary bifurcation, again delayed because of tunneling through the bifurcation, see Figure 2.7(right). Here the secondary branch has isotropy subgroup $S_6 \times S_2 \times S_{17}$ and the split changes to $p = 8, q = 17$ as two PODs belatedly change species.

In Figure 2.8(left) we have included a moderate level of noise. Noise triggers early instability of equilibria with small basins of attraction, and here leads to a secondary bifurcation that is not present in Figure 2.7. Figure 2.8(right) repeats this scenario but with a bigger range for λ. In this range, the noise triggers three consecutive secondary bifurcations, and the corresponding splits are $p = 4, q = 21$, then $p = 5, q = 20$, then $p = 6, q = 19$, and finally (in this figure) $p = 7, q = 18$.

The fossil record is too sparse to support the occurrence of secondary bifurcations of this kind, which affect short-term details of the speciation process but not its eventual outcome. However, it might be possible to detect secondary bifurcations in laboratory experiments on microorganisms in a suitably designed experiment.

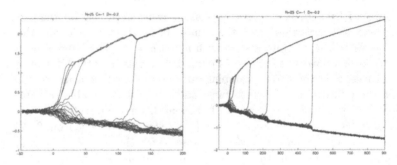

Figure 2.8: Numerical simulation of (2.10) in the presence of random noise with parameters $N = 25, C = -1, D = -0.2$. (Left) $\lambda \in [-50, 200]$; (right) $\lambda \in [-50, 900]$.

2.8 Steady-State Bifurcations in Reaction-Diffusion Systems

In this section we consider bifurcations from a constant equilibrium in a system of reaction-diffusion equations defined on a circle or equivalently on a line segment with periodic boundary conditions. This discussion extends the ideas in Section 1.8 concerning the single equation (1.17).

Let $u = (u_1, \ldots, u_m)$ be an m-vector of real-valued functions $u_j(x, t)$ that satisfy the reaction-diffusion system

$$u_t = D\Delta u + F(u, \lambda) \tag{2.16}$$

where the Laplacian Δ acts on u by acting individually on each component of u, D is a matrix of diffusion constants (usually assumed to be non-negative), and $F : \mathbf{R}^m \times \mathbf{R} \to \mathbf{R}^m$ is the reaction term. As mentioned, we assume that (2.16) satisfies *periodic boundary conditions* (PBC) on the interval $x \in [0, \ell]$ of length ℓ

$$\begin{aligned} u(0, t) &= u(\ell, t) \\ u_x(0, t) &= u_x(\ell, t) \end{aligned} \tag{2.17}$$

We assume that this system has a spatially homogeneous equilibrium for all λ; hence, we assume

$$F(0, \lambda) = 0$$

Finally, we assume that 0 is an asymptotically stable equilibium for $\lambda < 0$ that loses stability at $\lambda = 0$. To determine the ways in which this equilibrium can lose stability by a steady-state bifurcation, we must compute the eigenvalues of the linearized equations — and it is in this calculation that symmetry helps.

Proceeding exactly as in Section 1.8 we see that the steady-state nonlinear operator

$$\mathcal{P}u = D\Delta u + F(u, \lambda)$$

is $\mathbf{O}(2)$-equivariant where $\mathbf{SO}(2)$ acts on $x \in \mathbf{R}$ by translation modulo the spatial period ℓ and the reflection κ acts by $x \mapsto -x$. Indeed

$$\mathcal{P} : \mathcal{C}_\ell^2 \to \mathcal{C}_\ell^0$$

where $\mathcal{C}_\ell^0$ is the space of m-vectors of continuous ℓ-periodic functions and $\mathcal{C}_\ell^2$ is the subspace of twice continuously differentiable functions.

The isotypic decomposition theorem (Lemma 2.10 and Theorem 2.12 together) works just as well for compact Lie groups acting on (infinite-dimensional) function spaces. In particular, commuting linear operators must map isotypic components to isotypic components. We can use Fourier analysis to decompose C_ℓ^0 into isotypic components. By Fourier analysis we may write

$$u(x) = a_0 + a_1 e^{2\pi x i/\ell} + \cdots + a_k e^{2\pi k x i/\ell} + \cdots + \text{c.c.}$$

where $a_j \in \mathbf{C}^m$ and c.c. indicates complex conjugate. It follows that the subspaces

$$X_k = \{e^{2\pi k x i/\ell} a + \text{c.c.} : a \in \mathbf{C}^m\}$$

are the $\mathbf{O}(2)$-isotypic components of both C_ℓ^2 and C_ℓ^0, as we now show.

Each subspace X_k is the sum of m copies of the two-dimensional $\mathbf{O}(2)$ absolutely irreducible space

$$V_k = \{z e^{2\pi k x i/\ell} + \text{c.c.} : z \in \mathbf{C}\}$$

Note that when $k > 0$, $\theta \in \mathbf{SO}(2) = [0, \ell]$ acts on the complex coordinate z in V_k by

$$\theta z = e^{-2\pi k \theta i/\ell} z$$

It is easy to check that $\kappa x = -x$ induces the action

$$z \mapsto \bar{z}$$

Finally, it is straightforward to check that each of these two-dimensional representations of $\mathbf{O}(2)$ is absolutely irreducible and distinct (the kernel of the action on V_k is $\mathbf{Z}_k$ and isomophic representations have isomorphic kernels). Thus, the X_k are the desired isotypic components.

It follows by symmetry that the linearization of $\mathcal{P}$

$$Lu = D\Delta u + A(\lambda)u$$

where $A(\lambda) = (dF)_{(0,\lambda)}$ must map each subspace X_k into itself and that the eigenvalues of L are just the union of all of the eigenvalues of $L|X_k$ for $k = 0, 1, \ldots$. We could, of course, have verified this point by direct calculation — but we believe that it was worthwhile to have seen first why the calculation must succeed before actually performing that calculation.

Now, calculate

$$L(e^{2\pi k x i/\ell} a) = e^{2\pi k x i/\ell} \left(A(\lambda) - \left(\frac{2\pi k}{\ell} \right)^2 D \right) a$$

It follows that the eigenvalues of $L|X_k$ are the eigenvalues of the $m \times m$ matrix

$$L_k = A(\lambda) - \left(\frac{2\pi k}{\ell} \right)^2 D \tag{2.18}$$

A Single Reaction-Diffusion Equation. For a single reaction-diffusion equation we may rescale time so that $D = 1$. From (2.18) we see that the eigenvalues of the linearized equation at $\lambda = 0$ are

$$A(0) - \left(\frac{2\pi k}{\ell} \right)^2 \tag{2.19}$$

It follows that a steady-state bifurcation occurs when

$$A(0) = \left(\frac{2\pi k_0}{\ell} \right)^2$$

for some integer k_0. In this case the nonlinear theory augmented by the Equivariant Branching Lemma shows that generically a unique branch of solutions with $\mathbf{Z}_{k_0}$ symmetry and k_0 spatial oscillations on the interval $[0, \ell]$ (equivalently k_0 spatial oscillations on the circle) appear.

Note that these solutions must be linearly unstable when $k_0 > 0$, since the eigenvalues listed in (2.19) for all integers $k < k_0$ are positive. When the system consists of two or more equations, then bifurcation to stable $\mathbf{Z}_{k_0}$ symmetric solutions is possible, as we now show.

Systems of Two Reaction-Diffusion Equations. The Brusselator was designed as the simplest model consistent with chemical kinetics that exhibits oscillatory behavior like that seen in the Belousov-Zhabotinsky reaction (Prigogine and Lefever [438]). See Chapter 4. We use it here to explore steady-state bifurcations.

The Brusselator is a two-equation reaction-diffusion system in which the diffusion matrix is diagonal

$$D = \begin{bmatrix} D_1 & 0 \\ 0 & D_2 \end{bmatrix} \tag{2.20}$$

and the reaction term is

$$F(u) = \begin{bmatrix} \lambda - 1 & \alpha^2 \\ -\lambda & -\alpha^2 \end{bmatrix} u + \left(\frac{\lambda}{\alpha} u_1^2 + 2\alpha u_1 u_2 + u_1^2 u_2 \right) \begin{bmatrix} 1 \\ -1 \end{bmatrix} \tag{2.21}$$

This system was originally posed with Dirichlet boundary conditions where α and λ are positive constants representing concentrations of the two chemicals u_1, u_2 at the boundary.

As an example, we discuss here the Brusselator with periodic boundary conditions on the interval $[0, \ell]$ and set

$$D_1 = 1 \qquad D_2 = 2 \qquad \ell = 2\pi \tag{2.22}$$

It follows from (2.18) that the eigenvalues of the linear operator are the eigenvalues of the 2×2 matrices

$$L_k = \begin{bmatrix} \lambda - 1 - k^2 & \alpha^2 \\ -\lambda & -\alpha^2 - 2k^2 \end{bmatrix}$$

We now show that for any $k_0 > 1$ it is possible to choose α so that the first zero eigenvalue as λ increases from 0 occurs in in the matrix L_{k_0} and that all other eigenvalues of L have negative real part.

The eigenvalues of the 2×2 matrix L_k are determined by its determinant and trace. Thus we compute

$$\begin{aligned} \det(L_k) &= 2k^4 + (\alpha^2 + 2)k^2 + \alpha^2 - 2k^2\lambda \\ \operatorname{tr}(L_k) &= \lambda - (1 + \alpha^2 + 3k^2) \end{aligned}$$

The matrix L_k has a zero eigenvalue when its determinant vanishes; this occurs at the λ value

$$\lambda_k = k^2 + \frac{\alpha^2 + 2}{2} + \frac{\alpha^2}{2k^2}$$

Treat k as a real parameter — the curve of λ_k as a function of k is called a *dispersion curve*. It is easy to show that the minimum value λ^* of λ_k occurs at

$$k^2 = \frac{\alpha}{\sqrt{2}} \qquad \text{and} \qquad \lambda^* = 1 + \sqrt{2}\alpha + \frac{\alpha^2}{2}$$

So the first zero eigenvalue of the linearized PDE L occurs at k_0 when we set

$$\alpha = \sqrt{2}k_0^2$$

Finally, we show that all of the nonzero eigenvalues of L have negative real part. We verify this point by showing that $\det(L_k) > 0$ and $\operatorname{tr}(L_k) < 0$. Note that $\det(L_k)$ is decreasing in λ and positive at $\lambda = 0$. Hence $\det(L_k) > 0$ on the interval

$[0, \lambda^*]$ for all $k \neq k_0$, since the first zero eigenvalues of any L_k occurs when $k = k_0$. Next note that $\text{tr}(L_k)$ is linear in λ and negative at $\lambda = 0$. It is easy to show that $\text{tr}(L_k) < 0$ on the interval $[0, \lambda^*]$ for all k whenever

$$\lambda^* < 1 + \alpha^2$$

This inequality is valid as long as $\alpha > 2\sqrt{2}$, which in turn is valid whenever $k_0 > 1$. It can also be shown that $k_0 = 1$ is a possible first bifurcation for other values of the diffusion coefficients.

Pattern Formation. We use a variant of the calculations performed in the previous subsection to exhibit an often observed physical phenomenon — the spontaneous formation of regular patterns with a well-defined spatial wave length. The basic idea can now be viewed as rather simple, but it was quite a surprise when Alan Turing [506] first made this observation in his seminal paper in 1952.

Suppose that we compute the first instability of the spatially homogeneous solution 0 to a spatially periodic perturbation, that is, we look for eigenfunctions of L of the form $e^{\psi i x} a + \text{c.c.}$ for some spatial wave length $2\pi/\psi$. Turing observed that there is can be a critical wave length for which instability occurs first. We illustrate this point by using our previous calculations on the Brusselator.

There is an instability of the spatially homogeneous solution in the Brusselator when there is a zero eigenvalue of the linearization

$$L_\psi = \begin{bmatrix} \lambda - 1 - \psi^2 & \alpha^2 \\ -\lambda & -\alpha^2 - 2\psi^2 \end{bmatrix}$$

Note that in this calculation ψ need not be an intger. It follows from the previous calculations that the minimum value λ^* occurs at

$$\psi^2 = \frac{\alpha}{\sqrt{2}} \quad \text{and} \quad \lambda^* = 1 + \sqrt{2}\alpha + \frac{\alpha^2}{2}$$

Thus, for a fixed value of α, the first instability as λ increases occurs at λ^* with wave length $2\pi/\psi$ where

$$\psi = \sqrt{\frac{\alpha}{\sqrt{2}}}$$

Thus, bifurcation from a homogeneous state leads spontaneously to a spatially inhomogeneous state with a well-defined spatial wave length.

Applications of these ideas to PDEs can be found in Chapters 4 and 5.

Chapter 3

Time Periodicity and Spatio-Temporal Symmetry

After steady states, the next simplest type of dynamics is *periodic* behavior, in which the state of the system repeats after some fixed interval of time. That is, the solution $x(t)$ satisfies

$$x(t + T) = x(t) \quad \text{for all } t \in \mathbf{R}$$

where $0 < T \in R$ is the *period*. We also say that x is *T-periodic*. When discussing PDEs we often use the phrase 'time-periodic' to distinguish between such a state and one with spatially periodic patterning. This chapter focuses on periodic states in symmetric dynamical systems, a particularly rich area of equivariant dynamics.

The motivating application for this chapter is animal locomotion, a topic that we introduce in Section 3.1. It has long been observed that legged animals move using a variety of patterns, known as 'gaits'. These patterns are time-periodic, but they also have 'spatial' regularity corresponding to permutations of the legs. Thus, we are led to consider spatio-temporal patterns involving both space and time. The study of animal gaits illustrates the power of symmetry methods, because there is (as yet) no standard, agreed set of model equations in this area. (There are several widely used models, but none has the pedigree of, say, mechanical equations in the physical sciences.) We show that general symmetry considerations constrain the behavior that any symmetry-based model will permit.

In particular, we ask a basic existence question. Given a group Γ acting on $\mathbf{R}^n$, which spatio-temporal symmetries are *possible* in a Γ-equivariant system of ODEs? We answer this question completely, in terms of purely algebraic criteria, when Γ is a finite group: see Theorem 3.4. We apply the results to rings of identical 'cells' (or oscillators) and to models of animal locomotion. We also make a number of model-independent predictions about animal gaits, and review related experimental evidence. These predictions demonstrate that even without an explicit system of model equations, predictions made on grounds of symmetry can be testable.

Finally, we discuss multirhythms — periodic states in coupled cell systems in which symmetries force certain cells to be in resonance with other cells. We give examples of 2:1, 3:2, and 20:15:12 resonances of this kind. The 2:1 multirhythm has been observed in slime mold dynamics by Takamatsu *et al.* [500] and in CSTR's by Yoshimoto *et al.* [531].

3.1 Animal Gaits and Space-Time Symmetries

The gaits of quadrupeds (four-legged animals) provide an excellent introduction to the world of space time symmetries — or rhythms. It is well known that horses walk, trot, canter, and gallop (some horses can pace instead of trot), and that squirrels bound — but precisely what an animal does with its legs when performing one of these gaits is less well known. There is one feature that is common to all gaits: they are repetitive; that is, they are time-periodic.

In the pace, trot, and bound the animal's legs can be divided into two pairs — the legs in each pair move in synchrony, while legs in different pairs move with a half-period phase shift. The two pairs in a *bound* consist of the fore legs and the hind legs; the two pairs in a *pace* consist of the left legs and the right legs; and the two pairs in a *trot* consist of the the two diagonal pairs of legs. The quadruped *walk* has a more complicated cadence: each leg moves independently with a quarter-period phase shift in the order left hind, left fore, right hind, and right fore. As in the pace the left legs move and then the right legs move — but the left legs and the right legs do not move in unison.

The gallop (in fact there are two types of gallop) is a yet more complicated gait in which the animal 'leads' with one of its fore legs — its two fore legs (and its two hind legs, as well) execute different motions. Gallops are best described by the phases in the gait cycle in which the animal's legs touch the ground — for definiteness we begin the cycle with the left hind leg touching the ground.

In both types of gallop the animal then puts down its right hind leg a short time later. A half-period after putting down its left hind leg, it puts down its right fore leg in a *transverse gallop* and its left fore leg in a *rotary gallop*. The remaining leg hits the ground a half-period after the right hind leg.

We summarize the descriptions of these six gaits in Figure 3.1 by indicating the phases in the gait cycle when each given leg hits the ground. For definiteness, we start the gait cycle when the left hind leg hits the ground, while noting that gait analysts usually start the gait cycle when the left fore leg hits the ground. Numbers with asterisks are approximate. Phase differences of 0, $\frac{1}{4}$ and $\frac{1}{2}$ are (to within experimental error) exact.

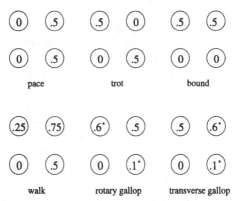

Figure 3.1: Six standard quadrupedal gaits.

Collins and Stewart [121, 122] and Schöner *et al.* [468] made the observation that each of these gaits can be distinguished by symmetry in the following sense. Spatio-temporal symmetries are permutations of the legs coupled with time shifts. So interchanging the two fore legs and the two hind legs of a bounding animal does not change the gait, while interchanging the two left legs and the two right legs leads to a half-period phase shift. In a walk permuting the legs in the order left hind to left fore to right hind to right fore leads to a quarter-period phase shift. We list the spatio-temporal symmetries of each of these gaits in Table 3.1.

Biologists often make the assumption that somewhere in the nervous system is a locomotor *central pattern generator* or CPG that produces the rhythms associated

Gait	Symmetries (leg permutation, phase shift)		
pace	$((1\,3)(2\,4), 0)$	$((1\,2)(3\,4), \frac{1}{2})$	$(1\,4)(2\,3), \frac{1}{2})$
trot	$((1\,3)(2\,4), \frac{1}{2})$	$((1\,2)(3\,4), \frac{1}{2})$	$(1\,4)(2\,3), 0)$
bound	$((1\,3)(2\,4), \frac{1}{2})$	$((1\,2)(3\,4), 0)$	$(1\,4)(2\,3), \frac{1}{2})$
walk	$((1\,3\,2\,4), \frac{1}{4})$	$((1\,2)(3\,4), \frac{1}{2})$	$((1\,4\,2\,3), \frac{3}{4})$
rotary gallop	$((1\,4)(2\,3), \frac{1}{2})$		
transverse gallop	$((1\,3)(2\,4), \frac{1}{2})$		

Table 3.1 Gait symmetries: 1 = left hind leg; 2 = right hind leg; 3 = left fore leg; 4 = right fore leg.

— but not in mammals. Nevertheless, suppose that we assume that there is a loco-motor CPG in quadrupeds — how can we model it? Since CPGs consist of neurons, neurons are often modeled by electrical circuits (for example, the Hodgkin-Huxley equations, see Keener and Sneyd [305]), and circuits are described by systems of ODEs, we may assume that our model is (a perhaps large dimensional) system of ODEs. We ask the question: What kind of structure does a system of ODEs need to have to produce all of these gaits — just by changing parameters within the model system? Not surprisingly, our answer is based on symmetry.

A favorite kind of model for CPGs is a coupled cell model (Kopell and Er-mentrout [322, 323], Schöner *et al.* [468], Rand *et al.* [442]). We imagine that for each leg there is a single group of neurons whose job is to signal that leg to move, and that the groups of neurons are otherwise identical. Moreover, we assume that the groups of neurons are coupled in some manner — and to simplify matters we assume that the kinds of coupling fall into a small number of identical types. A natural mathematical question now arises — even at this level of generality. Can couplings between these four groups of neurons be set up so that periodic solutions having the rhythms associated with each of these gaits exist? The answer is, per-haps surprisingly, no. To describe why, we must discuss a mathematically precise definition of spatio-temporal symmetries of time-periodic solutions.

In this chapter we prove a theorem that classifies the types of spatio-temporal symmetry that periodic solutions to systems of Γ-equivariant differential equations may have — at least when Γ is a finite group. See Theorem 3.4. This theorem is proved in phase space, but — as usual — its interesting applications are in physical space. We will use this theorem to analyze the structure of locomotor CPGs for quadrupeds in Section 3.5 and to introduce multirhythms in Section 3.6.

3.2 Symmetries of Periodic Solutions

Suppose that

$$\frac{\mathrm{d}x}{\mathrm{d}t} = f(x) \tag{3.1}$$

is a system of differential equations with $x \in \mathbf{R}^n$ and symmetry group Γ. Suppose that $x(t)$ is a T-periodic solution of (3.1) and that $\gamma \in \Gamma$. We discuss the ways in which γ can be a symmetry of $x(t)$; the main tool is the uniqueness theorem for solutions to the initial value problem for (3.1).

We know that $\gamma x(t)$ is another T-periodic solution of (3.1). Should the two trajectories intersect, then the common point of intersection would be the same initial point for the two solutions. Uniqueness of solutions then implies that the trajectories of $\gamma x(t)$ and $x(t)$ must be identical. So either the two trajectories are identical or they do not intersect.

Suppose that the two trajectories are identical. Then uniqueness of solutions implies that there exists $\theta \in \mathbf{S}^1 = [0, T]$ such that $\gamma x(t) = x(t - \theta)$, or

$$\gamma x(t + \theta) = x(t) \tag{3.2}$$

We call $(\gamma, \theta) \in \Gamma \times \mathbf{S}^1$ a *spatio-temporal* symmetry of the solution $x(t)$. A spatio-temporal symmetry of $x(t)$ for which $\theta = 0$ is called a *spatial symmetry*, since it fixes the point $x(t)$ at every moment of time. The group of all spatio-temporal symmetries of $x(t)$ is denoted

$$\Sigma_{x(t)} \subseteq \Gamma \times \mathbf{S}^1$$

Next we show how the symmetry group $\Sigma_{x(t)}$ can be identified with a pair of subgroups H and K of Γ and a homomorphism $\Theta : H \to \mathbf{S}^1$ with kernel K. Define

$$
\begin{aligned}
K &= \{\gamma \in \Gamma : \gamma x(t) = x(t) \quad \forall t\} \\
H &= \{\gamma \in \Gamma : \gamma\{x(t)\} = \{x(t)\}\}
\end{aligned}
\tag{3.3}
$$

The subgroup $K \subseteq \Sigma_{x(t)}$ is the group of spatial symmetries of $x(t)$ and the subgroup H consists of those symmetries that preserve the trajectory of $x(t)$ — in short, the spatial parts of spatio-temporal symmetries of $x(t)$. Indeed, the groups $H \subseteq \Gamma$ and $\Sigma_{x(t)} \subseteq \Gamma \times \mathbf{S}^1$ are isomorphic; the isomorphism is just the restriction to $\Sigma_{x(t)}$ of the projection of $\Gamma \times \mathbf{S}^1$ onto Γ. The group $\Sigma_{x(t)}$ can be written as

$$\Sigma^\Theta = \{(h, \Theta(h)) : h \in H\} \tag{3.4}$$

We call Σ^Θ a *twisted* subgroup of $\Gamma \times \mathbf{S}^1$.

Four Signals Do Not Suffice for Quadruped Locomotor CPGs. Suppose that we model the locomotor CPG of a quadruped by a system of four identical subsystems of differential equations that are coupled in some manner. That is, the state of the system of differential equations is determined by $x = (x_1, x_2, x_3, x_4) \in (\mathbf{R}^k)^4$. We imagine that each signal $x_j(t) \in \mathbf{R}^k$, derived from a periodic solution to this system, is sent to the leg indicated in Figure 3.2.

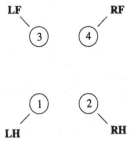

Figure 3.2: Schematic network for gaits in four-legged animals.

Suppose, in addition, that the system of differential equations has periodic solutions that correspond to walk, trot, and pace. From Table 3.1 we can make this correspondence precise. For example, if $((1\,3\,2\,4), \frac{1}{4})$ is a spatio-temporal symmetry of the 1-periodic solution $x(t)$, it follows that

$$x_2(t) = x_1\left(t + \frac{1}{2}\right) \qquad x_3(t) = x_1\left(t + \frac{1}{4}\right) \qquad x_4(t) = x_1\left(t + \frac{3}{4}\right)$$

and the solution sends the rhythm of a walk to the four legs.

We claim (following Golubitsky *et al.* [234] and Buono and Golubitsky [87]) that in such a network there must be a symmetry that transforms the trot into a pace. It follows that trot and pace are equally accessible in such a model. Consequently, no four signal system of this type is a good model for quadruped locomotor CPGs, since some animals trot but do not pace (squirrels) and some animals pace but do not trot (camels).

To verify this claim, note that the symmetry group of this system depends on the precise form of the coupling and is a subgroup $\Gamma \subseteq \mathbf{S}_4$. Since the walk exists, the four-cycle $(1324) \in \Gamma$. Supposing that the rhythms of trot and pace are generated by system symmetries, we see that $(12)(34) \in \Gamma$. Moreover, trot implies $(14)(23) \in \Gamma$ and pace implies $(13)(24) \in \Gamma$. Note that

$$
\begin{aligned}
(1324) \cdot (14)(23) \cdot (1324)^{-1} &= (13)(24) \\
(1324) \cdot (13)(24) \cdot (1324)^{-1} &= (14)(23) \\
(1324) \cdot (12)(34) \cdot (1324)^{-1} &= (12)(34)
\end{aligned}
$$

Thus, (1324) conjugates the generators of trot and pace symmetries and these solution types are conjugate. $\qquad\square$

What, then, is the architecture of a quadruped CPG? We wills how that it is possible to construct an essentially unique eight-subsystem model that produces solutions corresponding to each of the quadrupedal gaits. See Section 3.5.

Algebraic Restrictions on Spatio-Temporal Symmetries. There are algebraic restrictions on the pair H and K defined in (3.3) in order for them to correspond to symmetries of a periodic solution.

Lemma 3.1 *Let $x(t)$ be a periodic solution of (3.1) and let H and K be the subgroups of Γ defined in (3.3). Then*

(a) *K is an isotropy subgroup for the Γ-action.*
(b) *K is a normal subgroup of H and H/K is either cyclic or $\mathbf{S}^1$.*
(c) *$\dim \mathrm{Fix}(K) \geq 2$. If $\dim \mathrm{Fix}(K) = 2$, then either $H = K$ or $H = N(K)$.*

Proof. (a) follows from uniqueness of solutions. Let $x_0 = x(0)$ and let γ be in the isotropy subgroup of x_0. Then $\gamma x(t)$ is another solution to (3.1) with initial condition $\gamma x_0 = x_0$. It follows that $\gamma x(t) = x(t)$ for all t and that $\gamma \in K$. By definition, K is contained in the isotropy subgroup of x_0, so K is the isotropy subgroup of x_0.

(b) Let $\Theta : H \to \mathbf{S}^1$ be defined by $\Theta(\gamma) = \theta$ where θ is given in (3.2). We claim that Θ is a group homomorphism and $\ker \Theta = K$. It follows that K is a normal subgroup of H and that H/K is isomorphic to a closed subgroup of $\mathbf{S}^1$. Therefore either $H/K \cong \mathbf{Z}_m$ or $H/K \cong \mathbf{S}^1$. To verify the claim let γ and δ be in H. Uniqueness of solutions implies that $\gamma x(t) = x(t - \theta)$ and $\delta x(t) = x(t - \varphi)$. Thus

$$
\gamma \delta x(t) = \gamma x(t - \varphi) = x(t - \varphi - \theta)
$$

so $\Theta(\gamma\delta) = \Theta(\gamma) + \Theta(\delta)$. Also, if $\gamma x(t) = x(t - \theta)$, then $x(t) = \gamma^{-1} x(t - \theta)$ and $\gamma^{-1} x(t) = x(t - (-\theta))$. Therefore $\Theta(\gamma^{-1}) = -\Theta(\gamma)$. So Θ is a group homomorphism. Moreover, $\Theta(\gamma) = 0$ precisely when $\gamma \in K$, so K is the kernel of Θ.

(c) In fact, by definition, K is contained in the isotropy subgroup of every point on the trajectory of $x(t)$. Therefore, $x(t) \in \mathrm{Fix}(K)$ for all t. Since fixed-point subspaces are flow-invariant, $x(t)$ is a solution to the nonautonomous system (3.1) restricted to $\mathrm{Fix}(K)$. Since a nonautonomous system can have periodic solutions only when its dimension is at least two, $\dim \mathrm{Fix}(K) \geq 2$.

The final restriction occurs only when $\dim \text{Fix}(K) = 2$: A simple example illustrates the reason for this condition. Suppose that $\Gamma = \mathbf{Z}_4(\rho)$ acts by rotations on $\mathbf{R}^2$, $H = \mathbf{Z}_2(\rho^2)$, and $K = 1$. Observe that $H/K = \mathbf{Z}_2$ is cyclic, K is an isotropy subgroup, and $\dim \text{Fix}(K) = 2$; so all previous restrictions are satisfied. However, $N(K) = \mathbf{Z}_4$ so that $N(K) \gneqq H \gneqq K$. Suppose that $x(t)$ is a T-periodic solution whose spatio-temporal symmetry is $\rho^2 x(t) = x(t + \frac{T}{2})$. The trajectory of $x(t)$ must avoid the origin and, because ρ^2 is rotation by π, a simple topological argument shows that the trajectory must have the origin in its interior and $\rho\{x(t)\}$ must intersect $\{x(t)\}$. Hence $\rho\{x(t)\} = \{x(t)\}$ and $\rho x(t) = x(t \pm \frac{T}{4})$; that is, the spatio-temporal symmetry group is larger than H.

In effect, this is the only type of difficulty that can arise, as we now show. Suppose that $H \gneqq K$; then $H/K = \mathbf{Z}_m$ for some $m \geq 2$. We claim that H/K is generated by a rotation; the only other possibility is that $H/K = \mathbf{Z}_2(\tau)$ where τ acts as a reflection on $\text{Fix}(K)$. Since $\tau\{x(t)\} = \{x(t)\}$ it follows that $\{x(t)\} \cap \text{Fix}(\tau) \neq \emptyset$. Thus $\{x(t)\} \subseteq \text{Fix}(\tau)$, which is not possible. As in the example, H/K is generated by a rotation and $x(t)$ must contain the origin in its interior. If $\gamma \in N(K) \backslash H$, then $\gamma\{x(t)\}$ must intersect $\{x(t)\}$ and the spatio-temporal symmetry group of $x(t)$ must be larger than H. $\square$

Definition 3.2 When $H/K \cong \mathbf{Z}_m$ the periodic solution $x(t)$ is called either a *standing wave* or (usually for $m \geq 3$) a *discrete rotating wave*; and when $H/K \cong \mathbf{S}^1$ it is called a *rotating wave*.

Example 3.3 We can create a rotating wave in the plane in a straightforward way. The $\mathbf{SO}(2)$-equivariant differential equation

$$\begin{aligned} dx/dt &= (1 - (x^2 + y^2))x - y \\ dy/dt &= (1 - (x^2 + y^2))y + x \end{aligned}$$

has the unit circle as a stable periodic solution and is a rotating wave. If we perturb this system to the $\mathbf{Z}_3$-equivariant system

$$\begin{aligned} dx/dt &= (1 - (x^2 + y^2))x - y + 0.4(x^2 - y^2) \\ dy/dt &= (1 - (x^2 + y^2))y + x - 0.8xy \end{aligned}$$

then the perturbed periodic solution will have a trajectory with $\mathbf{Z}_3$ symmetry and will be a discrete rotating wave. See Figure 3.3. $\diamond$

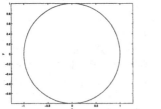 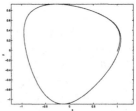

Figure 3.3: (Left) Planar rotating wave; (right) planar discrete rotating wave.

3.3 A Characterization of Possible Spatio-Temporal Symmetries

In Section 3.1 we posed the question: can a single network of identical coupled systems of differential equations produce the rhythms associated with with standard gaits? In Section 3.2 we used elementary group theory to show that the answer to this question is 'no' when there are exactly four subsystems. We also indicated that the answer to this question is 'yes' when there are eight subsystems. We now develop the mathematical tool that allows us to answer this question. This tool is a theorem that gives necessary and sufficient conditions for the existence of periodic solutions in systems of differential equations that are equivariant under some specified action of a given finite symmetry group.

The main theorem of this section, taken from Golubitsky *et al.* [87], is a characterization of the possible spatio-temporal symmetry of periodic solutions.

Let Γ be a finite group acting on $\mathbf{R}^n$ and let $x(t)$ be a periodic solution of a Γ-equivariant system of ODEs. Let K be the subgroup of spatial symmetries of $x(t)$ and let H be the subgroup of spatio-temporal symmetries defined in (3.3). Lemma 3.1 proves that H and K satisfy three algebraic conditions. We now show that there is one additional geometric restriction on H and K.

Define

$$L_K = \bigcup_{\gamma \notin K} \mathrm{Fix}(\gamma) \cap \mathrm{Fix}(K)$$

Since K is an isotropy subgroup (Lemma 3.1(a)), L_K is the union of proper subspaces of $\mathrm{Fix}(K)$. More precisely, suppose that $\mathrm{Fix}(\gamma) \supseteq \mathrm{Fix}(K)$. Then the isotropy subgroup of every point in $\mathrm{Fix}(K)$ contains both K and $\gamma \notin K$. Therefore, the isotropy subgroup of any point in $\mathrm{Fix}(K)$ is larger than K, and K is not an isotropy subgroup.

We claim that

$$H \text{ fixes a connected component of } \mathrm{Fix}(K) \backslash L_K. \tag{3.5}$$

To verify (3.5) we first show that any δ in the normalizer $N(K)$ permutes connected components of $\mathrm{Fix}(K) \backslash L_K$. Observe that

$$\delta(\mathrm{Fix}(\gamma) \cap \mathrm{Fix}(K)) = \mathrm{Fix}(\delta\gamma\delta^{-1}) \cap \mathrm{Fix}(\delta K\delta^{-1}) = \mathrm{Fix}(\delta\gamma\delta^{-1}) \cap \mathrm{Fix}(K)$$

Moreover, $\delta\gamma\delta^{-1} \notin K$. (If it were, then γ would be in $\delta^{-1}K\delta = K$, which it is not.) Therefore $\delta : L_K \to L_K$. Since δ is invertible, $\delta : \mathrm{Fix}(K) \backslash L_K \to \mathrm{Fix}(K) \backslash L_K$ and δ permutes the connected components of $\mathrm{Fix}(K) \backslash L_K$.

Since H/K is cyclic, we can choose an element $h \in H$ that projects onto a generator of H/K. We now show that h (and hence H) must fix one of the connected components of $\mathrm{Fix}(K) \backslash L_K$. Suppose that the trajectory of $x(t)$ intersects the flow-invariant subspace $\mathrm{Fix}(\gamma) \cap \mathrm{Fix}(K)$. Flow-invariance of $\mathrm{Fix}(\gamma)$ implies that γ is a spatial symmetry of the solution $x(t)$, and by definition $\gamma \in K$. Therefore the trajectory of $x(t)$ does not intersect L_K. Since h is a spatio-temporal symmetry of $x(t)$, it preserves the trajectory of $x(t)$. Therefore, h must map the connected component of $\mathrm{Fix}(K) \backslash L_K$ that contains the trajectory of $x(t)$ into itself, thus verifying (3.5).

Theorem 3.4 *Let Γ be a finite group acting on $\mathbf{R}^n$. There is a periodic solution to some Γ-equivariant system of ODEs on $\mathbf{R}^n$ with spatial symmetries K and spatio-temporal symmetries H if and only if*

    (a) *H/K is cyclic,*

    (b) *K is an isotropy subgroup,*

(c) $\dim \operatorname{Fix}(K) \geq 2$. If $\dim \operatorname{Fix}(K) = 2$, then either $H = K$ or $H = N(K)$,

(d) H fixes a connected component of $\operatorname{Fix}(K) \backslash L_K$.

Moreover, when these conditions hold, hyperbolic asymptotically stable limit cycles with the desired symmetry exist.

Proof. In the preceding discussion we have proved that the four conditions are necessary; now we prove that they are sufficient. We must prove the existence of a hyperbolic periodic solution with spatial symmetries K and spatio-temporal symmetries H. We sketch that proof here showing, in addition, that the periodic solution can be a stable limit cycle.

Choose a generator h of $H/K = \mathbf{Z}_m$. By assumption, H fixes a connected component C of $\operatorname{Fix}(K) \backslash L_K$. Recall that $N(K) \subseteq \Gamma$ is the largest subgroup that acts on $\operatorname{Fix}(K)$ and that elements in $N(K)$ permute the connected components of $\operatorname{Fix}(K) \backslash L_K$. Define

$$\hat{H} = \{\gamma \in N(K) : \gamma(C) = C\}$$

Two points need to be verified.

(i) There is a non-self-intersecting closed curve J in C that is mapped onto itself by h and such that no point on J is fixed by h. Moreover, $\gamma(J) \cap J = \emptyset$ for all $\gamma \in \hat{H} \backslash H$. If so, we can construct a C^∞ vector field f_1 on C for which J is a stable limit cycle and we can smooth f_1 so that it is zero near L_K and near $\gamma(J)$ for all $\gamma \in \hat{H} \backslash H$.

(ii) There is a smooth extension of f_1 to the whole of $\mathbf{R}^n$ that is Γ-equivariant.

Once these points are verified the proof may be completed as follows. Since K is an isotropy subgroup, the spatial symmetry subgroup of J is K. Since $h : J \to J$, the element h is a spatio-temporal symmetry of J and the spatio-temporal symmetry group of J is H. Proving that hyperbolic periodic solutions with spatial symmetry K and spatio-temporal symmetry H are robust (that is, they perturb to periodic solutions with the same symmetry subgroups) is straightforward. Since hyperbolicity implies that the perturbed periodic solution $V(t)$ is unique, it follows that $V(t) \in \operatorname{Fix}(K)$ for all t and that $hV(t)$ must be the same trajectory as $V(t)$. Since the number of temporal symmetries of $U(t)$ is m, it follows by continuity that the spatio-temporal symmetries of $V(t)$ form the subgroup H.

To verify (i) choose a point $x_1 \in C$ and form the group orbit $x_j = h^j x_1$ for $j = 2, \ldots, m$. Note that the points $x_j \in C$ since $h : C \to C$. Choose a smooth curve $\alpha(t)$ on a small neighborhood of $t = 0$ such that $\alpha(0) = x_1$. Use h to push the curve α to the curve $\beta(1+t) = h\alpha(t)$ where $\beta(1) = x_2$. Now extend these curves to a non-selfintersecting curve $J_1 : [0, 1] \to C$ so that $J_1(0) = x_1$, $J_1(1) = x_2$, and J_1 equals α near $t = 0$ and β near $t = 1$. This is possible since C is connected. Now let $J = \cup_j h^j(J_1)$. By construction $J \subseteq C$ is a smooth curve that is invariant under h. There are two difficulties: J can intersect itself and J might intersect $\gamma(J)$ where $\gamma \in \hat{H} \backslash H$. If $\dim \operatorname{Fix}(K) \geq 3$, then we can use transversality arguments (like those used to prove the Whitney Embedding Theorem, see Golubitsky and Guillemin [213]) to avoid these difficulties in our choice of J. If $\dim \operatorname{Fix}(K) = 2$, then there is a potential problem when constructing J; see Figure 3.4. Suppose h is rotation by $\frac{4\pi}{5}$, which generates the cyclic group $\mathbf{Z}_5$. Then self-intersections of J are unavoidable. If, however, we choose the generator to be $h_* = h^3$, which here is rotation through $\frac{2\pi}{5}$, then self-intersection can be avoided. So even when $\dim \operatorname{Fix}(K) = 2$ the construction of J is possible. Moreover, when $\dim \operatorname{Fix}(K) = 2$ assumption (c) states that either $H = N(K)$, in which case no restrictions come from $\hat{H}$, or $H = K$, in which case we choose a small curve J in $\mathbf{R}^2 \backslash L_K$ that does not intersect any of its images under $N(K)$.

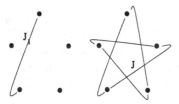

Figure 3.4: Unavoidable self-intersection in J with certain generators of H/K when $\dim \mathrm{Fix}(K) = 2$.

To verify (ii) we average f_1 over the group. First, extend $f_1 : C \to \mathrm{Fix}(K)$ to $f_1 : \mathrm{Fix}(K) \to \mathrm{Fix}(K)$ by setting $f_1 = 0$ outside an open neighborhood of J whose closure is inside C. Then let

$$f_2(x) = \frac{1}{|N(K)|} \sum_{\gamma \in N(K)} \gamma^{-1} f_1(\gamma x)$$

The vector field f_2 is $N(K)$-equivariant, zero near L_K, and has J as a stable limit cycle. Next extend f_2 to $f_2 : \mathbf{R}^n \to \mathbf{R}^n$ so that J is now a stable limit cycle in all of $\mathbf{R}^n$ and $f_2 = 0$ outside a neighborhood of J. In particular, $f_2 = 0$ near L_K. Now average again by setting

$$f_3(x) = \frac{1}{|\Gamma|} \sum_{\gamma \in \Gamma} \gamma^{-1} f_2(\gamma x)$$

The vector field f_3 is the desired extension. $\square$

Remark 3.5 Suppose that h is a generator of $\mathbf{Z}_m = H/K$. Is there a periodic solution such that $(h, \frac{1}{m})$ is a spatio-temporal symmetry? The answer may be no when $\dim \mathrm{Fix}(K) = 2$. In this case, we can always choose some generator $h' \in \mathbf{Z}_m = H/K$ so that $(h', \frac{1}{m})$ is a spatio-temporal symmetry; but we cannot necessarily choose *every* generator h. The essential obstacle is the one shown in Figure 3.4.

Theorem 3.6 *Let Γ be a finite group acting on $\mathbf{R}^n$. There is a T-periodic solution to some Γ-equivariant system of ODEs on $\mathbf{R}^n$ with spatial symmetries K and spatio-temporal symmetries H with*

$$hx(t) = x\left(t - \frac{T}{m}\right)$$

for some fixed $h \in N(K)$ if and only if

    (a) *$H/K \cong \mathbf{Z}_m$ is cyclic, $m \geq 2$, and $h \in H$ projects onto a generator of H/K,*
    (b) *K is an isotropy subgroup,*
    (c) *$\dim \mathrm{Fix}(K) \geq 2$. If $\dim \mathrm{Fix}(K) = 2$, then $H = N(K)$ and h acts on $\mathrm{Fix}(K)$ by rotation through $\pm \frac{2\pi}{m}$,*
    (d) *H fixes a connected component of $\mathrm{Fix}(K) \backslash L_K$.*

Proof. This theorem follows from Theorem 3.4 by noting that the only obstruction to the existence of a periodic solution with a specified spatiotemporal symmetry is the one described in Remark 3.5 and Figure 3.4. $\square$

Corollary 3.7 *For pairs (H, K) satisfying conditions (a)-(d) of Theorem 3.4, the property of having periodic solutions with spatial symmetries K and spatio-temporal symmetries H is robust in Γ-equivariant systems of ODEs on $\mathbf{R}^n$.*

Proof. Suppose that a hyperbolic periodic solution $x(t)$ has the symmetry group pair $K \subseteq H$. Hyperbolicity implies that any sufficiently small Γ-equivariant perturbation of the ODE system has a unique periodic solution $y(t)$ near $x(t)$ and that $y(t)$ is hyperbolic. We claim that the spatial symmetry group of $y(t)$ is K and the spatio-temporal symmetry group of $y(t)$ is H. We begin with the spatial symmetries.

Since the periodic solution $x(t)$ is also a hyperbolic solution to the system of ODEs restricted to $\text{Fix}(K)$, it follows that there is a periodic solution to the system restricted to $\text{Fix}(K)$ near $x(t)$. Uniqueness of $y(t)$ implies that $y(t)$ is in $\text{Fix}(K)$. The connected components of $\text{Fix}(K)\backslash L_K$ are open, and the trajectory of $x(t)$ lies in one of these connected components, so for sufficiently small perturbations, $y(t)$ must be in the same connected component as $x(t)$. Thus the spatial symmetry subgroup of $y(t)$ is K.

Next we discuss the spatio-temporal symmetries. Let $h \in H$ project onto a generator of H/K and let $\theta \in \mathbf{S}^1$ satisfy $hx(t + \theta) = x(t)$. Since h maps the trajectory $x(t)$ to itself, $hy(t + \theta)$ must be a periodic solution to the perturbed system that is near $x(t)$. Uniqueness of the perturbed periodic solution implies that $hy(t + \theta) = y(t)$. So the spatio-temporal symmetries of $y(t)$ are identical to those of $x(t)$. $\qquad\qquad\square$

3.4 Rings of Cells

For the purposes of this section, we define a system of N identical coupled cells to be a system of differential equations of the form

$$\frac{\mathrm{d}x_j}{\mathrm{d}t} = f(x_j) + \sum_{i \to j} \alpha_{ij} h(x_i, x_j) \qquad (3.6)$$

where $x_j \in \mathbf{R}^k$ is the state of the jth cell, $f : \mathbf{R}^k \to \mathbf{R}^k$ is the internal dynamics of the cells, $h : \mathbf{R}^k \times \mathbf{R}^k \to \mathbf{R}^k$ is the coupling function, α_{ij} is an $k \times k$ matrix of coupling strengths, and the summation is taken over all cells i that are actually coupled to cell j (the notation $i \mapsto j$). CPG models, for example, are often assumed to have the form of a coupled cell system. The symmetry group of a coupled cell system is a subgroup Γ of the permutation group $\mathbf{S}_N$ consisting of those permutations of the cells that preserve the coupling. The state space of a coupled cell system is $W = (\mathbf{R}^N)^k$. A more general definition of a coupled cell system, based on possession of such a symmetry group, can be found in Golubitsky and Stewart [233] and Section 3.7

The case when the internal dynamics of a coupled cell system is $k \geq 2$ motivates the following corollary to Theorem 3.4.

Corollary 3.8 *Let Γ be a finite group acting on V and suppose that $W = V^k$ for some $k \geq 2$. Then there is a hyperbolic periodic solution to some Γ-equivariant system of ODEs on $\mathbf{R}^n$ with spatial symmetries K and spatio-temporal symmetries H if and only if*

(a) *H/K is cyclic*
(b) *K is an isotropy subgroup*
(c) *If $\dim \text{Fix}(K) = 2$, then either $H = K$ or $H = N(K)$*

Proof. We must show that $\dim \text{Fix}(K) \geq 2$ and $\text{codim}\,\text{Fix}(\gamma) \cap \text{Fix}(K) \geq 2$ for each $\gamma \notin K$. Then $W \backslash L_K$ is connected and Theorem 3.4 applies. Since K is an isotropy subgroup, it fixes a point $x = (v_1, \ldots, v_k) \in V^k$. It follows that K fixes each v_j and that K is an isotropy subgroup for the action of Γ on V. Therefore, $\dim \text{Fix}(K) \geq k \geq 2$. In a similar way, the codimension of $\text{Fix}(\gamma) \cap \text{Fix}(K)$ in V must be positive and therefore the codimension of $\text{Fix}(\gamma) \cap \text{Fix}(K)$ in W is at least $k \geq 2$. $\qquad\qquad\qquad\qquad\qquad\qquad\qquad\qquad\qquad\qquad\qquad\qquad\square$

In the remainder of this section we use Corollary 3.8 to discuss the types of periodic solutions that can appear in rings of coupled cells. We begin by giving a complete classification for unidirectional rings (which are much simpler than bidirectional rings) and we simplify the discussion by assuming that the internal dynamics is at least two-dimensional ($n \geq 2$).

A *unidirectional ring* is a coupled cell system with N cells and $\mathbf{Z}_N$ symmetry. A *bidirectional ring* is a coupled cell system with N cells and $\mathbf{D}_N$ symmetry. In either case we may number the cells so that the N-cycle $(1\ 2\ \cdots\ N)$ moves cell 1 to 2 to 3, etc. Rings with *nearest neighbor coupling* are those for which couplings exist only between cells i and $i + 1 \pmod{N}$.

Unidirectional Rings.
Unidirectional Ring of Four Cells. Let the action of $\Gamma = \mathbf{Z}_4$ on $V = \mathbf{R}^4$ by cyclic permutation of the coordinates and let $\mathbf{R}^n = V^k$ for $k > 1$. Then Γ-equivariant systems of differential equations can have robust spatio-temporal symmetry pairs of six types: $(K, H) = (\mathbf{Z}_4, \mathbf{Z}_4), (\mathbf{Z}_4, \mathbf{Z}_2), (\mathbf{Z}_4, 1), (\mathbf{Z}_2, 1), (\mathbf{Z}_2, \mathbf{Z}_2)$, and $(1, 1)$. Note that $\dim \text{Fix}(1) = 4k > 2$; so Corollary 3.8(c) is not relevant here.

Let $X(t) = (x_1(t), x_2(t), x_3(t), x_4(t))$ be a periodic solution with period (normalized to) 1, where $x_j \in \mathbf{R}^k$. Then these symmetry pairs force the periodic solutions to have the form shown in (3.7).

$$
\begin{array}{ll}
(\mathbf{Z}_4, \mathbf{Z}_4) & (x(t), x(t), x(t), x(t)) \\
(\mathbf{Z}_4, \mathbf{Z}_2) & \left(x(t), x\left(t + \tfrac{1}{2}\right), x(t), x\left(t + \tfrac{1}{2}\right)\right) \\
(\mathbf{Z}_4, 1) & \left(x(t), x\left(t + \tfrac{1}{4}\right), x\left(t + \tfrac{1}{2}\right), x\left(t + \tfrac{3}{4}\right)\right) \\
(\mathbf{Z}_2, 1) & \left(x(t), y(t), x\left(t + \tfrac{1}{2}\right), y\left(t + \tfrac{1}{2}\right)\right) \\
(\mathbf{Z}_2, \mathbf{Z}_2) & (x(t), y(t), x(t), y(t)) \\
(1, 1) & (x(t), y(t), z(t), u(t))
\end{array}
\tag{3.7}
$$

Unidirectional ring of N Cells. We now consider periodic solutions in a general unidirectional ring. The only subgroups of $\mathbf{Z}_N$ are the groups $\mathbf{Z}_h$ where $h | N$ and each of these is an isotropy subgroup. Thus the only possible pairs of (H, K) are $(\mathbf{Z}_h, \mathbf{Z}_k)$ where $k | h$ and the quotient group is cyclic. The pattern of oscillation in solutions corresponding to this pair is described as follows. Separate the N cells into $\frac{N}{h}$ intertwined blocks with a block consisting of cells j, $j + \frac{N}{h}$, $j + 2\frac{N}{h}$, etc. Within each block all of the x_j are identical up to a phase shift. Assume that ℓ is coprime with $\frac{h}{k}$ and, for simplicity, assume that the periodic solution is 1-periodic. Then, within each block $x_{j+\frac{N}{h}}(t) = x_j(t + \ell\frac{k}{h})$. The periodic solutions with the most symmetry are those for which $h = N$. For such solutions all cells produce the same waveform and when $\ell = 1$ the solutions form a (discrete) rotating wave.

Bidirectional Rings. The periodic solution types that can exist in bidirectional rings are quite different from those that can exist in unidirectional rings. To reduce the combinatorial complexity, we discuss only the cases $N = 3$ and $N = 4$.

Bidirectional Rings of Three Cells. The subgroups of D_3 are 1, D_1, Z_3, and D_3; all but Z_3 are isotropy subgroups. The possible pairs (H, K) are $(1, 1)$, $(D_1, 1)$, $(Z_3, 1)$, (D_1, D_1), and (D_3, D_3). See Table 3.2.

H	K	Pattern of oscillation
D_3	D_3	$(x(t), x(t), x(t))$
D_1	D_1	$(x(t), x(t), y(t))$
Z_3	1	$(x(t), x(t + \frac{1}{3}), x(t + \frac{2}{3}))$
D_1	1	$(x(t), x(t + \frac{1}{2}), \hat{y}(t))$
1	1	$(x(t), y(t), z(t))$

Table 3.2 Oscillation in three cell bidirectional rings: $\hat{y}(t + \frac{1}{2}) = \hat{y}(t)$.

There are two periodic solutions that have nontrivial spatio-temporal symmetry: $(D_1, 1)$ and $(Z_3, 1)$. The symmetry associated to $(D_1, 1)$ is: interchange cells 1 and 2 and phase shift by half a period. Thus, the third cell is fixed by the symmetry and $\hat{y}(t)$ is forced to satisfy $\hat{y}(t) = \hat{y}(t + \frac{1}{2})$; that is, $\hat{y}(t)$ oscillates with half the period of the other cells. The pair $(Z_3, 1)$ forces the three cells to oscillate with the same wave-form but with a one-third period phase shift between cells. These solutions are discrete rotating waves. Examples are given by:

$$\frac{dx_i}{dt} = F(x_i) + K(2x_i - x_{i-1} - x_{i+1}) \qquad i = 1, 2, 3$$

where

$$F(x, y) = \left(\begin{bmatrix} -4 & 1 \\ -1 & -4 \end{bmatrix} - 2K - 5(x^2 + y^2) \right) \begin{bmatrix} x \\ y \end{bmatrix} - 50 \begin{bmatrix} -y \\ x \end{bmatrix}$$

and

$$K = \lambda \begin{bmatrix} -4 & 2 \\ -2 & -4 \end{bmatrix}$$

with $\lambda = 1.1$ and $\lambda = 1.05$, respectively. See Figure 3.5.

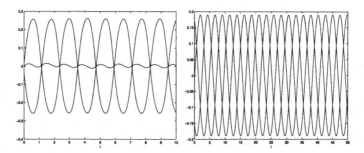

Figure 3.5: (Left) $\lambda = 1.1$: cells 1 and 2 oscillate a half-period out of phase while cell 3 oscillates at twice the frequency. (Right) $\lambda = 1.05$: discrete rotating wave.

Bidirectional Ring of Four Cells. A ring of four cells with D_4 symmetry is sketched in Figure 3.6 along with the lattice of subgroups up to conjugacy. The descriptions of the groups 1, Z_4, and D_4 are clear. The group D_1^p is generated by a reflection across a diagonal, say $(2\ 4)$ and the group D_1^s is generated by reflection across horizontal or vertical, say $(1\ 2)(3\ 4)$. The group D_2^p is generated by reflections across both diagonals and the group $\mathbf{D}_2^s$ is generated by the horizontal and vertical reflections. Finally the group $\mathbf{Z}_2$ is generated by rotation by π, that is, by $(1\ 3)(2\ 4)$.

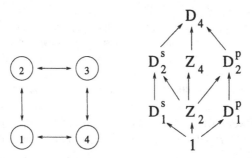

Figure 3.6: (Left) Unidirectional ring of four cells; (right) lattice of subgroups of $\mathbf{D}_4$.

A short calculation shows that there are five isotropy subgroups for $\mathbf{D}_4$ acting on the four cells: $\mathbf{D}_4$, $\mathbf{D}_2^p$, $\mathbf{D}_1^p$, $\mathbf{D}_1^s$, and 1. There are eleven pairs of subgroups H and K for which H/K is cyclic and K is an isotropy subgroup. These pairs are listed in Table 3.3. (Note that there are only six solution types in a unidirectional ring with four cells.)

H	K	Pattern of oscillation
$\mathbf{D}_4$	$\mathbf{D}_4$	$(x(t), x(t), x(t), x(t))$
$\mathbf{D}_2^p$	$\mathbf{D}_2^p$	$(x(t), y(t), x(t), y(t))$
$\mathbf{D}_4$	$\mathbf{D}_2^p$	$(x(t), x(t+\frac{1}{2}), x(t), x(t+\frac{1}{2}))$
$\mathbf{D}_1^p$	$\mathbf{D}_1^p$	$(x(t), y(t), z(t), y(t))$
$\mathbf{D}_2^p$	$\mathbf{D}_1^p$	$(x(t), \hat{y}(t), x(t+\frac{1}{2}), \hat{y}(t))$
$\mathbf{D}_1^s$	$\mathbf{D}_1^s$	$(x(t), x(t), y(t), y(t))$
$\mathbf{D}_2^s$	$\mathbf{D}_1^s$	$(x(t), x(t), x(t+\frac{1}{2}), x(t+\frac{1}{2}))$
1	1	$(x(t), y(t), z(t), u(t))$
$\mathbf{D}_1^p$	1	$(x(t), \hat{y}(t), x(t+\frac{1}{2}), \hat{u}(t))$
$\mathbf{D}_1^s$	1	$(x(t), x(t+\frac{1}{2}), y(t), y(t+\frac{1}{2}))$
$\mathbf{Z}_2$	1	$(x(t), y(t), x(t+\frac{1}{2}), y(t+\frac{1}{2}))$
$\mathbf{Z}_4$	1	$(x(t), x(t\pm\frac{1}{4}), x(t+\frac{1}{2}), x(t\mp\frac{1}{4}))$

Table 3.3 Patterns of oscillation in bidirectional rings of four cells: $\hat{y}(t+\frac{1}{2}) = \hat{y}(t)$. Thus, $\hat{y}(t)$ oscillates with twice the frequency of $x(t)$. Similarly for $\hat{u}(t)$.

A Mathematical Note on Existence. The examples in this section lead to an important mathematical point. We have classified the possible patterns of oscillation that are permitted by symmetry for each ring of coupled cells. But we do not know that these solutions actually exist in rings of cells because a cell system, as currently defined, is not a general equivariant system. In our coupled cell systems we have assumed that all coupling is *point to point coupling* — the effect of cell i on cell j does not depend on cell k. The point is further emphasized when we restrict say to nearest neighbor coupling. At this time there is no general theorem, like Theorem 3.4, that asserts existence of periodic solutions in coupled cell systems. In certain cases, existence can be proved by bifurcation techniques — in this case Hopf bifurcation. This approach is discussed in Section 4.8.

3.5 An Eight-Cell Locomotor CPG Model

In this section we discuss a schematic CPG network, introduced by Golubitsky *et al.* [234], and shown in Figure 3.7. The most notable feature of this network is that it has twice as many cells as the animal has legs. For expository purposes we assume that cells $1, \ldots, 4$ determine the timing of leg movements, and refer to the remaining four cells as 'hidden'. We also follow Golubitsky *et al.* [235] and show how the model-independent mathematical analysis of the structure of this CPG network can still lead to testable predictions about the structure of gaits.

A Sketch of the Network Derivation. The structure of the quadrupedal CPG network shown in Figure 3.7(right) can be deduced from six assumptions:

(a) The abstracted CPG network is composed of identical cells, and the signal from each cell goes to one leg.
(b) Different gaits are generated by the same network, and switching between gaits arises from changes in parameters (such as coupling strengths).
(c) The locomotor CPG has the same architecture for all quadrupeds.
(d) The network can generate the rhythms of *walk, trot,* and *pace.*
(e) Trot and pace are dynamically independent — they exist in different regions of parameter space.
(f) The network generates only simple rhythmic patterns observed in quadrupedal locomotion.

Assumption (a) introduces symmetry into model CPGs and accounts for the observed symmetries of many gaits. Different CPGs could control the rhythms of each gait; then there would have to be a controller that signals each CPG when it should be active, and there is no evidence for such controllers. It seems more likely that (b) is valid. Similarly, there is an evolutionary advantage to have a single architecture valid for all quadruped locomotor CPGs, thus justifying (c).

Since virtually all quadrupeds walk and either trot or pace, the network must produce these gaits, and (d) is appropriate. Since camels pace but do not trot, and squirrels trot but do not pace, (e) must be valid. Finally, (f) is assumed in order to create the simplest CPG model.

It is shown in Buono and Golubitsky [87] that any CPG satisfying (a)-(f) must consist of eight identical cells, whose interconnections have the same *symmetry* as Figure 3.7(right). See also Golubitsky *et al.* [234, 235]. There are two types of symmetry in the network: contralateral symmetry κ that interchanges cells on the left with cells on the right and ipsilateral symmetry ω that cyclically and simultaneously permutes cells on both left and right. Thus the symmetry group of the eight-cell quadruped CPG is $\Gamma = \mathbf{Z}_2(\kappa) \times \mathbf{Z}_4(\omega)$.

The eight-cell network in Figure 3.7(right) is essentially the only one that satisfies (a)-(f). For a proof see Golubitsky *et al.* [234, 235] or Buono and Golubitsky [87]. We sketch the argument. Walk implies that the network must have a four-cycle symmetry, and either trot or pace implies that the network must have a two-cycle symmetry. These statements follow from (a) and (d). Assumption (b) implies that the network must possess both of these symmetries. The four-cycle implies that the number of cells is a multiple of four. If the network has four cells, then the two-cycle and four-cycle cannot commute and this forces trot and pace to be conjugate solutions, contradicting (e). If the network has more than eight cells, then unnatural phase shift symmetries would be allowed, contrary to (f). Thus there

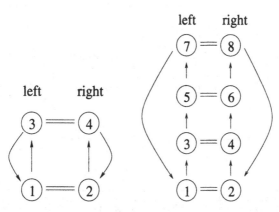

Figure 3.7: (Left) four-cell network for bipedal locomotor CPG; (right) eight-cell network for quadrupeds. Double lines indicate contralateral coupling; single lines indicate ipsilateral coupling. Direction of ipsilateral coupling is indicated by arrows; contralateral coupling is bidirectional.

	walk	jump	trot	pace	bound	pronk
	$\pm\frac{3}{4}$ $\pm\frac{1}{4}$		$\frac{1}{2}$ 0	0 $\frac{1}{2}$	$\frac{1}{2}$ $\frac{1}{2}$	0 0
	$\frac{1}{2}$ 0	$\pm\frac{3}{4}$ $\pm\frac{3}{4}$	0 $\frac{1}{2}$	0 $\frac{1}{2}$	0 0	0 0
LF RF	$\pm\frac{1}{4}$ $\pm\frac{3}{4}$	0 0	$\frac{1}{2}$ 0	0 $\frac{1}{2}$	$\frac{1}{2}$ $\frac{1}{2}$	0 0
LH RH	0 $\frac{1}{2}$	$\pm\frac{1}{4}$ $\pm\frac{1}{4}$	0 $\frac{1}{2}$	0 $\frac{1}{2}$	0 0	0 0
Subgroup K	$\mathbf{Z}_2(\kappa\omega^2)$	$\mathbf{Z}_2(\kappa)$	$\mathbf{Z}_4(\kappa\omega)$	$\mathbf{Z}_4(\omega)$	$\mathbf{D}_2(\kappa,\omega^2)$	$\mathbf{Z}_2\times\mathbf{Z}_4$

Table 3.4 Phase shifts for primary gaits in the eight-cell network.

are eight cells. In an eight-cell network, (e) implies that the four-cycle and two-cycle commute, and this leads to the network whose symmetry type is indicated in Figure 3.7(right).

Primary versus Secondary Gaits. Table 3.4 lists the most symmetric rhythms; those for which $H = \mathbf{Z}_2 \times \mathbf{Z}_4$. We call such gaits *primary*. They have the property that all signals $x_j(t)$ $(j = 1,\ldots,8)$ are identical up to a phase shift. To classify primary gaits we just need to find all subgroups K such that $(\mathbf{Z}_2 \times \mathbf{Z}_4)/K$ is cyclic. This is a straightforward group theory calculation; there are six such subgroups and they are listed in Table 3.4. These periodic solutions correspond to standard quadrupedal gaits with the exception of the 'jump'. For concrete networks, the values of the coupling terms in the network dynamics determine which rhythms occur [85, 87, 86].

This model also produces a variety of *secondary* gaits where the signals to different legs are different waveforms. These secondary gaits include the two forms of gallops [234, 85, 86].

The lattice of subgroups of $\mathbf{Z}_4 \times \mathbf{Z}_2$ is given in Table 3.5 where ω is the generator of $\mathbf{Z}_4$ and κ is the generator of $\mathbf{Z}_2$.

The secondary gaits are those that have two independent wave forms, and we have indicated the fourteen secondary gaits listed in this table by a *.

Some results from numerical simulation of this gait model (when the internal dynamics consists of a nonlinear system of differential equations known as the Morris-Lecar equations) by Buono [85, 86] are shown in Figure 3.8. Note that in

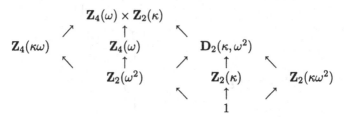

Table 3.5 Lattice of subgroups of $\Gamma = \mathbf{Z}_4 \times \mathbf{Z}_2$.

H	K	name	twist	$x_2(t)$	$x_3(t)$	$x_4(t)$
Γ	$\mathbf{Z}_4(\omega) \times \mathbf{Z}_2(\kappa)$	pronk	0	$x_1(t)$	$x_1(t)$	$x_1(t)$
	$\mathbf{Z}_4(\omega)$	rack	2	$x_1(t+\frac{1}{2})$	$x_1(t)$	$x_1(t+\frac{1}{2})$
	$\mathbf{D}_2(\kappa,\omega^2)$	bound	2	$x_1(t)$	$x_1(t+\frac{1}{2})$	$x_1(t+\frac{1}{2})$
	$\mathbf{Z}_4(\kappa\omega)$	trot	2	$x_1(t+\frac{1}{2})$	$x_1(t+\frac{1}{2})$	$x_1(t)$
	$\mathbf{Z}_2(\kappa)$	jump	4^+	$x_1(t)$	$x_1(t+\frac{1}{4})$	$x_1(t+\frac{1}{4})$
			4^-	$x_1(t)$	$x_1(t-\frac{1}{4})$	$x_1(t-\frac{1}{4})$
	$\mathbf{Z}_2(\kappa\omega^2)$	walk	4^+	$x_1(t+\frac{1}{2})$	$x_1(t+\frac{1}{4})$	$x_1(t+\frac{3}{4})$
			4^-	$x_1(t-\frac{1}{2})$	$x_1(t-\frac{1}{4})$	$x_1(t-\frac{3}{4})$
$\mathbf{Z}_4(\kappa\omega)$	$\mathbf{Z}_4(\kappa\omega)^*$	loping trot	0	$x_2(t)$	$x_2(t)$	$x_1(t)$
	$\mathbf{Z}_2(\omega^2)^*$	rotary gallop	2	$x_2(t)$	$x_2(t+\frac{1}{2})$	$x_1(t+\frac{1}{2})$
	1^*	rotary	4^+	$x_2(t)$	$x_2(t+\frac{1}{4})$	$x_1(t+\frac{1}{4})$
		richocheting jump	4^-	$x_2(t)$	$x_2(t-\frac{1}{4})$	$x_1(t-\frac{1}{4})$
$\mathbf{Z}_4(\omega)$	$\mathbf{Z}_4(\omega)^*$	loping rack	0	$x_2(t)$	$x_1(t)$	$x_2(t)$
	$\mathbf{Z}_2(\omega^2)^*$	transverse gallop	2	$x_2(t)$	$x_1(t+\frac{1}{2})$	$x_2(t+\frac{1}{2})$
	1^*	transverse	4^+	$x_2(t)$	$x_1(t+\frac{1}{4})$	$x_2(t+\frac{1}{4})$
		richocheting jump	4^-	$x_2(t)$	$x_1(t-\frac{1}{4})$	$x_2(t-\frac{1}{4})$
$\mathbf{D}_2(\kappa,\omega^2)$	$\mathbf{D}_2(\kappa,\omega^2)^*$	loping bound	0	$x_1(t)$	$x_3(t)$	$x_3(t)$
	$\mathbf{Z}_2(\omega^2)^*$	scuttle	2	$x_1(t+\frac{1}{2})$	$x_3(t)$	$x_3(t+\frac{1}{2})$
	$\mathbf{Z}_2(\kappa\omega^2)^*$	scuttle	2	$x_1(t+\frac{1}{2})$	$x_3(t)$	$x_3(t+\frac{1}{2})$
	$\mathbf{Z}_2(\kappa)^*$	loping bound	2	$x_1(t)$	$x_3(t)$	$x_3(t)$
$\mathbf{Z}_2(\omega^2)$	$\mathbf{Z}_2(\omega^2)$		0	$x_2(t)$	$x_3(t)$	$x_4(t)$
	1		2	$x_2(t)$	$x_3(t)$	$x_4(t)$
$\mathbf{Z}_2(\kappa)$	$\mathbf{Z}_2(\kappa)^*$	loping bound	0	$x_1(t)$	$x_3(t)$	$x_3(t)$
	1^*	scuttle	2	$x_1(t+\frac{1}{2})$	$x_3(t)$	$x_3(t+\frac{1}{2})$
$\mathbf{Z}_2(\kappa\omega^2)$	$\mathbf{Z}_2(\kappa\omega^2)$		0	$x_2(t)$	$x_3(t)$	$x_4(t)$
	1		2	$x_2(t)$	$x_3(t)$	$x_4(t)$
1	1		0	$x_2(t)$	$x_3(t)$	$x_4(t)$

Table 3.6 Pairs of subgroups H and K when $\Gamma = \mathbf{Z}_4 \times \mathbf{Z}_2$.

the walk the wave motion is translated by quarter-periods from cell 1 to cell 3 to cell 2 to cell 4, as desired. In the transverse gallop the cells divide into pairs — cells 1 and 3 have the same wave form a half-period out of phase, as do cells 2 and 4; but the wave forms in cells 1 and 2 are different. This grouping is characteristic of a secondary gait.

Predictions from the Model. Even though this model prescribes only the symmetries of the CPG, and not the differential equations, it still leads to several predictions. By a *prediction* we mean a consequence obtained from the model that was not an assumption made in deriving the model.

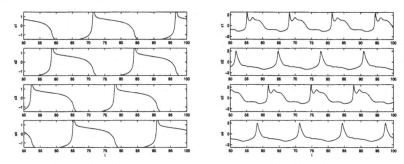

Figure 3.8: Simulations of eight-cell network: (left) walk; (right) transverse gallop.

Prediction 1: Observable Differences Between Primary and Secondary Gaits. The *duty factor* of a leg in a particular gait is the percentage of time during the gait cycle that the leg is on the ground. Our model suggests that since the signals sent to fore legs in primary gaits are identical and in secondary gaits are unequal, the duty factors of the fore legs should be equal in primary gaits and unequal in secondary gaits. Indeed, the duty factors of fore legs of a walking horse are equal (Hildebrand [274]) and those of a galloping horse are different (Leach and Sprigings [354], Deuel and Lawrence [151]).

Prediction 2: The Existence of the Jump Gait. Table 3.4 includes a non-standard primary gait, the jump, which can be described as 'fore feet hit ground, then hind feet hit ground, then three beats later fore feet hit ground'. We observed a gait with that pattern at the Houston Livestock Show and Rodeo. Figure 3.9 shows four, equal time interval, video frames of a bucking bronco. The timing of the footfalls is close to 0 and 1/4 of the period of this rhythmic motion. Additionally, the *primitive ricocheting jump* of a Norway rat and an Asia Minor gerbil has the cadence of the jump, Gambaryan [206].

Prediction 3: The Role of Hidden Cells in Bipeds. Finally, we specialize the network to bipeds where hidden cells seem unnecessary, Collins and Stewart [121]. For evolutionary reasons, we expect bipeds not to break the pattern hypothesized for quadrupeds, thus resulting in the four-cell locomotor CPG architecture illustrated in Figure 3.7 and the four primary bipedal gaits listed in Table 3.7. It is perhaps a surprise that symmetry considerations lead to the prediction that there should be two different human gaits in which swapping legs leads to a half-period phase shift in the gait cycle.

slow hop	fast hop	walk	run
0 0	$\frac{1}{2}$ $\frac{1}{2}$	$\frac{1}{2}$ 0	0 $\frac{1}{2}$
0 0	0 0	0 $\frac{1}{2}$	0 $\frac{1}{2}$

Table 3.7 Bipedal gaits for a four-cell CPG network.

The supposition made in Golubitsky *et al.* [235] is that the two hidden cells in bipedal CPGs do play an active role in gaits by controlling timings of different muscle groups. Thus, muscle groups may reveal the presence of two distinct gaits in which the legs move half a period out-of-phase. More precisely, lower leg muscles should be activated synchronously in one gait and asynchronously in the other — because of the phases of the hidden cells.

Figure 3.9: Approximate quarter cycles of bareback bronc jump at Houston Livestock Show and Rodeo.

Human walk and run support this prediction (Mann [369], Mann *et al.* [370]). During walking, gastrocnemius (GA), an ankle plantarflexor, and tibialis anterior (TA), an ankle dorsiflexor, are activated out-of-phase, whereas during running, GA and TA are co-activated during significant portions of the gait cycle.

Cell to Leg Assignments. From this discussion we conclude that there is a physiological need for having each leg receive signals from two different cells: each joint is driven by two muscles, a flexor and an extensor, and the activity in these muscles needs to be coordinated. This observation allows us to specify the network architecture even more exactly. Previously, we have discussed how the observed quadrupedal gaits uniquely specify the symmetry of the eight-cell network; now we show that there is a unique assignment of the leg to which each cell sends its signal.

Suppose that Φ : cells → legs is a possible assignment. We assume that each leg is assigned two cells and, without loss of generality, we can assume that $\Phi(1) = LH$, that is, one of the two cells that is assigned to the left hind leg is cell 1. Based on the cell to muscle group interpretation of a CPG a reasonable assumption about Φ states that if the signals from cells i and j are sent to the same leg, then for any network symmetry γ the signals from cells γi and γj are also sent to the same leg. In symbols:

$$\text{if } \Phi(i) = \Phi(j), \text{ then } \Phi(\gamma i) = \Phi(\gamma j). \tag{3.8}$$

With this assumption it was shown in [87] that there are precisely two different assignments of cells to legs.

Theorem 3.9 *Assume that the cell to leg assignment Φ satisfies $\Phi(1) = LH$ and (3.8). Then Φ is one of the two cell to leg assignments given in Table 3.8.*

$$\begin{vmatrix} 7 & 8 \\ 5 & 6 \\ \\ 3 & 4 \\ 1 & 2 \end{vmatrix} \rightarrow \begin{vmatrix} LF & RF \\ LH & RH \\ \\ LF & RF \\ LH & RH \end{vmatrix} \quad \text{and} \quad \begin{vmatrix} 7 & 8 \\ 5 & 6 \\ \\ 3 & 4 \\ 1 & 2 \end{vmatrix} \rightarrow \begin{vmatrix} RF & LF \\ RH & LH \\ \\ LF & RF \\ LH & RH \end{vmatrix}$$

Table 3.8 Cell to leg assignments in Theorem 3.9.

Proof. The four-cycle permutation ω accounts for the quarter-period phase shift of the walk and the transposition κ accounts for the half-period phase shift between left and right in the trot and pace. Transitivity of the group $\mathbf{Z}_4 \times \mathbf{Z}_2$ implies that ω cyclically permutes four cells, which we number 1,3,5,7.

Since κ and ω commute and $\kappa \neq \omega^2$, there exists a fifth cell, labeled 2, such that κ interchanges cell 1 with cell 2. It follows that

$$\omega = (1\ 3\ 5\ 7)(2\ 4\ 6\ 8) \quad \text{and} \quad \kappa = (1\ 2)(3\ 4)(5\ 6)(7\ 8).$$

We assumed that the signal from cell 1 goes to the left hind leg. Since κ interchanges left with right in the trot, the signal from cell 2 must be sent to the right hind leg. Similarly, since ω is responsible for the quarter-period phase shift in the walk, the signal from cell 3 must be sent to the left fore leg and then κ implies that the signal from cell 4 is sent to the right fore leg.

In principle, the signal from cell 5 can be sent to any of the four legs. It is straightforward to check that if the signal from cell 5 is sent to either the left hind or the right hind leg, then (3.8) forces the leg assignments given in Table 3.8.

If the signal from cell 5 is sent to either the left fore or the right fore leg, then (3.8) forces a contradiction. First, if that signal is sent to the left fore leg, then since ω maps cell 3 to cell 5 it must map left fore leg cells to left fore leg cells by (3.8). Therefore, the signal from cell 7 (which is mapped by ω from cell 5) must also be sent to the left fore leg. Moreover, applying ω once again shows that the signal from cell 1 must also be sent to the left fore leg, contradicting the assumption that the signal from cell 1 is sent to the left hind leg. Second, suppose that signal is sent to the right fore leg. Since ω maps cells 3 to 5 and 2 to 4, it follows that ω maps cells assigned to the left fore leg (3) and the right hind leg (2) to cells assigned to the right fore leg (5 and 4), which contradicts assumption (3.8). □

We call the first network listed in Table 3.8 the *zig-zag network* and the second the *criss-cross network*. These networks are illustrated in Figure 3.10. They have the property that for primary gaits the cells that send signals to the same leg are either in-phase or a half-period out-of-phase. In the zig-zag network, the cell signals to the same leg are out-of-phase in walk and jump, while the cell signals are in-phase for the remaining primary gaits; in the criss-cross network, the situation is reversed.

Thus, unlike bipeds, the hidden phases of quadrupedal primary gaits can be deduced from the observable half-network and each primary gait corresponds either to run-like synchronous muscle activation or walk-like asynchronous activation. Pearson [429] presents data showing that that the flexor and extensor muscles in a walking cat fire out-of-phase (see Figure 3.11), from which we conclude that the cat walk is a walk-like gait and that the zig-zag network is the simplest architecture for describing quadrupedal locomotor CPGs.

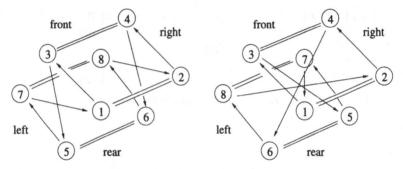

Figure 3.10: Schematic eight-cell networks for gaits in four-legged animals consistent with (3.8). (L) The zig-zag network. (R) The criss-cross network.

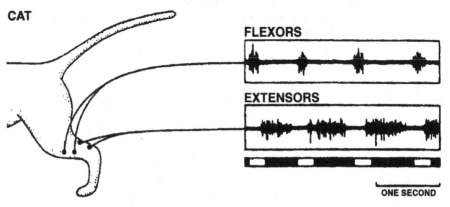

Figure 3.11: Times series of flexor and extensor muscles in a walking cat. Adapted from Pearson [429] p. 81.

This work suggests that the symmetries of animal gaits can be used to infer a class of CPG architectures. Moreover, the symmetries of these architectures can be used to form new insights and make novel predictions about animal gaits, which are supported by available information.

3.6 Multifrequency Oscillations

Coupled cell dynamics can lead to situations where different cells are forced by symmetry to oscillate at different frequencies (Golubitsky and Stewart [228], Golubitsky et al. [237], Armbruster and Chossat [7]). As we have seen in bidirectional rings, certain cells can be forced to oscillate at twice the frequency of other cells — but the range of possibilities is much more complicated.

The basic principle is simple (though combinatorial bells and whistles can be added). Let γ be an m-cycle that is a spatio-temporal symmetry of a coupled cell system having corresponding phase shift $\frac{1}{m}$. Suppose, in addition, that γ cyclicly permutes cells $1, \ldots, m$ and fixes cell $m+1$. Then cell $m+1$ must oscillate m times as quickly as cell 1 — with one caveat that we will return to in a moment. A simple example of a four-cell system that illustrates this point is given in Figure 3.12. In this coupled cell system the discrete rotating wave solution $(\mathbf{Z}_3, \mathbf{1})$ forces cell 4 to oscillate at three times the frequencies of the other three cells.

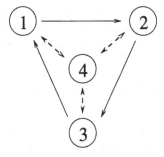

Figure 3.12: Unidirectional ring of three cells with a center cell.

Example 3.10 Multirhythms have been observed in biological systems. Takamatsu *et al.* [500] performed experiments using a ring of three coupled identical cells, in which the dynamics of each cell is of biological origin. The cells contain the plasmodial slime mold *Physarum polycephalum*. Slime molds are aggregations of amoebas, and they are known to exhibit chemical oscillations, for example in the concentration of calcium ions or of adenosine triphosphate. Such concentrations can affect the movement of the amoebas, and here the variable observed was the thickness of the slime film.

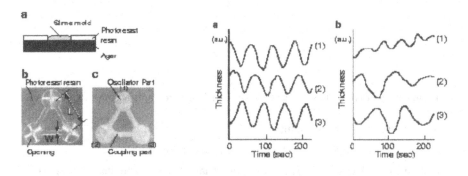

Figure 3.13: (Left) Three identical coupled cells, each containing slime mold [500]. (Right) Time series of thickness of slime layer in each of the three cells. (a) discrete rotating wave with one third period phase shifts. (b) The second and third traces are half a period out of phase with each other, and the first has twice the frequency. Courtesy of Reiko Tanaka.

The experimental aparatus is illustrated in Figure 3.13 (left): it comprises three chambers cut into a microfabricated structure formed from photo-resist resin, linked by three channels (which produce the coupling). The symmetry group of the apparatus is $\mathbf{S}_3$, and a number of possible patterns of periodic oscillation can be predicted on symmetry grounds. For example, spatio-temporal symmetry $(\mathbf{Z}_3, 1)$ corresponds to successive phase shifts of one third of a period in the three time series, Figure 3.13 (right)(a). The pattern of greatest interest in the context of multirhythms has spatio-temporal symmetry $(\mathbf{Z}_2, 1)$, where $\mathbf{Z}_2$ is generated by the transposition $\rho = (2\,3)$. The temporal effect of ρ is a half-period phase shift. As a result, cells 2 and 3 oscillate half a period out of phase with each other, and — the key feature — cell 1 oscillates with twice the frequency of cells 2 and 3. Figure 3.13 (right)(b) shows dynamics of this kind.

The same paper also reports a number of different oscillation patterns in rings of four and five slime mold cells. Several of these patterns can be identified with periodic states having spatio-temporal symmetries that are expected on the general theoretical grounds discussed in this chapter. Other states are more puzzling and have yet to be characterized.   ◇

We now return to the caveat: suppose two different cycles with nontrivial temporal symmetries exist. Then, they can force two different frequency relations between the cells — and it is quite curious how these two frequency restrictions are resolved into one relation, as we now show.

Consider a five-cell system consisting of two rings — one with three cells and one with two cells — as shown in Figure 3.14. The symmetry group of this system is $\Gamma = \mathbf{Z}_3 \times \mathbf{Z}_2 \cong \mathbf{Z}_6$. Note that the internal dynamics of cells 4 and 5 do not have to be the same as that of cells 1, 2, and 3 (indeed, they do not even have to have the same dimensions).

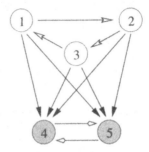

Figure 3.14: Five cell system made of a ring of three and a ring of two.

Suppose that a 1-periodic solution

$$X(t) = (x_1(t), x_2(t), x_3(t), y_1(t), y_2(t))$$

to this coupled cell system exists. Suppose that this solution has two spatio-temporal symmetries $((1\ 2\ 3), \frac{1}{3})$ and $(4\ 5), \frac{1}{2})$. The first symmetry forces the x_j to be a discrete rotating wave with (nominally) the frequency of the y_i equal to three times the frequency of the x_j. The second symmetry forces the y_i to be a half-period out-of-phase and the x_j to oscillate at twice the frequency of the y_i. This apparent nonsense is resolved as follows. The product of the two symmetries is

$$\gamma = ((1\ 2\ 3)(4\ 5), \tfrac{1}{6})$$

explicitly exhibiting the isomorphism $\mathbf{Z}_3 \times \mathbf{Z}_2 \cong \mathbf{Z}_6$. Thus $X(t)$ actually has the form

$$X(t) = (x(t), x(t + \tfrac{1}{3}), x(t + \tfrac{2}{3}), y(t), y(t + \tfrac{1}{2}))$$

where three times the frequency of x is twice the frequency of y.

Does such a solution actually exist? Corollary 3.8 states that it does — at least if all nonlinearities consistent with $\mathbf{Z}_6$ symmetry are permitted to be present. The difficulty is to find a solution corresponding to the pair $(\mathbf{Z}_6, 1)$ in the coupled cell system context.

The difficulty is compounded by the following fact, which we explain in the next chapter: no such solution is supported by a primary Hopf bifurcation in this coupled-cell system. The reason is that in Hopf bifurcation the available representations of the symmetry group $\mathbf{Z}_6 \cong \mathbf{Z}_3 \times \mathbf{Z}_2$ are sums of irreducible components of the permutation representation on $\mathbf{R}^{5k}$, where k is the dimension of the state space of a single cell. However, there does exist a more complicated bifurcation scenario that contains such a representation: primary Hopf bifurcation to a $\mathbf{Z}_3$ discrete rotating wave, followed by a secondary Hopf bifurcation using the nontrivial $\mathbf{Z}_2$ representation. We therefore seek a 3:2 resonant solution arising from such a scenario. Let $x_1, x_2, x_3 \in \mathbf{R}$ be the state variables for the ring of three cells and let $y_1 = (y_1^1, y_2^1), y_2 = (y_1^2, y_2^2) \in \mathbf{R}^2$ be the state variables for the ring of two cells. Consider the system of ODEs

$$
\begin{aligned}
\dot{x}_1 &= -x_1 - x_1^3 + 2(x_1 - x_2) + D(y_1 + y_2) + 3((y_2^1)^2 + (y_2^2)^2) \\
\dot{x}_2 &= -x_2 - x_2^3 + 2(x_2 - x_3) + D(y_1 + y_2) + 3((y_2^1)^2 + (y_2^2)^2) \\
\dot{x}_3 &= -x_3 - x_3^3 + 2(x_3 - x_1) + D(y_1 + y_2) + 3((y_2^1)^2 + (y_2^2)^2) \qquad (3.9) \\
\dot{y}_1 &= B_1 y_1 - |y_1|^2 y_1 + B_2 y_2 + 0.4(x_1^2 + x_2^2 + x_3^2)C \\
\dot{y}_2 &= B_1 y_2 - |y_2|^2 y_2 + B_2 y_1 + 0.4(x_1^2 + x_2^2 + x_3^2)C
\end{aligned}
$$

where

$$
B_1 = \begin{bmatrix} -\frac{1}{2} & 1 \\ -1 & -\frac{1}{2} \end{bmatrix} \qquad B_1 = \begin{bmatrix} -1 & -1 \\ 1 & -1 \end{bmatrix} \qquad D = (0.20, -0.11) \qquad C = \begin{bmatrix} 0.10 \\ 0.22 \end{bmatrix}
$$

Starting at the initial condition

$$
x_1^0 = 1.78 \quad x_2^0 = -0.85 \quad x_3^0 = -0.08 \quad y_1^0 = (-0.16, 0.79) \quad y_2^0 = (0.32, -0.47)
$$

We obtain Figures 3.15 and 3.16.

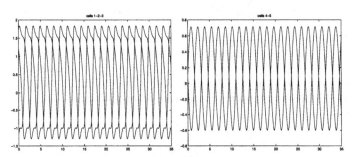

Figure 3.15: Integration of (3.9). (Left) Cells 1-2-3 out of phase by one-third period; (right) cells 4-5 out of phase by one-half period.

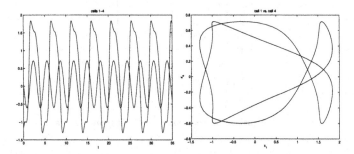

Figure 3.16: Integration of (3.9). (Left) Time series of cells 1 and 4 indicating that triple the frequency of cell 4 equals double the frequency of cell 1; (right) Plot of cell 1 versus cell 4 showing a closed curve that indicates a time-periodic solution.

Higher Resonances Forced by Symmetry. The examples so far are merely the tip of an iceberg. To illustrate the possibilities, we create a coupled cell system consisting of three subsystems pairwise connected by all-to-all connections, where the subsystems consist of rings of three, four, and five cells. In total there are twelve cells; we assume that the internal dynamics in each cell is one-dimensional. The symmetry group of the coupled cell network is $\mathbf{Z}_3 \times \mathbf{Z}_4 \times \mathbf{Z}_5$, which is isomorphic to $\mathbf{Z}_{60}$. In such a system we can expect to find a periodic solution with $\mathbf{Z}_{60}$ spatio-temporal symmetry. If $\mathbf{Z}_{60} = \langle \alpha \rangle$ where $\alpha = (1\,2\,3)(4\,5\,6\,7)(8\,9\,10\,11\,12)$, then (among others) we can expect to find a solution with spatio-temporal symmetry group

$$\tilde{\mathbf{Z}}_{60} = \langle (\alpha, \frac{1}{60}) \rangle$$

For such a solution, each ring of 3, 4 or 5 cells will exhibit a rotating wave, with temporal resonances between rings whose frequencies are in the ratio 20 : 15 : 12 — or, more simply, whose periods are in the ratio 3 : 4 : 5.

Such a solution cannot arise by primary Hopf bifurcation (see Chapter 4 Section 4.8). However, it is consistent with theorem 3.4, as we now check. We have $H = \mathbf{Z}_{60} = \langle \alpha \rangle$ acting on $\mathbf{R}^3 \oplus \mathbf{R}^4 \oplus \mathbf{R}^5 = \mathbf{R}^{12}$, and $K = 1$. Condition (a) clearly holds. Condition (b) holds since any generic point has isotropy 1. For (c) we compute dim Fix$(K) = 12 \geq 2$. Since in fact $12 > 2$, condition (d) is irrelevant. Finally we consider (e). What is the set L_K? We have

$$L_K = \bigcup_{\gamma \notin K} \text{Fix}(\gamma) \cap \text{Fix}(K) = \bigcup_{\gamma \neq 1} \text{Fix}(\gamma)$$

We claim that when $\gamma \neq 1$ the space Fix(γ) always has codimension ≥ 2, so the complement of L_K is connected and (e) is trivially valid. The elements $\gamma \neq 1$ are $\gamma = \alpha^m$ for $1 \leq m \leq 59$. Let $\mu = \gcd(m, 60)$, and let C_j denote some j-cycle. Then we easily compute Table 3.9. The largest dimension for Fix(γ) is 10, and $12 - 10 = 2$. So this completes the verification of conditions (a)-(e).

μ	α^m	dim Fix
1	$C_3 C_4 C_5$	3
2	$C_3 C_2 C_5$	4
3	$C_4 C_5$	5
4	$C_3 C_5$	6
5	$C_3 C_4$	7
6	$C_2 C_5$	6
10	$C_3 C_2$	8
12	C_5	7
15	C_4	9
20	C_3	10
30	C_2	10

Table 3.9 Dimensions of fixed-point spaces of elements $\gamma \neq 1$.

We now find such a solution numerically in a coupled cell system. Let $x = (x_1, x_2, x_3)$, $y = (y_1, \ldots, y_4)$, and $z = (z_1, \ldots, z_5)$ be coordinates for the cell system.

The differential equations for this cell system are:

$$\dot{x}_1 = 1.1x_1 - x_1^3 - 2.35x_2 + 0.1|y|^2 + 0.1|z|^2$$
$$\dot{x}_2 = 1.1x_2 - x_2^3 - 2.35x_3 + 0.1|y|^2 + 0.1|z|^2$$
$$\dot{x}_3 = 1.1x_3 - x_3^3 - 2.35x_1 + 0.1|y|^2 + 0.1|z|^2$$

$$\dot{y}_1 = 1.4y_1 - y_1^3 - 2y_2 - 3y_3 - 0.4|x|^2 + 0.5(x_1x_2 + x_2x_3 + x_3x_1)$$
$$\dot{y}_2 = 1.4y_2 - y_2^3 - 2y_3 - 3y_4 - 0.4|x|^2 + 0.5(x_1x_2 + x_2x_3 + x_3x_1)$$
$$\dot{y}_3 = 1.4y_3 - y_3^3 - 2y_4 - 3y_1 - 0.4|x|^2 + 0.5(x_1x_2 + x_2x_3 + x_3x_1)$$
$$\dot{y}_4 = 1.4y_4 - y_4^3 - 2y_1 - 3y_2 - 0.4|x|^2 + 0.5(x_1x_2 + x_2x_3 + x_3x_1)$$

$$\dot{z}_1 = z_1 - z_1^3 - 2.6z_2 - 0.5z_3 + 0.2|x|^2 + 0.5(x_1x_2 + x_2x_3 + x_3x_1)$$
$$\dot{z}_1 = z_2 - z_2^3 - 2.6z_3 - 0.5z_4 + 0.2|x|^2 + 0.5(x_1x_2 + x_2x_3 + x_3x_1)$$
$$\dot{z}_1 = z_3 - z_3^3 - 2.6z_4 - 0.5z_5 + 0.2|x|^2 + 0.5(x_1x_2 + x_2x_3 + x_3x_1)$$
$$\dot{z}_1 = z_4 - z_4^3 - 2.6z_5 - 0.5z_1 + 0.2|x|^2 + 0.5(x_1x_2 + x_2x_3 + x_3x_1)$$
$$\dot{z}_1 = z_5 - z_5^3 - 2.6z_1 - 0.5z_2 + 0.2|x|^2 + 0.5(x_1x_2 + x_2x_3 + x_3x_1)$$

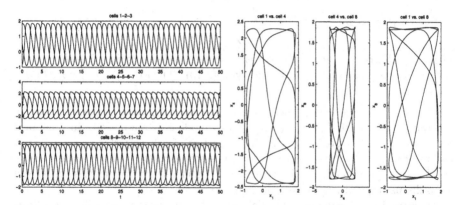

Figure 3.17: $3 : 4 : 5$ resonance solution in cell system: (left) time series showing three rotating waves; (right) pairwise plots of cells i versus j for i, j taken from $1, 4, 8$ illustrating that solution is time-periodic.

3.7 A General Definition of a Coupled Cell Network

In Section 3.4 we introduced the concept of a coupled cell system. In this section we give a formal definition of a coupled cell system.

Formal Definition of a Coupled Cell System. First, we generalize the working definition in Section 3.4, which defined a coupled cell system in terms of its equations. This approach is too restrictive for many purposes: what really matters is that the state space of a coupled cell system is a Cartesian product, equipped with canonical projections onto its factors, and each factor is the state space of one cell. The symmetries of the system permute the cells, and the central object of study is this permutation action. We therefore begin by giving a formal definition of a coupled cell system that is intimately related to its symmetries.

Let $\mathcal{N} = \{1, \ldots, N\}$ and let P_j be a manifold for $j \in \mathcal{N}$.

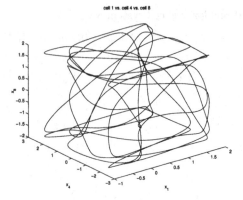

Figure 3.18: Three-dimensional plots of cells 1, 4, 8 illustrating that solution is time-periodic.

Definition 3.11 A *coupled cell system* is a dynamical system

$$\frac{\mathrm{d}x}{\mathrm{d}t} = F(x) \qquad (3.10)$$

defined on the space

$$P = P_1 \times \cdots \times P_N$$

where the P_j are the *cells* of the system and the projections $\pi_j : P \to P_j$ are the *cell projections*. Let $x(t)$ be a trajectory of (3.10). Then the jth *cell trajectory* is $x_j(t) = \pi_j(x(t))$.   ◇

There is one point that distinguishes a general coupled cell system from a general system of ODEs and that point is a modeling assumption. In a coupled cell system we interested primarily in properties of cell trajectories $x_j(t)$ — not just in properties of the trajectory $x(t)$. It is this focus that allows us to identify *synchrony* — two identical cell trajectories — and relative frequencies (as we did when discussing multirhythms in Section 3.6).

Abstractly, this completes the definition, but we need to be able to do three things: interpret a coupled cell system in terms of its individual cells and how they are coupled, decompose the dynamical system into different levels of coupling, and impose symmetry constraints. That is, we need to set up links between the abstract concept and the intuitive one employed in areas of application. The basic idea is that F can be decomposed as a sum of terms, which correspond to various types of coupling. A formal treatment can be found in Golubitsky and Stewart [233]; we summarize the main points here.

The vector field F can be written (in an essentially unique way) as a sum of terms that depend on none of the x_j, on just one of them, on just two of them, and so on. Let Φ_k be the terms that depend on exactly k of the x_j. Each Φ_k can be further decomposed according to which x_j actually occur (that is, the value of $\Phi_k(x)$ is not independent of x_j). The constant part of F is Φ_0. We can write $\Phi_1 = f_1 + \cdots + f_N$ where f_j depends only on x_j. Then the ith component of f_i defines the internal dynamics of cell i. In a similar manner (the details require a little care) we can define the coupling from cell i to cell j. When the system has 'point to point' coupling, as in (3.6), this takes care of the whole of F, but in general there might be 'three-cell' coupling terms involving three different x_j, and so on. Such terms can also be given a canonical meaning.

Associated with a coupled cell system is a 'decorated directed graph' (more generally a labelled oriented simplicial complex) whose nodes correspond to the N cells of the system and whose edges (or higher-dimensional simplices) correspond to various types of coupling. An edge from node j to node i exists if and only if F_i contains terms that depend only on x_i and x_j, and so on. The resulting graph (or complex) provides a schematic description of which cells influence which — but not of what these influences actually are.

Suppose that a group $\Gamma \subseteq \mathbf{S}_N$ permutes the nodes in $\mathcal{N}$. Nodes are said to be *identical* (or to have the same *type*) if they lie in the same Γ-orbit. Edges are said to be *identical* (or to have the same *type*) if they lie in the same Γ-orbit, where Γ is now acting on pairs (i, j) with $i \neq j$. In practice we draw nodes of the same type with the same kind of symbol (circle, box,...) and we draw edges of the same type with the same kind of arrow (single head, two heads, double shaft,...).

The animal gaits model in Figure 3.7(right) involves two different types of arrows: one for ipsilateral coupling and the other for contralateral coupling.

The key idea we wish to explore is symmetry of a coupled cell network. The symmetry group of the network is some subgroup $\Gamma \subseteq \mathbf{S}_N$. Recall that cell j has phase space P_j, and suppose that whenever $i = \gamma(j)$ for $\gamma \in \Gamma$ there is an isomorphism between P_i and P_j. That is, symmetrically related cells have the same internal phase space. We can identify P_i and P_j. Now Γ acts on $P = P_1 \times \cdots \times P_N$ permuting coordinates:

$$\gamma.(y_1, \ldots, y_N) = (y_{\gamma^{-1}(1)}, \ldots, y_{\gamma^{-1}(N)}) \qquad (y_j \in Y)$$

so that the following definition now makes sense:

Definition 3.12 A coupled cell system is said to be *symmetric* under Γ if F is Γ-equivariant. By extension we also say that the coupled cell system is Γ-equivariant.

$\diamond$

The equivariance assumption implies that the internal dynamics of nodes of the same type are identical. Similarly, coupling terms corresponding to edges of the same type are identical, and the same goes for multi-cell coupling terms. In particular:

Definition 3.13 A cell complex has *identical cells* if Γ acts transitively on the nodes.

$\diamond$

We end our discussion with three simple examples of rings of three identical cells. The general unidirectional ring, as pictured in Figure 3.19(left), corresponds to a system

$$\begin{aligned} \dot{x}_1 &= f(x_1, x_3) \\ \dot{x}_2 &= f(x_2, x_1) \\ \dot{x}_3 &= f(x_3, x_2) \end{aligned} \qquad (3.11)$$

The figure indicates that the output of cell 2 does not affect cell 1 and that requirement is fulfilled in system (3.11). This coupled cell system has $\mathbf{Z}_3$ symmetry.

The general bidirectional ring, as pictured in Figure 3.19(right), corresponds a system

$$\begin{aligned} \dot{x}_1 &= f(x_1, x_2, x_3) \\ \dot{x}_2 &= f(x_2, x_3, x_1) \\ \dot{x}_3 &= f(x_3, x_1, x_2) \end{aligned} \qquad (3.12)$$

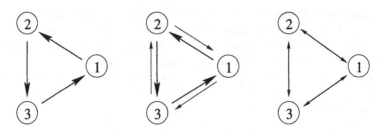

Figure 3.19: Three identical cells. (Left) unidirectional ring with $\mathbf{Z}_3$ symmetry; (middle) asymmetric bidirectional ring with $\mathbf{Z}_3$ symmetry; (right) bidirectional ring with $\mathbf{D}_3$ symmetry.

where $f(x, y, z) = f(x, z, y)$. The figure indicates that the outputs of cell 3 and cell 2 affect cell 1 equally and that requirement is fulfilled in system (3.12). This coupled cell system has $\mathbf{D}_3$ symmetry.

The general asymmetric bidirectional ring, as pictured in Figure 3.19(middle), corresponds system (3.12) where $f(x, y, z) \neq f(x, z, y)$. The figure indicates that the outputs of cell 3 and cell 2 affect cell 1, but not equally, and that requirement is fulfilled. This coupled cell system has $\mathbf{Z}_3$ symmetry.

Symmetry and Modelling. Many times coupled cell systems are used as models in a schematic sense: the exact form that model equations may have is unknown. All that is known is which cells have equal influences on other cells. The examples on speciation and animal gaits fall into this category. In these cases, it is the symmetry of the coupled cell system that is the important modelling assumption, not the detailed equations for the cells.

For example, in the animal gaits locomotor CPG model, the cells themselves may represent individual neurons or, as is more likely, collections of neurons. Should the internal dynamics of each cell be modelled by a single Hodgkin-Huxley system, or for simplicity by Morris-Lecar or Fitzhugh-Nagumo equations, or more realistically by a collection of Hodgkin-Huxley systems? Should the cell coupling be modelled by nearest-neighbor point-to-point coupling or more realistically by couplings that include dependence on all cells? In many cases, such issues are secondary because there is no well-established physical or biological reason to make any particular choice.

In this sense, the most important modelling assumption for the locomotor CPG model is the symmetry assumption; that is, the coupled cell system has $\Gamma = \mathbf{Z}_2 \times \mathbf{Z}_4$ symmetry. In these circumstances, the only *a priori* assumption on the form of the coupled cell system that we should make is Γ-equivariance. That is, we need to study Γ-symmetric coupled cell systems defined on the state space $(\mathbf{R}^k)^8$. In the $\mathbf{Z}_3$-symmetric examples pictured in Figure 3.19, it is arguable that the cell system in the middle is usually to be preferred to the cell system on the left.

Chapter 4

Hopf Bifurcation with Symmetry

In the previous chapter we studied the symmetry properties of time-periodic states of equivariant dynamical systems. We did not enquire how such states might arise. In this chapter we develop the theory of one of the most widespread routes to time-periodicity: Hopf bifurcation. The physical characteristics of Hopf bifurcation are the loss of stability of a steady state, as a parameter is varied, leading to a bifurcation to small-amplitude periodic states with 'finite period'. That is, the limiting period as the amplitude tends to zero (immediately after bifurcation) is finite and nonzero.

An excellent candidate for a system that can be modeled as a Hopf bifurcation occurs in the Belousov-Zhabotinskii chemical reaction. In 1958 Belousov discovered that if a mixture of citric acid and sulfuric acid is dissolved in water with potassium bromate and a cerium salt, then the color of the mixture oscillates periodically over time. Interest in this phenomenon increased when Zhabotinskii [534] found that related reactions produced spatial patterns. In a modern version of this celebrated experiment, the following chemicals are dissolved in water and mixed in a beaker in suitable proportions:

- Sodium bromate and sulfuric acid
- Sodium bromide
- Malonic acid

See Cohen and Stewart [116] for the complete recipe. The mixture turns yellow, and as it loses bromine it turns straw-colored or colorless. At this stage, a color indicator is added: 1, 10 phenanthroline ferrous complex. This turns blue or red respectively depending on whether the reaction is oxidizing or reducing. The mixture is stirred and poured into a petri dish, where it appears to remain in a spatially uniform equilibrium state. After a time, however — between 1 and 10 minutes, usually — a pattern of concentric expanding blue and red rings appears, known as *target patterns*. if one of the rings is carefully broken, say by dragging a wire through it, then the pattern breaks up into a series of rotating *spirals*, often linked in pairs. See Figure 4.1.

Each of these patterns is time-periodic, and each of these patterns has symmetry. Target patterns have spatial symmetry: at each instant of time the pattern is invariant under $O(2)$. Spirals have a more interesting spatio-temporal symmetry. At any instant of time the spiral has trivial symmetry in space. However, spirals rotate 'rigidly' — that is, without changing their form — so time-translation is equivalent to some rotation in $SO(2)$. Indeed the angle of rotation is proportional to the time translation: that is, the speed of rotation of the pattern is constant. The concept of spatio-temporal symmetry is central to equivariant Hopf bifurcation, indeed to the general phenomenology of time-periodicity in equivariant ODEs, as we saw in the previous chapter.

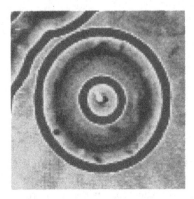 

Figure 4.1: Belousov-Zhabotinskii reaction. (Left) Target patterns; (right) Spiral waves. Pictures courtesy of A.L. Lin and H.L. Swinney.

The Belousov-Zhabotinskii reaction has spawned a huge literature. Similar patterns occur in many planar systems, and related, more complex patterns occur in three dimensions. We survey some of the more complicated temporal patterns of spiral waves in Section 6.7. A complete explanation turns out to be complicated, and some of the associated pattern-forming phenomena are best understood by taking the symmetry group to be the Euclidean group $\mathbf{E}(2)$ acting on the infinite plane. This group is non-compact, which causes technical difficulties but opens the way to some important phenomena. In this chapter we avoid the issue of non-compactness by restricting the domain of the system to a disk. In this setting, we provide a partial explanation of target patterns and spirals in this and related systems, using a model that is symmetric under $\mathbf{O}(2)$ acting on the disk.

The history of Hopf bifurcation goes back to 1942, with the classic theorem of Hopf [282]. This theorem provides a sufficient condition for an equilibrium of an ODE to undergo a local bifurcation to a branch of time-periodic solutions. The main requirement is that the linearization about the equilibrium state should possess purely imaginary eigenvalues $\pm\omega i$, $\omega \neq 0$, which cross the imaginary axis with nonzero speed as a bifurcation parameter is varied. There is also a technical condition, requiring this eigenvalue to be non-resonant. That is, no other eigenvalues exist that are integer multiples of ωi, so that in particular ωi is a simple eigenvalue. For modern proofs see Hale [262], Marsden and McCracken [375], and Golubitsky and Schaeffer [226].

When the ODE has symmetry, eigenvalues of the linearization are often forced to be multiple, and the hypotheses of the Hopf Theorem are seldom valid. In this chapter we derive an equivariant analogue of the Hopf Theorem, proving the existence of branches of symmetry-breaking time-periodic solutions for certain *spatiotemporal* isotropy subgroups. These isotropy subgroups are said to be **C**-axial, and there is a close analogy with axial subgroups in steady-state bifurcation. The Equivariant Hopf Theorem is thus very similar to the Equivariant Branching Lemma.

The proof of the Equivariant Hopf Theorem is obtained using an idea that goes back to Hale [262], namely: restate the conditions for the existence of a bifurcating branch of time-periodic solutions as conditions for the existence of a branch of zeros on an infinite-dimensional Banach space, known as 'loop space'. The phase-shift symmetry of periodic solutions translates into an action of the circle group $\mathbf{S}^1$ on loop space. If the original ODE has symmetry group Γ, then the operator reformulation has symmetry group $\Gamma \times \mathbf{S}^1$. Liapunov-Schmidt reduction then turns the problem of Γ-equivariant Hopf bifurcation into what is, in effect, 'steady-state' $\Gamma \times \mathbf{S}^1$-equivariant bifurcation on the real eigenspace for $\pm \omega i$. If $\Sigma \subseteq \Gamma \times \mathbf{S}^1$ has a 2-dimensional fixed-point space, then equivariance can be used to prove the generic existence of a branch of periodic solutions whose symmetry group contains Σ. It is this condition that defines a **C**-axial subgroup.

We begin this chapter by setting up the necessary linear analysis, obtaining conditions under which a Γ-equivariant ODE can have a linearization with purely imaginary eigenvalues. This leads to the notion of a Γ-simple representation, which plays the same role in Hopf bifurcation as an absolutely irreducible representation in steady-state bifurcation. Next we state the Equivariant Hopf Theorem and outline its proof by Liapunov-Schmidt reduction. We then introduce an alternative to Liapunov-Schmidt reduction, the idea of Poincaré-Birkhoff normal form.

The remaining sections deal with various issues that arise from the Equivariant Hopf Theorem. We consider some technical questions related to the application of the Equivariant Hopf Theorem to coupled cell networks, and point out a number of potential pitfalls. We end the chapter by expanding on what is perhaps the most interesting feature of equivariant Hopf bifurcation: the new $\mathbf{S}^1$ symmetry that arises — not from the Γ symmetry of the ODE, but from the type of solution being sought. We describe a number of generalizations and related 'dynamic' symmetries, including period-q points of mappings and the celebrated Feigenbaum period-doubling cascade in the logistic map. We point out some open questions in this area, which seems to have considerable potential for future research.

4.1 Linear Analysis

Suppose we have a Γ-equivariant system of equations

$$\frac{\mathrm{d}x}{\mathrm{d}t} = f(x, \lambda) \tag{4.1}$$

where $f : \mathbf{R}^n \times \mathbf{R} \to \mathbf{R}^n$ and $f(x_0, \lambda_0) = 0$, which undergoes a local bifurcation at (x_0, λ_0).

By the Implicit Function Theorem, local bifurcation of steady states cannot occur when the Jacobian $(\mathrm{d}f)_{x_0,\lambda_0}$ is nonsingular. Therefore a necessary (and 'morally sufficient') condition for local bifurcation of steady states is that the Jacobian should be singular, that is, have at least one zero eigenvalue. It turns out that local bifurcation to a time-periodic state cannot occur unless $(\mathrm{d}f)_{x_0,\lambda_0}$ has at least one nonzero purely imaginary eigenvalue. The intuition here is that the linearized equation must have a periodic trajectory. Both types of local bifurcation can be defined in a unified manner:

Definition 4.1 A *local bifurcation* occurs when $(\mathrm{d}f)_{x_0,\lambda_0}$ has an eigenvalue on the imaginary axis. $\diamond$

The two principal types of local bifurcation can be distinguished by the nature of the critical eigenvalues:

(1) *Steady-state bifurcation*: a zero eigenvalue.
(2) *Hopf bifurcation*: a pair of purely imaginary eigenvalues.

We also mention the concept of a *mode interaction*, where different critical eigenvalues occur for the same value of λ_0. There are three kinds of mode interaction: steady-state/steady-state, steady-state/Hopf, and Hopf/Hopf.

The first step is to work out when a linearization of equivariant equations can have non-real (and in particular purely imaginary) eigenvalues.

Lemma 4.2 *Suppose* $(df)_{x_0,\lambda_0}$ *has a non-real eigenvalue. Then there exists an irreducible representation* $V \subseteq \mathbf{R}^n$ *such that either*

(a) $V \oplus V \subseteq \mathbf{R}^n$ *and* Γ *acts absolutely irreducibly on* V, *or*

(b) Γ *acts non-absolutely irreducibly on* V.

Proof. Suppose neither is true. Then we can write

$$\mathbf{R}^n = V_1 \oplus \cdots \oplus V_s$$

where the V_j are absolutely irreducible and distinct. Standard properties of isotypic components (Lemma 2.10) imply that $(df)_{x_0,\lambda_0}(V_j) \subseteq V_j$. Moreover, since Γ acts absolutely irreducibly on V_j, we have

$$(df)_{x_0,\lambda_0}|_{V_j} = c_j I_{V_j}$$

Therefore all eigenvalues are real. $\square$

It is not hard to show that if Lemma 4.2(a) or (b) holds, then not only can df have non-real eigenvalues — it can have nonzero purely imaginary eigenvalues. This leads to a central concept for equivariant Hopf bifurcation:

Definition 4.3 The vector space W is Γ-*simple* if either:

(a) $W = V \oplus V$ where V is an absolutely irreducible representation of Γ, or

(b) Γ acts irreducibly but not absolutely irreducibly on W. $\diamond$

Example 4.4 The standard actions of $\mathbf{SO}(2)$ on $\mathbf{R}^2$ and of $\mathbf{Z}_q$ ($q \geq 3$) on $\mathbf{R}^2$ are irreducible but not absolutely irreducible. However $\mathbf{O}(2)$ and $\mathbf{D}_q$ act absolutely irreducibly.

We now consider Hopf bifurcation from a group-invariant equilibrium x_0, that is, one that satisfies the condition $\gamma x_0 = x_0$ for all $\gamma \in \Gamma$. Recall from Theorem 1.27 that at steady-state bifurcations, generically $\ker(df)_{x_0,\lambda_0}$ is an absolutely irreducible representation of Γ. The corresponding result for Hopf bifurcation with symmetry is:

Theorem 4.5 *Generically, at a Hopf bifurcation, the action of* Γ *on the center subspace is* Γ-*simple.*

Proof. See Golubitsky *et al.* [237] Chapter XVI Proposition 1.4. $\square$

This theorem implies that (away from a mode interaction) when studying Hopf bifurcation in a one-parameter family of ODEs, we may assume that Γ acts Γ-simply on $\mathbf{R}^{2m}$.

Remark 4.6 Note that $\mathrm{Fix}_{\mathbf{R}^{2m}}(\Gamma) = \{0\}$ when Γ acts Γ-simply on $\mathbf{R}^{2m}$. It follows that $f(0) = 0$ for any Γ-equivariant f; thus f always has a trivial solution.

For the remainder of this chapter we assume for simplicity that $(x_0, \lambda_0) = (0,0)$.

Lemma 4.7 *Assume that* $\mathbf{R}^n$ *is* Γ-*simple and* $f : \mathbf{R}^n \times \mathbf{R} \to \mathbf{R}^n$ *is a* Γ-*equivariant mapping, such that* $(df)_{0,0}$ *has eigenvalues* $\pm \omega i$ *with* $\omega > 0$. *Then the following conditions hold:*

(a) *The eigenvalues of* $(df)_{0,\lambda}$ *are* $\sigma(\lambda) \pm i\rho(\lambda)$, *each of multiplicity* $m = n/2$, *and* σ *and* ρ *depend smoothly on* λ *with* $\sigma(0) = 0$ *and* $\rho(0) = \omega$.

(b) *There exists a linear invertible mapping $S : \mathbf{R}^n \to \mathbf{R}^n$ that commutes with Γ such that*

$$(df)_{0,0} = \omega S J S^{-1}$$

where

$$J = \begin{bmatrix} 0 & -I_m \\ I_m & 0 \end{bmatrix}$$

Proof. Both statements are straightforward exercises in representation theory. For details see Golubitsky *et al.* [237] Chapter XVI Section 1 Lemma 1.5 p.265. $\square$

By rescaling time we can also ensure that $\omega = 1$, so we may assume from now on that

$$(df)_{0,0} = J \tag{4.2}$$

The group-theoretic approach to spatio-temporal symmetries in Hopf bifurcation depends on one simple idea: the linearization (4.2) provides us with an action of $\mathbf{S}^1$ on $\mathbf{R}^n$, namely,

$$\theta x = e^{-i\theta J} x \tag{4.3}$$

This action is crucial to the proof of the existence of periodic solutions in the Equivariant Hopf Theorem, and it is closely related to the phase shift issues discussed in the previous chapter. The $\mathbf{S}^1$-action (4.3) commutes with the action of Γ, because J commutes with Γ. In fact, $\mathbf{S}^1$ symmetry appears in the Equivariant Hopf Theorem in three ways:

- Symmetry of periodic solutions.
- Liapunov-Schmidt reduction proof of existence.
- Stability by center manifold reduction and Poincaré-Birkhoff normal form.

To understand the Equivariant Hopf Theorem, it is necessary to address each of the above roles.

4.2 The Equivariant Hopf Theorem

The Equivariant Branching Lemma asserts the existence of branches of steady states corresponding to axial subgroups — those with one-dimensional fixed-point subspaces. The Equivariant Hopf Theorem asserts the existence of branches of periodic solutions corresponding to **C**-axial subgroups, in the following sense:

Definition 4.8 Let W be a Γ-simple representation and let $\mathbf{S}^1$ act on W as in (4.3). A subgroup $\Sigma \subseteq \Gamma \times \mathbf{S}^1$ is **C**-*axial* if Σ is an isotropy subgroup and

$$\dim \text{Fix}(\Sigma) = 2$$

$\diamond$

Theorem 4.9 (Equivariant Hopf Theorem) *Let a compact Lie group Γ act Γ-simply, orthogonally, and nontrivially on $\mathbf{R}^{2m}$. Assume that*

(a) $f : \mathbf{R}^{2m} \times \mathbf{R} \to \mathbf{R}^{2m}$ *is Γ-equivariant. Then $f(0,\lambda) = 0$ and $(df)_{0,\lambda}$ has eigenvalues $\sigma(\lambda) \pm i\rho(\lambda)$ each of multiplicity m.*
(b) $\sigma(0) = 0$ *and $\rho(0) = 1$.*
(c) $\sigma'(0) \neq 0$ — *the eigenvalue crossing condition.*
(d) $\Sigma \subseteq \Gamma \times \mathbf{S}^1$ *is a **C**-axial subgroup.*

Then there exists a unique branch of periodic solutions with period $\approx 2\pi$ emanating from the origin, with spatio-temporal symmetries Σ.

We require one lemma before sketching the proof of Theorem 4.9.

Lemma 4.10 *Let* Γ *act on* $\mathbf{R}^n$ *and let* $\Sigma \subset \Gamma$ *be a subgroup. Suppose that* γ *is in the normalizer* N_Σ, *that is* $\gamma\Sigma\gamma^{-1} = \Sigma$. *Then* $\gamma \operatorname{Fix}(\Sigma) = \operatorname{Fix}(\Sigma)$.

Proof. Let $x \in \operatorname{Fix}(\Sigma)$ and $\sigma \in \Sigma$. We want to show that γx is also in $\operatorname{Fix}(\Sigma)$. By assumption $\gamma^{-1}\sigma\gamma \in \Sigma$. Therefore, $\gamma^{-1}\sigma\gamma x = x$ which implies $\sigma(\gamma x) = \gamma x$, as desired. $\square$

Sketch of Proof of Theorem 4.9. Details of the proof are given in Golubitsky *et al.* [237] Chapter XVI Theorem 4.1, p. 275. We outline the main steps.

The main idea is to use Liapunov-Schmidt Reduction to prove the existence of periodic solutions. To do this, we set up a mapping (in infinite dimensions) whose zeros correspond to periodic solutions of the original system of differential equation (4.1).

We view the system (4.1) in operator form (see (4.4) below) and look for periodic solutions with period approximately equal to 2π. To ensure this, we assume that $x(t)$ has period

$$\frac{2\pi}{1+\tau} \approx 2\pi$$

where $\tau \approx 0$, and define

$$u(s) = x\left(\frac{s}{1+\tau}\right)$$

Then $u(s)$ is a 2π-periodic solution of

$$(1+\tau)\frac{du}{ds} - f(u, \lambda) = 0$$

Let $C_{2\pi}$ be the space of continuous functions $\mathbf{S}^1 \to \mathbf{R}^{2m}$, and let $C_{2\pi}^1 \subseteq C_{2\pi}$ consist of those functions that are C^1 differentiable. We call $C_{2\pi}$ and $C_{2\pi}^1$ *loop spaces*. Define the operator $\mathcal{N} : C_{2\pi}^1 \times \mathbf{R} \times \mathbf{R} \to C_{2\pi}$ by

$$\mathcal{N}(u, \lambda, \tau) = (1+\tau)\frac{du}{ds} - f(u, \lambda) \qquad (4.4)$$

Then zeros of $\mathcal{N}$ correspond to periodic solutions of (4.1) with period near 2π. Clearly $\mathcal{N}$ is Γ-equivariant; it is also $\mathbf{S}^1$-equivariant where the action of θ on $C_{2\pi}$ is given by

$$\theta u(s) = u(s - \theta) \qquad (4.5)$$

Since the actions of Γ and $\mathbf{S}^1$ commute, $\mathcal{N}$ is $\Gamma \times \mathbf{S}^1$-equivariant.

To solve $\mathcal{N} = 0$ by Liapunov-Schmidt reduction we compute $\ker(d\mathcal{N})_{0,0,0}$ and use the Implicit Function Theorem to reduce the system to $g : \ker(d\mathcal{N}) \times \mathbf{R} \times \mathbf{R} \to \ker(d\mathcal{N})$. We have:

$$(d\mathcal{N})_{0,0,0}v = \frac{dv}{ds} - (df)_{0,0}v = \frac{dv}{ds} - Jv$$

where J is as in Lemma 4.7. We may identify $V \oplus V = \mathbf{R}^{2m}$ with $\mathbf{C}^m$ by identifying $(v_1, v_2) \in V \oplus V$ with $v_1 + iv_2 \in \mathbf{C}^m$. Then $\ker(d\mathcal{N})_{0,0,0}$ is identified with $\mathbf{C}^m$ by viewing the independent solutions of

$$\frac{dv}{ds} - Jv = 0$$

as $\operatorname{Re}(e^{is}v_0)$ and $\operatorname{Im}(e^{is}v_0)$ where $v_0 \in \mathbf{C}^m$. Thus

$$g : \mathbf{R}^{2m} \times \mathbf{R} \times \mathbf{R} \to \mathbf{R}^{2m}$$

g is $\Gamma \times \mathbf{S}^1$-equivariant, and the zeros of g are in one-to-one correspondence with $\frac{2\pi}{1+\tau}$-periodic solutions of the original system.

In addition, it follows from (4.5) that the action of $\mathbf{S}^1$ on $C_{2\pi}$ induces the standard action of $\mathbf{S}^1$ on $\mathbf{R}^{2m} \cong \mathbf{C}^m$. Functions v in $\ker(\mathrm{d}\mathcal{N})_{0,0,0}$ have the form $v(s) = e^{is}v_0$ for some $v_0 \in \mathbf{C}^{2m}$. Then (4.5) implies that $\theta v = e^{-i\theta}v$.

The $\Gamma \times \mathbf{S}^1$-equivariance of g implies that $g : \mathrm{Fix}(\Sigma) \times \mathbf{R} \times \mathbf{R} \to \mathrm{Fix}(\Sigma)$ and the assumption that $\mathrm{Fix}(\Sigma)$ is two-dimensional allows us to identify $\mathrm{Fix}(\Sigma)$ with $\mathbf{C}$. Since $\mathbf{S}^1 \subset N(\Sigma)$, Lemma 4.10 implies that $g|\,\mathrm{Fix}(\Sigma) \times \mathbf{R} \times \mathbf{R}$ commutes with $\mathbf{S}^1$. Moreover, any smooth map $f : \mathbf{C} \to \mathbf{C}$ that commutes with $\mathbf{S}^1$ has the form

$$f(z) = P(|z|^2)z, \quad P \in \mathbf{C}$$

See Golubitsky and Schaeffer [226] Chapter VIII Lemma 2.5. Therefore g has the form

$$g(z, \lambda, \tau) = P(|z|^2, \lambda, \tau)z$$

for some smooth P, and we can write this map in real coordinates as

$$g(z, \lambda, \tau) = p(|z|^2, \lambda, \tau)z + q(|z|^2, \lambda, \tau)iz$$

We need to solve $g = 0$. This requires either $z = 0$ or $p = q = 0$. A key step in the proof of the standard Hopf theorem is to show that $q_\tau(0) \neq 0$. The same argument holds here, so by the Implicit Function Theorem

$$\tau = \tau(|z|^2, \lambda)$$

solves $q(|z|^2, \lambda, \tau) = 0$. Hence $g = 0$ if and only if $z = 0$ or

$$a(|z|^2, \lambda) \equiv P(|z|^2, \lambda, \tau(|z|^2, \lambda)) = 0$$

Again, standard computations in the proof of Hopf bifurcation, Golubitsky *et al.* [237] Chapter XVI Lemma 4.3, shows that

$$a_\lambda(0) = \sigma'(0)$$

where the eigenvalues of $\mathrm{d}f|_{0,\lambda}$ are $\sigma(\lambda) \pm i\omega(\lambda)$. The Implicit Function Theorem again implies the existence of a unique branch of periodic solutions in $\mathrm{Fix}(\Sigma)$. $\square$

There are several useful ways to write the $\mathbf{S}^1$-action. We can consider $\mathbf{R}^{2m}$ to be identified with $\mathbf{C}^m = V \otimes \mathbf{C}$, so the action of $\Gamma \times \mathbf{S}^1$ can be written as

$$(\gamma, \theta)(v \otimes z) = (\gamma v) \otimes (e^{i\theta}z)$$

Alternatively we may identify $\ker(\mathrm{d}\mathcal{N})_{0,0}$ with $V \oplus V$ and form the $m \times 2$ matrix $[v_1|v_2]$ where $(v_1, v_2) \in V \oplus V$. Then γ acts as $m \times m$ matrix

$$\gamma[v_1|v_2] \tag{4.6}$$

and θ acts as right multiplication by the 2×2 rotation matrix

$$\theta[v_1|v_2] = [v_1|v_2]R_\theta. \tag{4.7}$$

The only fixed point of θ is the origin.

Example 4.11 Let $\Gamma = \mathbf{Z}_2(\kappa)$. The irreducible representations of Γ are one-dimensional, and there are two such representations, one trivial and one nontrivial. We consider the nontrivial representation in which $\Gamma = \mathbf{Z}_2(\kappa)$ acts on $\mathbf{R}$ by $\kappa v = -v$. Now consider $\mathbf{Z}_2 \times \mathbf{S}^1$ acting on $\mathbf{R}^2$. Since (κ, π) acts trivially, every 2π-periodic solution $x(t)$ must satisfy

$$\kappa x(t) = x(t - \pi)$$

Thus reflection corresponds to a half-period phase-shift. $\diamond$

Example 4.12 Consider $\Gamma = \mathbf{O}(2)$ with $\dim V = 2$. Using (4.6) and (4.7), the action of $\mathbf{O}(2) \times \mathbf{S}^1$ on $\mathbf{R}^4 \cong \mathbf{C}^2$ is

$$\phi(w_1, w_2) = (e^{i\phi}w_1, e^{i\phi}w_2)$$
$$\kappa(w_1, w_2) = (\bar{w}_1, \bar{w}_2)$$
$$\theta(w_1, w_2) = (\cos(\theta)w_1 - \sin(\theta)w_2, \sin(\theta)w_1 + \cos(\theta)w_2)$$

We transform to nicer coordinates by setting

$$w(z_1, z_2) = (\bar{z}_1 + z_2, i(\bar{z}_1 - z_2))$$
$$z(w_1, w_2) = \tfrac{1}{2}(\overline{w_1 - iw_2}, w_1 + iw_2)$$

In these new coordinates the action of $\mathbf{O}(2) \times \mathbf{S}^1$ is

$$
\begin{array}{rcl}
\phi(z_1, z_2) & = & (e^{-i\phi}z_1, e^{i\phi}z_2) \\
\kappa(z_1, z_2) & = & (z_2, z_1) \\
\theta(z_1, z_2) & = & (e^{i\theta}z_1, e^{i\theta}z_2)
\end{array}
\qquad (4.8)
$$

Computations show that there are precisely two $\mathbf{C}$-axial subgroups (up to conjugacy): $\mathbf{Z}_2(\kappa) \oplus \mathbf{Z}_2(\pi, \pi)$ and $\widetilde{\mathbf{SO}(2)} = \{(\theta, \theta) \in \mathbf{SO}(2) \times \mathbf{S}^1\}$. Their fixed-point spaces are

$$
\begin{array}{rcl}
\mathrm{Fix}(\mathbf{Z}_2(\kappa)) & = & \{(z, z) : z \in \mathbf{C}\} \\
\mathrm{Fix}(\widetilde{\mathbf{SO}(2)}) & = & \{(z, 0) : z \in \mathbf{C}\}
\end{array}
$$

Both fixed-point subspaces are two-dimensional, so the Equivariant Hopf Theorem implies that two periodic solutions exist:

Standing waves: The solution has two symmetries: it is invariant under one reflection at all instants of time, and spatial rotation through angle π has the same effect as a half-period phase shift.

Rotating waves: The solution's trajectory is invariant under rotation, and satisfies $u(s) = R_s u(0)$. That is, time evolution is the same as spatial rotation.   ◇

Example 4.13 Consider $\mathbf{D}_3$ acting $\mathbf{D}_3$-simply on $\mathbf{C}^2$. Since there is only one two-dimensional irreducible representation of $\mathbf{D}_3$ we can write the action of $\mathbf{D}_3 \times \mathbf{S}^1$ on $\mathbf{C}^2$ by restricting the action of $\mathbf{O}(2) \times \mathbf{S}^1$ from the previous example. Thus the action of $\mathbf{D}_3 \times \mathbf{S}^1$ is

$$\phi(z_1, z_2) = (e^{-i\phi}z_1, e^{i\phi}z_2)$$
$$\kappa(z_1, z_2) = (z_2, z_1)$$
$$\theta(z_1, z_2) = (e^{i\theta}z_1, e^{i\theta}z_2)$$

where $\phi = \frac{2\pi}{3}$ and $\theta \in \mathbf{S}^1$. This action is one of the actions we expect to find for three identical all-to-all coupled oscillators. In this case, we have three $\mathbf{C}$-axial subgroups: $\mathbf{Z}_2((\kappa, 0))$, $\widetilde{\mathbf{Z}_3} = <(\phi, \phi)>$, and $\mathbf{Z}_2(\kappa, \pi)$, and

$$
\begin{array}{rcl}
\mathrm{Fix}(\mathbf{Z}_2(\kappa)) & = & \{(z, z) : z \in \mathbf{C}\} \\
\mathrm{Fix}(\widetilde{\mathbf{Z}_3}) & = & \{(z, 0) : z \in \mathbf{C}\} \\
\mathrm{Fix}(\mathbf{Z}_2(\kappa, \pi)) & = & \{(z, -z) : z \in \mathbf{C}\}
\end{array}
$$

Each fixed-point subspace is two-dimensional, so the Equivariant Hopf Theorem implies that there exist three branches of solutions, corresponding to the following pairs (H, K): $(\mathbf{D}_1, \mathbf{D}_1)$, $(\mathbf{Z}_3, \mathbf{1})$, and $\mathbf{D}_1, \mathbf{1})$. See Table 3.2.   ◇

4.3 Poincaré-Birkhoff Normal Form

The Poincaré-Birkhoff normal form provides a method for using the linearization of a vector field at an equilibrium to successively simplify higher order terms in that vector field. See Golubitsky *et al.* [237] or Broer [78] for details — some of which are sketched here. Consider the system

$$\frac{dy}{dt} = f(y) = Ly + f_2(y) + \cdots + f_k(y) + \cdots \tag{4.9}$$

where $f(0) = 0$, f_j is a polynomial map homogeneous of degree k, and all eigenvalues of L lie on the imaginary axis. Let $\mathcal{P}_k$ be the space of all homogeneous polynomial maps from $\mathbf{R}^n$ to $\mathbf{R}^n$ of degree k. Define $\mathrm{ad}_L : \mathcal{P}_k \to \mathcal{P}_k$ by

$$\mathrm{ad}_L(P_k)(y) = (LP_k)(y) - (dP_k)_y Ly$$

Choose a subspace $\mathcal{G}_k \subseteq \mathcal{P}_k$ such that

$$\mathcal{P}_k = \mathcal{G}_k \oplus \mathrm{Im}\,\mathrm{ad}_L$$

Theorem 4.14 (Poincaré-Birkhoff Normal Form Theorem) *For each k there exists a polynomial change of coordinates $y \mapsto x$ of degree k such that in the new coordinates (4.9) becomes*

$$\frac{dx}{dt} = Lx + g_2(x) + \cdots + g_k(x) + r_{k+1}(x)$$

where $g_j \in \mathcal{G}_j$ and r_{k+1} is of degree $\geq k + 1$.

There is a natural choice for the $\mathcal{G}_k$, in which we make use of the one-parameter group $\exp(sL^T) \subseteq \mathbf{GL}(n)$. Then

$$\mathbf{S} = \overline{\exp(sL^T)}$$

is a Lie group which is either $\mathbf{T}^k$ if L is semisimple or $\mathbf{T}^k \times \mathbf{R}$ if L has a nilpotent part. (Here the bar indicates topological closure.)

The next theorem is due to Elphick *et al.* [170]:

Theorem 4.15 *Let $\mathcal{P}_k(\mathbf{S})$ be the vector space of $\mathbf{S}$-equivariant homogeneous polynomials of degree k. Then*

$$\mathcal{P}_k = \mathcal{P}_k(\mathbf{S}) \oplus \mathrm{Im}\,\mathrm{ad}_L .$$

Theorem 4.16 *Let $f : \mathbf{R}^n \to \mathbf{R}^n$ be Γ-equivariant and let $L = (df)_0$. Choose $k \geq 1$. Then there exists a Γ-equivariant polynomial change of coordinates of degree k such that in the new coordinates $y' = f(y)$ transforms to*

$$x' = Lx + g_2(x) + \cdots + g_k(x) + r_{k+1}(x)$$

where g_j is $\Gamma \times \mathbf{S}$-equivariant and r_{k+1} is of degree $\geq k + 1$ and is Γ-equivariant.

Thus, up to order $k+1$, we can assume that the normal form is $\Gamma \times \mathbf{S}$-equivariant. This theorem is proved by averaging the standard Poincaré-Birkhoff normal form over the group Γ.

Example 4.17 In ordinary Hopf bifurcation

$$L = \begin{bmatrix} 0 & -1 \\ 1 & 0 \end{bmatrix}$$

so

$$\mathbf{S} = e^{sL^T} = \begin{bmatrix} \cos s & \sin s \\ -\sin s & \cos s \end{bmatrix}$$

Thus $\mathbf{S} = \mathbf{S}^1$. For equivariant Hopf,

$$L = \begin{bmatrix} 0 & -I \\ I & 0 \end{bmatrix} \equiv J$$

so

$$\mathbf{S} = e^{sL^T} = \begin{bmatrix} \cos sI & \sin sI \\ -\sin sI & \cos sI \end{bmatrix}$$

Since $L = J = (\mathrm{d}f)_0$ commutes with $\mathbf{S}^1$, the truncated normal form $Lx + g_2(x) + \cdots + g_k(x)$ is $\Gamma \times \mathbf{S}^1$-equivariant. $\diamond$

Periodic Solutions of Normal Form Equations are Circles. Suppose that $x(t)$ is a 2π-periodic solution to the truncated normal form equation with initial condition $x(0) = x_0$. We claim that $\mathbf{S}^1$-symmetry of the normal form implies that

$$x(s) = (I, -s)x_0$$

where I is the identity: that is, the solution is a *circle*.

The proof of this remark follows directly from the proof of the Liapunov-Schmidt reduction. Since the truncated normal form is $\Gamma \times \mathbf{S}^1$-equivariant, the Liapunov-Schmidt reduced function is $\Gamma \times \mathbf{S}^1 \times \mathbf{S}^1$-equivariant, where the two actions of $\mathbf{S}^1$ are *identical*. Hence $(1, \theta, -\theta) \in \Gamma \times \mathbf{S}^1 \times \mathbf{S}^1$ acts trivially on all 2π-periodic solutions, that is,

$$\theta x(s + \theta) = x(s) \tag{4.10}$$

or, by setting $\theta = -s$

$$x(s) = (-s)x_0 = e^{(-s)J^t}x_0 = e^{sJ}x_0 \tag{4.11}$$

Liapunov-Schmidt Reduction of Normal Form Vector Fields. Suppose we have a Γ-equivariant vector field $f : \mathbf{R}^{2m} \to \mathbf{R}^{2m}$ in normal form for Hopf bifurcation. Then

$$f(x) = Jx + g(x)$$

is $\Gamma \times \mathbf{S}^1$-equivariant, where

$$J = \begin{bmatrix} 0 & -I \\ I & 0 \end{bmatrix}$$

The action of $\Gamma \times \mathbf{S}^1$ is

$$\gamma(x_1, x_2) = (\gamma x_1, \gamma x_2)$$

$$\theta(x_1, x_2) = e^{-\theta J} \begin{bmatrix} x_1 \\ x_2 \end{bmatrix}$$

Recall from (4.11) that in normal form 2π-periodic solutions can be written as

$$x(s) = e^{sJ}x_0 \tag{4.12}$$

Thus finding periodic solutions is equivalent to solving

$$\frac{\mathrm{d}x}{\mathrm{d}t} = f(x) \tag{4.13}$$

where the solutions $x(s)$ are of the form (4.12). Substituting (4.12) into (4.13) we find that

$$Je^{sJ}x_0 = \frac{\mathrm{d}x}{\mathrm{d}s} = f(x) = f(e^{sJ}x_0) = e^{sJ}x_0 f(x_0)$$

where the rightmost equality follows from the $\mathbf{S}^1$-equivariance of the normal form. Hence 2π-periodic solutions are found by solving

$$g(x_0) \equiv f(x_0) - Jx_0 = 0$$

The stability of the bifurcating solutions can be determined from the eigenvalues of dg. The verification of this fact requires solving the so-called Floquet equation — which can be done explicitly because of Poincaré-Birkhoff normal form. See Golubitsky *et al.* [237] Chapter XVI Section 6.

4.4 **O**(2) **Phase-Amplitude Equations**

In Poincaré-Birkhoff normal form, truncated at sufficiently high order to remove terms that break the $\mathbf{S}^1$ symmetry, $\mathbf{O}(2)$ symmetry-breaking Hopf bifurcation reduces to the study of the ODE

$$
\begin{aligned}
dz_1/dt &= (p+iq)z_1 + (r+is)(|z_2|^2 - |z_1|^2)z_1 \\
dz_2/dt &= (p+iq)z_2 - (r+is)(|z_2|^2 - |z_1|^2)z_2
\end{aligned}
\tag{4.14}
$$

where p, q, r, s are smooth functions of $|z_2|^2 + |z_1|^2$ and $(|z_2|^2 - |z_1|^2)^2$. See [237], Proposition XIV, 2.1, p. 331.

An important property of (4.14) is that it can be decoupled into phase-amplitude equations. Let

$$
\begin{aligned}
z_1 &= xe^{i\phi_1} & z_1\bar{z} &= x^2 \\
z_2 &= ye^{i\phi_2} & z_2\bar{z}_2 &= y^2
\end{aligned}
$$

and substitute. Computing t-derivatives of x, y, ϕ_1, ϕ_2 we find that the amplitude equations are

$$
\frac{dx}{dt} = (p+\delta r)x
$$

$$
\frac{dy}{dt} = (p-\delta r)y
$$

and the phase equations are

$$
\frac{d\phi_1}{dt} = q + \delta s
$$

$$
\frac{d\phi_2}{dt} = q - \delta s
$$

Here $\delta = y^2 - x^2$ and p, q, r and s are functions of $N = x^2 + y^2$ and $\Delta = \delta^2$. The phase equations represent a flow on a 2-torus. Thus, an equilibrium of the amplitude equations corresponds to either an equilibrium point, a periodic orbit, or an invariant 2-torus in the full system (4.14). We also have that stability of equilibria of the amplitude equations corresponds to stability of equilibria, periodic solutions or invariant 2-tori of the original system and hyperbolicity of equilibria of the amplitude equations to normal hyperbolicity of solutions to the original system.

Note that the amplitude equations have $\mathbf{D}_4$ symmetry since they are equivariant with respect to the actions

$$
(x, y) \mapsto (y, x)
$$

$$
(x, y) \mapsto (-x, y)
$$

Remark 4.18 It can be shown that the general amplitude equations are equivalent to the general $\mathbf{D}_4$-equivariant mapping

$$
g(z) = p(z\bar{z}, z^4 + \bar{z}^4)z + q(z\bar{z}, z^4 + \bar{z}^4)\bar{z}^3
$$

Thus, it is possible to study degenerate Hopf bifurcations with $\mathbf{O}(2)$ symmetry by studying degenerate steady-state bifurcations with $\mathbf{D}_4$ symmetry. See Golubitsky *et al.* [237] Chapter XVII.   ◇

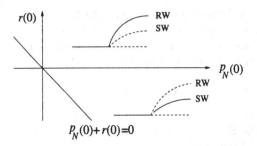

Figure 4.2: Stable solutions in $\mathbf{O}(2)$ Hopf bifurcation. SW $= \mathbf{Z}_2(\kappa)$ standing wave, RW $= \widetilde{\mathbf{SO}(2)}$ rotating wave.

Now assume there exists a bifurcation parameter λ and solve for the zeros of the amplitude equations

$$(p + \delta r)x = 0$$
$$(p - \delta r)y = 0$$

where $p = p(N, \Delta, \lambda)$ and $r = r(N, \Delta, \lambda)$. There are three possibilities

(1) $x = y = 0$. This is the trivial solution, an $\mathbf{O}(2)$-invariant equilibrium.
(2) $x = 0, y \neq 0$. In this case

$$p(y^2, y^4, \lambda) + r(y^2, y^4, \lambda)y^2 = 0$$

Expanding in a Taylor series gives

$$0 = p_\lambda(0)\lambda + (p_N(0) + r(0))y^2 + \text{h.o.t.}$$

So if $p_\lambda(0) \neq 0$ and $p_N(0) + r(0) \neq 0$ there exists a branch of solutions, the rotating waves.

(3) $x \neq 0 \neq y$. There are two subcases.
   (a) $p = 0 = \delta$. Here we have $x^2 = y^2$ and so

   $$O = p(2x^2, 0, \lambda) = p_\lambda(0)\lambda + p_N(0)x^2 + \text{h.o.t.}$$

   Thus if $p_\lambda(0) \neq 0$ and $p_N(0) \neq 0$ there exists a branch of solutions, the standing waves.
   (b) $p = 0 = r$. Generically, we have $r \neq 0$; so there are no solutions of this type.

Therefore the only solutions are standing and rotating waves.

Theorem 4.19 *Generically, only rotating and standing waves appear near an* $\mathbf{O}(2)$ *symmetry-breaking Hopf bifurcation. Moreover, in order for either* $\widetilde{\mathbf{SO}(2)}$ *or* $\mathbf{Z}_2(\kappa)$ *periodic solutions to be stable, both must be supercritical. Then precisely one is stable (depending on* sgn$(r(0))$*).*

Figure 4.2 illustrates the regions of parameter space in which the standing waves and rotating waves exist and are stable.

4.5 Traveling Waves and Standing Waves

We discussed $\mathbf{O}(2)$ steady-state bifurcation in the context of reaction-diffusion systems on a line with periodic boundary conditions (PBC) in Section 2.8. We can now extend that discussion to include $\mathbf{O}(2)$ Hopf bifurcation. Recall the form of a reaction-diffusion system (2.16)

$$u_t = D\Delta u + F(u, \lambda)$$

where $u = (u_1, \dots, u_m)$ be an m-vector of real-valued functions of x and t, D is an $m \times m$ matrix of diffusion coefficients and F is the reaction term. On the interval $[0, \ell]$ PBC have the form

$$
\begin{aligned}
u(0, t) &= u(\ell, t) \\
u_x(0, t) &= u_x(\ell, t)
\end{aligned}
$$

We assume that this system has a spatially homogeneous equilibrium for all λ, that is,

$$F(0, \lambda) = 0$$

We showed in Section 2.8 that

$$X_k = \{e^{2\pi k x i/\ell} a + \text{c.c.} : a \in \mathbf{C}^m\}$$

are the $\mathbf{O}(2)$-isotypic components of the spaces of periodic functions C_ℓ^2 and C_ℓ^0 on which the PDE are defined. It follows that the linearized system is

$$Lu = D\Delta u + A(\lambda)u$$

where $A(\lambda) = (dF)_{(0,\lambda)}$. Symmetry forces L to map each subspace X_k into itself; hence, the eigenvalues of L are just the union of all of the eigenvalues of $L_k = L|X_k$ for $k = 0, 1, \dots$. A calculation summarized in (2.18) showed that the matrix form of L_k is

$$L_k = A(\lambda) - \left(\frac{2\pi k}{\ell}\right)^2 D$$

We discussed the real eigenvalues of L_k and how they lead to $\mathbf{O}(2)$ steady-state bifurcation in Section 2.8. We now discuss the purely imaginary eigenvalues of L_k and how they lead to $\mathbf{O}(2)$ Hopf bifurcation.

Suppose that $\sigma \pm i\omega$ are eigenvalues of L_k for $k > 0$ and that

$$L_k(w(x)) = (\sigma - i\omega)w(x)$$

Then

$$u(x, t) = e^{i\omega t}w(x) + \text{c.c.}$$

is a solution to the linear system

$$u_t = D\Delta u + A(\lambda)u$$

Recall that $w(x)$ has the form $e^{2\pi k i x/\ell}a$ and to linear level the functions $u(x, t)$ approximate the time-periodic spatially-periodic solutions obtained by Hopf bifurcation.

To simplify the discussion we assume that $k = 1$. As we discussed in Sections 4.2 and 4.4 there are two branches of time-periodic solutions emanating from an $\mathbf{O}(2)$ Hopf bifurcation — rotating waves and standing waves. In $\mathbf{SO}(2) \times \mathbf{S}^1$ rotating waves have spatio-temporal symmetries (θ, θ) and standing waves have symmetries $(\kappa, 0), (\pi, \pi)$ where κ is a specific reflection symmetry. On the line rotations act by translation (modulo the spatial period ℓ) and $\kappa x = -x$.

Suppose that $u(x, t)$ is a rotating wave. It follows from symmetry that $u(x, t) = w(x - t)$ (time evolution is the same as spatial translation), that is, symmetry forces u to be a *traveling wave*. On the other hand, suppose that $u(x, t)$ is a standing wave. Then $u(-x, t) \equiv u(x, t)$ and $u_x(0, t) = 0$. Thus the wave form $u(x, t)$ has for fixed t

a maximum or a minimum at $x = 0$; it is a wave that appears to be pinned to the origin, a standing wave.

Hopf Bifurcation in the Brusselator. Hopf bifurcation can be found in the Brusselator model discussed in Section 2.8. We proceed assuming that the reader is familiar with the calculations in that section.

When $m = 2$ (two equations) we can search for purely imaginary eigenvalues of L_k by finding parameter values where $\mathrm{tr}(L_k) = 0$ and $\det(L_k) > 0$. It is easy to check that no Hopf bifurcation occurs in the Brusselator (2.20), (2.21) when the diffusion constants are set as in (2.22), that is, $D_1 = 1$ and $D_2 = 2$. Hopf bifurcation does occur in the Brusselator for $k = 1$ if we set $D_1 = 2$ and $D_1 = 1$. In this instance,

$$L_1 = \begin{bmatrix} \lambda - 3 & \alpha^2 \\ -\lambda & -\alpha^2 - 1 \end{bmatrix}$$

and

$$\det(L_1) = -\lambda + 3\alpha^2 + 3$$
$$\mathrm{tr}(L_1) = \lambda - (4 + \alpha^2)$$

It follows that $\mathrm{tr}(L_1) = 0$ when $\lambda = 4 + \alpha^2$ and at such a point $\det(L_1) = 2\alpha^2 - 1$. If $\alpha = 1$, then $\det(L_1) > 0$ and a Hopf bifurcation occurs at $\lambda = 5$.

Note that we have not found a point where $\mathbf{O}(2)$ symmetry-breaking Hopf bifurcation occurs from a stable equilibrium.

4.6 Spiral Waves and Target Patterns

We now return to the application that motivates this chapter: pattern formation in the Belousov-Zhabotinskii reaction. It is now widely recognized that time-periodic spiral waves and target patterns are universal features of pattern-forming systems in the plane (Murray [413], Cross and Hohenberg [134]). Since their discovery in the Belousov-Zhabotinskii reaction, spirals have been observed in biological systems, for example in the growth pattern of slime molds (Newell [416]), in heart muscle (Winfree [522]), in catalysis (Eiswirth *et al.* [174]), in fluid systems such as convection (Bodenschatz *et al.* [69]) and the Faraday system (Kiyashko *et al.* [314]), and even in vibrated sand (Umbanhowar *et al.* [507]). In some of these cases the spirals form spontaneously; in others a finite amplitude perturbation is necessary to initiate their formation. In some cases the shape of the container, a circular disk for example, is instrumental in facilitating spiral formation; in others spirals form in large aspect ratio systems by a process that is essentially unaffected by the shape of the container. In all of these systems, spiral waves are rotating waves (time evolution is the same as spatial rotation about their centers) and target patterns are circularly symmetric for each moment of time.

The formation of spirals (and to a lesser extent target patterns) has been studied from numerous points of view — numerical simulation, matched asymptotic expansions, phase equations, and various phenomenological models. Most easily treated are certain PDEs known as λ-ω systems; when truncated at third order in the amplitude these systems can be written in the form of an (isotropic) complex Ginzburg-Landau equation in the plane. For these PDEs Kopell and Howard [324] establish the existence of both target and spiral solutions. More recently, Scheel [465, 466] proved the existence of solutions of reaction-diffusion equations in the infinite plane that possess the main characteristics of spirals near infinity. Scheel studies Hopf bifurcation to solutions in the form of appropriately defined (and possibly many-armed) spiral waves, and in particular constructs a finite-dimensional manifold that contains all small rotating waves close to the homogeneous equilibrium. He also relates his results to earlier work on $\lambda - \omega$ systems.

We follow [214] and study the formation of spirals from the viewpoint of equivariant bifurcation theory, treating both spirals and target patterns as general, universal phenomena with many common features. To do this we formulate the onset of the instability as a bifurcation problem on a disk. We begin with the observation that with Robin (otherwise called mixed) boundary conditions reaction-diffusion equations on a disk can undergo a Hopf bifurcation from the trivial state, whose associated eigenfunctions have a prominent spiral character. They take the form of complex Fourier-Bessel functions, and their amplitude grows rapidly towards the boundary. Consequently we refer to them as *wall* modes. In contrast, with Neumann or Dirichlet boundary conditions, the eigenfunctions take the form of *body* modes filling the interior of the domain but lacking the expected spiral character. For certain parameter values these body modes may, in fact, possess a roughly spiral character, but with 'dislocations' at which the strands of the spiral may split or join.

Spiral waves in the form of wall modes have been found in single reaction-diffusion equations on a disk with *spiral* boundary conditions (Dellnitz *et al.* [145]) and in binary fluid convection in a cylinder with realistic boundary conditions at the sidewall (Mercader *et al.* [397]).

In the next subsection we describe the level curves of complex Fourier-Bessel eigenfunctions and show that these are suitable for describing both spirals and target patterns. In Section 4.7 we show, using a mixture of analytical and numerical calculations, that solutions of this type are produced naturally as a result of a Hopf bifurcation in systems of reaction-diffusion equations with Robin boundary conditions. We also show that, unlike the situation for Neumann boundary conditions (Auchmuty [35]), there are few restrictions if any on the (azimuthal) mode number of the primary bifurcation, or on the sequence of mode numbers that appear in successive bifurcations from the trivial state.

Eigenfunction Patterns in a Disk. What kinds of pattern should we expect to be generated as solutions of a Euclidean-invariant system of PDEs, such as a reaction-diffusion equation, posed on a circular disk? A partial answer to this question is found by considering solutions that bifurcate from a trivial constant solution. The pattern in these solutions is dominated by the pattern in the eigenfunctions of the associated linearized system of PDEs, at least near the primary bifurcation.

We consider reaction-diffusion systems whose domain Ω is a circular disk of radius R, defined by

$$U_t = D\Delta U + F(U) \tag{4.15}$$

Here $U = (U_1, \ldots, U_\ell)^t$ is a ℓ-vector of functions, D is an $\ell \times \ell$ matrix of diffusion coefficients, and F is an ℓ-dimensional smooth mapping. The linearized system is

$$U_t = D\Delta U + AU \tag{4.16}$$

where A, the linearization of F at the origin, is an $\ell \times \ell$ reaction matrix. As we discuss in more detail in Section 4.7, the eigenfunctions of (4.16) depend on boundary conditions and are linear combinations of Fourier-Bessel functions. To be specific, let (r, θ) denote polar coordinates on Ω and let J_m be the (complex)

Bessel function of the first kind of order m, for some nonnegative integer m. Then a Fourier-Bessel function has the form

$$f(r, \theta, t) = \text{Re}\left[z\, e^{i\omega t + im\theta} J_m(qr)\right] \tag{4.17}$$

where $q \in \mathbf{C}$ is some nonzero constant and $z \in \mathbf{C}$ is a constant scalar. The possibility that q may be *complex* permits eigenfunctions with spiral geometry.

Patterns from $\mathbf{O}(2)$ *Bifurcation.* We assume that the *pattern* associated with a planar function $g : \mathbf{R}^2 \to \mathbf{R}$ is given by the level contours of g. Typically, the pattern associated with a solution is the pattern of some *observable*, some real-valued function of the (components of the) solution. See Golubitsky *et al.* [219] and Section 6.6. For example, one common observable is the projection onto one component of the solution vector. The observed pattern is then the pattern of a linear combination of ℓ Fourier-Bessel functions of the form (4.17). To gain a feeling for the types of pattern that can form in these systems, we consider the level contours of (4.17) for some appropriate choices of the resulting eigenfunctions.

We imagine producing patterned solutions via bifurcations obtained through variation of a parameter (typically, a parameter in the reaction matrix A). When the problem is posed on a disk, the primary bifurcation from a spatially uniform equilibrium is either a steady-state bifurcation with $\mathbf{O}(2)$ symmetry or a Hopf bifurcation with $\mathbf{O}(2)$ symmetry. When the critical eigenvalues are simple, the resulting states are $\mathbf{SO}(2)$-invariant, and the patterns of contours consist of concentric circles about the origin. When the critical eigenvalues are double, steady-state bifurcation produces a pitchfork of revolution, and the bifurcating solutions are always invariant under a reflection. In addition, the action of $\mathbf{SO}(2)$ on the eigenspace has kernel $\mathbf{Z}_m$ for some integer $m \geq 1$, and the bifurcating solution is then invariant under m different reflections. Hopf bifurcation with double critical eigenvalues produces a *rotating wave* and a *standing wave*. In addition, the spatial $\mathbf{Z}_m$ symmetry is present in both solutions.

Patterns from Steady-State Bifurcation. We now discuss the exact pattern associated to eigenfunctions corresponding to different $\mathbf{O}(2)$ bifurcations. Which of these eigenfunctions are relevant depends on boundary conditions. In the most familiar cases of Dirichlet boundary conditions ($U = 0$ on $\partial\Omega$) or Neumann boundary conditions ($U_r = 0$ on $\partial\Omega$, where the subscript denotes the partial derivative with respect to r), the complex constant q is forced to be real, leading to real-valued Bessel functions $J_m(qr)$ with a real argument. Such eigenfunctions arise, for example, in the vibration of a circular drum, see for example Courant and Hilbert [124]. On the other hand, as we show in Section 4.7, Robin (or mixed) boundary conditions ($(U_j)_r + \beta_j U_j = 0$ on $\partial\Omega$) can lead to eigenfunctions with complex q. Then $J_m(qr)$ becomes a complex-valued function on the line $\{qr : r \in \mathbf{R}\}$ in the complex plane. Complex q have also been observed when using spiral boundary conditions ($U_r = KU_\theta$ on $\partial\Omega$), see Dellnitz *et al.* [145], and in oscillatory convection with realistic lateral boundary conditions, Mercader *et al.* [397].

After scaling, we may assume $|q| = 1$. The patterns associated with $q = \pm 1$ are well known. When $m \geq 1$, the time-independent states have patterns with m radial nodal lines and a number of concentric nodal circles (Figure 4.3(left)). The position of the nodal circles relative to the boundary depends on the details of the boundary conditions. The amplitude of the eigenfunction decays like $r^{-1/2}$, so all contours except the zero contour are bounded. In the corresponding Hopf bifurcation, the instantaneous time contours are as in Figure 4.3(left); in the rotating wave the contours rotate at a uniform speed about the origin and in the standing wave the nodal lines are fixed while the remaining contours oscillate up and down periodically. Note that one cannot distinguish between these two possibilities because the snapshot retains an m-fold reflection symmetry. However, if the pattern of Figure 4.3(left) represents a rotating wave, we expect this symmetry to be broken once nonlinear terms are added.

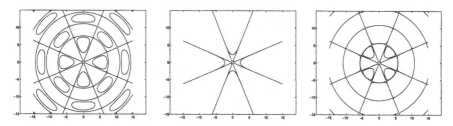

Figure 4.3: Contour plots of (4.17) with $m = 4$, $\omega = 0$; two contours 0 and 0.1 are shown. (Left) $q = 1$; (middle) $q = i$; (right) $q = e^{i\pi/4}$.

Another exceptional case arises when $q = \pm i$. Now there are just m radial nodal lines (Figure 4.3(middle)). Solutions of this type cannot occur with either Neumann or Dirichlet boundary conditions. This is because one of the two independent eigenfunctions is $K_m(|q|r)$ which is singular at $r = 0$ and must therefore be discarded while the other, $I_m(|q|r)$, has no (real) zeros. However, with identical Robin boundary conditions on all species, solutions of this type are possible although there is at most one, see Friedlander and Siegmann [201]. This is in contrast to the case of real q for which there is a countable number of solutions.

In general, however, q is complex (so in particular $q \neq \pm 1, \pm i$ can occur) and the structure of the possible eigenfunctions does not appear to be widely known. Presumably this is because the traditional boundary conditions employed in many problems are either Dirichlet or Neumann. For q that is neither real nor imaginary, and $m > 0$, the functions $J_m(qr)$ are neither real nor purely imaginary. As already discussed, in steady-state bifurcation theory the nonlinear theory picks an eigenfunction that is invariant under a reflection — say reflection across the real axis. Such a function has the form

$$\text{Re}[(e^{im\theta} + e^{-im\theta})J_m(qr)] \tag{4.18}$$

The level contours of (4.18) are shown in Figure 4.3(right).

Patterns from Hopf Bifurcation. The contours of the rotating wave in the corresponding Hopf bifurcation are the contours of (4.17) at an instant of time — an m-armed spiral (Figure 4.4). In time, these contours rotate rigidly with uniform angular velocity, Mercader *et al.* [397]. The amplitude of these complex Bessel functions $J_m(qr)$ grows exponentially with r instead of decreasing algebraically, as in the case of q real. The standard asymptotic formula for Bessel functions, discussed in the next subsection, shows that the spiral is asymptotically Archimedean (spaced equally in the radial direction) for large r. In practice this approximately equal spacing also remains valid for quite small r.

The standing wave in this Hopf bifurcation has nodal lines: a typical case is shown Figure 4.3(right). The nonzero contours oscillate periodically in time and the nodal lines remain fixed. However, the phasing of this oscillation is such that the circular sections of the contours drift outward from the origin in a wave-like motion, see [397]. When $m = 0$ the eigenfunction (4.17) has no θ-dependence, and

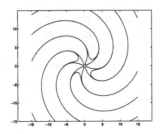

Figure 4.4: Contour plot of (4.17) with $m = 4$, $q = e^{i\pi/4}$, $t = 0$. Two contours 0 and 0.1 are shown.

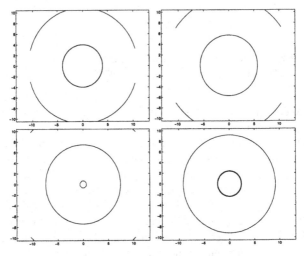

Figure 4.5: Contour plot of (4.17) with $m = 0$, $q = e^{0.35i\pi}$, $\omega = 1$, $t = 0, \frac{\pi}{4}, \frac{\pi}{2}, \frac{3\pi}{4}$. Two contours 0, 0.1 are shown.

the level contours are concentric circles about the origin, as expected. Figure 4.5 shows a time sequence of the level contours showing these contours propagating radially as in a target pattern. Again the amplitude of the eigenfunction grows exponentially in r.

Asymptotics of Bessel Functions. We now briefly describe how to employ the standard asymptotic expansion of Bessel functions, Whittaker and Watson [517], to verify the above statements analytically. Recall that we are dealing with eigenfunctions of the form

$$f(z, \theta, t) = \text{Re}[e^{i\omega t + im\theta} J_m(z)]$$

where $z = re^{i\psi}$. For large $|z|$ the asymptotics of Bessel functions imply that

$$f(z, \theta, t) \sim \sqrt{\frac{1}{2\pi r}} e^{r\sin\psi} \cos\left(\omega t + m\theta - \frac{\psi}{2} - r\cos\psi + \frac{m\pi}{2} + \frac{\pi}{4}\right) \qquad (4.19)$$

when $\sin\psi > 0$, and

$$f(z, \theta, t) \sim \sqrt{\frac{1}{2\pi r}} e^{r|\sin\psi|} \cos\left(\omega t + m\theta - \frac{\psi}{2} + r\cos\psi - \frac{m\pi}{2} + \frac{\pi}{4}\right) \qquad (4.20)$$

when $\sin\psi < 0$.

The zero-sets of the asymptotic eigenfunctions are easily determined. For simplicity we assume $\sin\psi > 0$. The case $\sin\psi < 0$ is similar, with a few minor changes.

The zero set of (4.19) comprises those points (r, θ) in polar coordinates for which

$$\omega t + m\theta - \frac{\psi}{2} - r\cos\psi + \frac{m\pi}{2} + \frac{\pi}{4} = (2s+1)\frac{\pi}{2} \qquad (4.21)$$

for integer s. It is easy to see that when $m > 0$ and $\psi \neq (2k+1)\frac{\pi}{2}$ for integer k, this zero-set is a rotating m-armed set of Archimedean spiral pairs whose 'pitch' $p = \frac{2m\pi}{\cos\psi}$. By 'Archimedean spiral pair' we mean that the zero-set is comprised of m curves, each consisting of two separate Archimedean spirals which interlace with each other and meet at the origin (see Figure 4.4); each member of this pair moves radially by a distance p as the spiral makes one full turn round the origin. The structure of double spirals is natural because we are considering the zero contour, and we expect to find spiral regions in which the function is positive, separated from spiral regions in which it is negative.

However, when $\psi = (2k+1)\frac{\pi}{2}$, the function $J_m(e^{i\psi})$ is either real or purely imaginary, depending on m. In this case for even m (4.19) is replaced by

$$f(z, \theta, t) \sim \sqrt{\frac{1}{2\pi r}} e^r \cos(\omega t + m\theta) \cos\left((m-k)\frac{\pi}{2}\right) \qquad (4.22)$$

when $\sin\psi > 0$, with a similar expression involving $\sin(\omega t + m\theta)$ for odd m. Thus at given t the only nodes are at $\theta = (2s+1)\frac{\pi}{2m} - \frac{\omega t}{m}$ and there is no outward (or inward) propagation.

The case $m = 0$ is exceptional. Here we get target patterns with no fixed nodal lines, provided again $\psi \neq (2k+1)\frac{\pi}{2}$.

4.7 **O**(2) Hopf Bifurcation in Reaction-Diffusion Equations

We study the primary bifurcation in the system (4.16), written in the form

$$U_t = D\Delta U + (B + \sigma I)U \qquad (4.23)$$

using the method of Goldstein *et al.* [212]. Here $\sigma \in \mathbf{R}$ is the bifurcation parameter, $U = U(r, \theta, t)$, and the linear problem (4.23) is to be solved subject to Robin boundary conditions (RBC)

$$(U_j)_r + \beta_j U_j = 0 \quad \text{on} \quad r = R \qquad (4.24)$$

In the following we assume that the β_j are all distinct; this assumption excludes the case of Neumann or Dirichlet boundary conditions on U. We look for Hopf bifurcations and make the *ansatz*

$$U(r, \theta, t) = e^{i\omega t} V(r, \theta)$$

Substituting this assumption into (4.23) yields the eigenvalue problem

$$D\Delta V + (B + \lambda I)V = 0 \qquad (4.25)$$

where $\lambda = \sigma - i\omega$. To solve this problem we assume that V is an eigenfunction of the Laplacian,

$$\Delta V = -k^2 V$$

Then (4.25) reduces to

$$(B + \lambda I - k^2 D)V = 0 \qquad (4.26)$$

so that (nontrivial) solutions exist provided

$$\det(B + \lambda I - k^2 D) = 0 \qquad (4.27)$$

Equation (4.27) is a polynomial of degree ℓ in k^2. Thus there are ℓ roots $k_1^2, \ldots, k_\ell^2$, assumed to be distinct, which are functions of the complex quantity λ. Let $V_j \equiv (v_j^1, \ldots, v_j^\ell)$ be the nullvector of (4.26) corresponding to k_j^2.

We now choose a fixed value of m. Using separation of variables we can write the solutions to (4.25) as

$$V(r, \theta) = \mathrm{Re}\left[a_1 J_m(k_1 r)e^{im\theta}V_1 + \cdots + a_\ell J_m(k_\ell r)e^{im\theta}V_\ell\right]$$

where $a_1, \ldots, a_\ell$ are complex constants. Applying Robin boundary conditions to V on the disk of radius R for all t implies that the real parts of the ℓ expressions

$$\left\{a_1 v_1^1(k_1 J'_m(k_1 R) + \beta_1 J_m(k_1 R)) + \cdots + a_\ell v_\ell^1(k_\ell J'_m(k_\ell R) + \beta_1 J_m(k_\ell R))\right\}e^{i(\omega t + m\theta)}$$

$$\vdots$$

$$\left\{a_1 v_1^\ell(k_1 J'_m(k_1 R) + \beta_\ell J_m(k_1 R)) + \cdots + a_\ell v_\ell^\ell(k_\ell J'_m(k_\ell R) + \beta_\ell J_m(k_\ell R))\right\}e^{i(\omega t + m\theta)}$$

vanish. These equations hold for all θ precisely when

$$a_1 v_1^1(k_1 J'_m(k_1 R) + \beta_1 J_m(k_1 R)) + \cdots + a_\ell v_\ell^1(k_\ell J'_m(k_\ell R) + \beta_1 J_m(k_\ell R)) = 0$$

$$\vdots$$

$$a_1 v_1^\ell(k_1 J'_m(k_1 R) + \beta_\ell J_m(k_1 R)) + \cdots + a_\ell v_\ell^\ell(k_\ell J'_m(k_\ell R) + \beta_\ell J_m(k_\ell R)) = 0$$

There is a nontrivial solution to these complex equations for $a_1, \ldots, a_\ell$ precisely when

$$\det \begin{bmatrix} v_1^1(k_1 J'_m(k_1 R) + \beta_1 J_m(k_1 R)) & \cdots & v_\ell^1(k_\ell J'_m(k_\ell R) + \beta_1 J_m(k_\ell R)) \\ \vdots & & \vdots \\ v_1^\ell(k_1 J'_m(k_1 R) + \beta_\ell J_m(k_1 R)) & \cdots & v_\ell^\ell(k_\ell J'_m(k_\ell R) + \beta_\ell J_m(k_\ell R)) \end{bmatrix} = 0$$

$$(4.28)$$

Note that when the β_j are distinct we expect the k_j to be distinct also.

Systems of Two Equations. When $\ell = 1$ it is easy to show that all bifurcations are steady-state and the corresponding eigenfunctions are body modes. This is no longer the case when $\ell \geq 2$. In the following we examine the case $\ell = 2$ and write

$$B = \begin{bmatrix} a & b \\ c & d \end{bmatrix}$$

In this case $bc < 0$ is a necessary condition for the presence of a Hopf bifurcation; if $bc > 0$ all bifurcations are necessarily steady regardless of the values of β_1, β_2. To find explicit Hopf bifurcations with eigenfunctions in the form of wall modes we define

$$z_1 = k_1 R \qquad z_2 = k_2 R \qquad \tilde{\beta}_1 = \beta_1 R \qquad \tilde{\beta}_2 = \beta_2 R$$

and write equation (4.28) in the form

$$v_1^1 v_2^2(J'_m(z_1)z_1 + \tilde{\beta}_1 J_m(z_1))(J'_m(z_2)z_2 + \tilde{\beta}_2 J_m(z_2))$$
$$-v_2^1 v_1^2(J'_m(z_2)z_2 + \tilde{\beta}_1 J_m(z_2))(J'_m(z_1)z_1 + \tilde{\beta}_2 J_m(z_1)) = 0$$

$$(4.29)$$

In general this is a complex equation for σ and ω. Bessel functions satisfy the identity

$$zJ'_m(z) = zJ_{m-1}(z) - mJ_m(z)$$

This identity is useful even when $m = 0$ because $J_{-1} = -J_1$. Then (4.29) implies that

$$v_1^1 v_2^2(J_{m-1}(z_1)z_1 + (\tilde{\beta}_1 - m)J_m(z_1))(J_{m-1}(z_2)z_2 + (\tilde{\beta}_2 - m)J_m(z_2))$$
$$-v_2^1 v_1^2(J_{m-1}(z_2)z_2 + (\tilde{\beta}_1 - m)J_m(z_2))(J_{m-1}(z_1)z_1 + (\tilde{\beta}_2 - m)J_m(z_1)) = 0$$

$$(4.30)$$

Numerical Results in Two Dimensions. In this section we describe solution of equation (4.30) for a particular choice of the matrices D and B and the coefficients β_1, β_2. We solve this equation numerically using Matlab. First we used a graphical method to locate parameter values at which the primary bifurcation is to a nonzero mode number, and then we used Matlab's PDE toolbox to solve the linearized equation (4.25) numerically and compute eigenfunctions. The two methods yield results in close agreement.

To locate suitable parameter values we sketched — for the range of mode numbers $0 \leq m \leq 5$ — the curves along which the real and imaginary parts of the left-hand side of (4.30) vanish, as σ and ω vary within suitable ranges. We experimented with parameter values until the first bifurcation was to a nonzero mode number. This occurs, for example, at parameter values

$$D = \begin{bmatrix} 0.01 & 0 \\ 0 & 0.015 \end{bmatrix} \qquad B = \begin{bmatrix} 0.5 & 1 \\ -1 & 0 \end{bmatrix}$$

with

$$R = 1 \qquad \beta_1 = 10 \qquad \beta_2 = 0.01$$

See Figure 4.6.

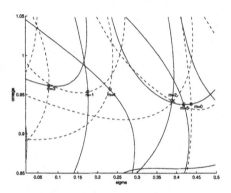

Figure 4.6: Solutions to (4.30) when $m = 0, \ldots, 5$ given by intersections of the zeros of the real (solid curves) and imaginary parts (dashed curves) of (4.30). Open circles indicate intersection points and are labelled by mode number.

In this example the mode numbers of the primary bifurcations occur in the order 3, 1, 4, 2, 5, 0 as σ increases. By definition, the eigenfunctions corresponding to nonzero mode numbers have the symmetry of a rotating wave but lack reflectional symmetry. In our terminology they are therefore 'spiral', although as already discussed they are 'good spirals' only when they are in addition wall modes. Despite the appearance of some of the figures, this is in fact so for all the $m \neq 0$ eigenfunctions illustrated, even for $m = 3$, 1, and 4. Thus in this example the primary bifurcation is to a 3-armed ($m = 3$) spiral, while the bifurcation to a target pattern ($m = 0$) occurs only much later. This behavior is very different from what happens with Neumann boundary conditions (and positive definite D) for which the first instability is always $m = 0$.

Solutions of the PDE (4.25) for these parameter values using the PDE toolbox confirm the eigenvalues and mode numbers obtained with greater accuracy from (4.30) and shown in Figure 4.6. Repeated refinements of the mesh were used to check convergence of the PDE calculation. Table 4.1 lists the resulting values of λ, ω and the associated values of m. Figures 4.7-4.8 show the corresponding

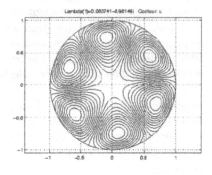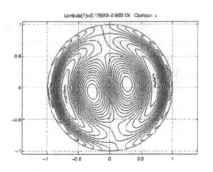

Figure 4.7: Contours of numerically computed eigenfunctions for (4.30). (Left) 1st eigenvalue, for which $m = 3$; (right) 2nd eigenvalue, for which $m = 1$.

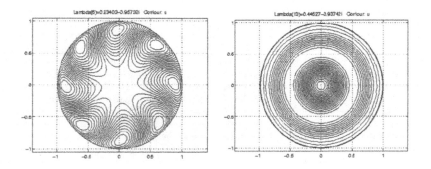

Figure 4.8: Contours of numerically computed eigenfunctions for (4.30). (Left) 3rd eigenvalue, for which $m = 4$ (right) 6th eigenvalue, for which $m = 0$.

eigenfunctions in the form of contour plots. The experiment by Hartmann *et al.*[267] on the NO+CO reaction on a small circular Pt(100) catalyst shows a spiral whose Karhunen-Loève decomposition reveals the presence of $m = 1, 2, 3$ modes with structure remarkably like that shown in these plots. Such states are also found in the model of excitable media introduced by Barkley [267].

real part σ	imaginary part ω	mode number m
0.080	0.961	3
0.179	0.953	1
0.234	0.957	4
0.396	0.941	2
0.415	0.951	5
0.445	0.937	0

Table 4.1 Numerical data for the first six eigenmodes obtained from the PDE.

Since $\omega \neq 0$ the roots k^2 of (4.27) are necessarily complex and the eigenfunctions take the form of wall modes. Note that the requirement $\omega \neq 0$ is not necessary: it is possible for k^2 to be complex even when $\omega = 0$ since the quadratic equation for k^2 may not have real roots. However, the point is that for Robin boundary conditions with $\beta_1 \neq \beta_2$ this is inevitable, and in this sense model-independent. When $\beta_1 = \beta_2 \equiv \beta$ (this case includes both Neumann and Dirichlet boundary conditions) the solution takes the form of a *single* Bessel function with k^2 real. There is a countable number of solutions with $k^2 > 0$ (body modes) and (if $\beta \neq 0, \infty$) at most one solution with $k^2 < 0$ (a wall mode), see Friedlander and Siegmann [201].

Nonlinear Theory. Center manifold reduction can now be used to establish the presence of nonlinear spirals (rotating waves) and target patterns (standing waves), as in the analysis of the Hopf bifurcation with **O**(2) symmetry. As we have seen, when q is neither real nor purely imaginary the spirals consist of waves that travel outwards and at the same time rotate azimuthally. The targets are an equal-amplitude superposition of clockwise and counterclockwise rotating m-armed spirals and so have m-fold reflection symmetry; as a result they do not rotate, although the waves do continue to propagate radially outwards.

Solutions to nonlinear equations, obtained via bifurcation analysis, resemble appropriate linear eigenfunctions. This is because local bifurcation theorems guarantee only solutions with sufficiently small amplitude. However, when the linear eigenfunction involves a complex Bessel function, the amplitude of the eigenfunction increases exponentially from the center to the boundary, forcing the domain of validity of the (weakly) nonlinear theory to be much smaller than might otherwise be expected. This increase of amplitude also leads to interpretational difficulties in physical space, since exponentially growing spiral states do not resemble the spirals of approximately uniform amplitude that are observed in chemical or fluid systems.

The view of spirals that emerges suggests that in large domains the region in parameter space where linear or weakly nonlinear theory provides a global description of target or spiral patterns is very small — in fact, so small as to be practically undetectable. In a somewhat larger parameter range (though still small) the eigenfunctions (wall modes) of the linear theory describe only the inner part of the core, while the visible part of these patterns, separated from the core by a front, is fully nonlinear and hence inaccessible to local bifurcation theory. See the discussion in Golubitsky *et al.* [214].

In our approach these boundary conditions provide a convenient, but physically well motivated, procedure for generating a Hopf bifurcation from the trivial state with a nonzero spatial wave number and exponentially growing eigenfunctions. In fact, solutions of this type may be expected whenever the linear problem is non-selfadjoint, a situation that we believe to be generic for reaction-diffusion systems of sufficient complexity.

What about spirals on the infinite plane? The limit $R \to \infty$ is mathematically problematic. Formally, our results suggest that the region in which the weakly nonlinear theory applies shrinks to zero. The instability is then to a mode that grows exponentially in r. The solution therefore becomes fully nonlinear immediately at onset with no intermediate weakly nonlinear regime in which the solution resembles the linear eigenfunction over the whole domain. There has been much recent work on bifurcation from spiral waves in the infinite plane, which presumes the existence of spiral waves; this work relies on Euclidean symmetry. See Section 6.7.

4.8 Hopf Bifurcation in Coupled Cell Networks

In this section we show how the general results on Hopf bifurcation in symmetric systems of ODEs can be realized in coupled cell systems. A number of issues concerning the interpretation of the abstract symmetry results arise, and we discuss these in some detail.

Interpretation of Permutational Symmetries. We clarify two fundamental issues in Hopf bifurcation for coupled cell systems. These issues center around the role of permutational symmetries in describing the symmetry of the network, and the sometimes subtle differences between an abstract group, the same group represented as a group of linear transformations, and the same group represented as a group of permutations. The Equivariant Hopf Theorem is normally stated in terms of a group of linear transformations — that is, a group representation. However, in coupled cell networks the interpretation of a symmetry group depends on its structure as a permutation group. The section is divided into three parts. First, we discuss the fact that not all representations occur as permutation representations for a given network. Second, we show by example that couple cell networks support two different presentations of each nonabsolutely irreducible permutation representations, and finally we discuss this last point more abstractly in terms of representations on loop spaces — the spaces relevant to the proof of the Equivariant Hopf Theorem given in Section 4.2.

Representations in a Coupled Cell Network at Hopf Bifurcation. The first issue is the non-occurrence of certain types of primary Hopf bifurcations, because of the absence of the corresponding Γ-simple representation.

Consider a system of N coupled cells, and let Γ be the global symmetry group of the network, considered as a permutation group. Suppose that the internal dynamics of each cell has state space R^ℓ. Let ρ be the permutation action of Γ on $\mathbf{R}^N$. Suppose that $\ell = 1$ and write the phase space of the coupled cell system as:

$$\mathbf{R}^N = V_1 \oplus \cdots \oplus V_k \tag{4.31}$$

where the V_j are isotypic components of this permutation action of the permutation group Γ. Moreover, for general ℓ, we can write the phase space as

$$(\mathbf{R}^\ell)^N = V_1^\ell \oplus \cdots \oplus V_k^\ell$$

where the V_j^ℓ are the isotypic components of the phase space for this coupled cell network. In particular, the only irreducible representations of Γ that occur in the coupled cell network are those that occur when $\ell = 1$ and not all irreducible representations of Γ may occur in a given cell network. We call the irreducible representations that occur in the isotypic components listed in (4.31) *permutation representations of ρ.*

Theorem 4.20 *Hopf bifurcation in a coupled cell network can be supported by the permutation representations V of Γ when*

- *V is a non-absolutely irreducible permuatation representation of ρ.*
- *V is an absolutely irreducible permutation representation of ρ when $\ell \geq 2$.*
- *V is an absolutely irreducible representation of ρ that occurs with multiplicity ≥ 2 when $\ell = 1$.*

Proof. The representations supported by Hopf bifurcation are Γ-simple.   □

Examples of Permutation Representations. Next we compare and contrast two examples of coupled cell networks, each having symmetry group abstractly isomorphic to $\mathbf{Z}_6$ but with distinct permutation representations.

An Example with All $\mathbf{Z}_6$ Irreducible Representations. Here the network comprises six identical cells arranged in a ring with unidirectional coupling (Figure 4.9). The appropriate permutation representation is $\mathbf{Z}_6 \subseteq \mathbf{S}_6$ where $\mathbf{Z}_6$ is generated by the 6-cycle (1 2 3 4 5 6).

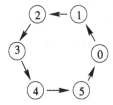

Figure 4.9: Six identical cells arranged in a ring.

There are four distinct irreducible real representations of $\mathbf{Z}_6$:

$$
\begin{aligned}
\rho_0 \text{ on } \mathbf{R} &: \quad \rho_0(\sigma)(x) = x \\
\rho_1 \text{ on } \mathbf{C} &: \quad \rho_0(\sigma)(x) = \zeta x \\
\rho_2 \text{ on } \mathbf{C} &: \quad \rho_0(\sigma)(x) = \zeta^2 x \\
\rho_3 \text{ on } \mathbf{R} &: \quad \rho_0(\sigma)(x) = -x
\end{aligned}
$$

where $\zeta = e^{2\pi i/6}$. Of these, ρ_0, ρ_3 are absolutely irreducible, whereas ρ_1, ρ_2 are nonabsolutely irreducible of complex type. In this example the permutation representation of $\mathbf{Z}_6$ on $\mathbf{R}^6$ splits as $\rho_0 \oplus \rho_1 \oplus \rho_2 \oplus \rho_3$. Thus all irreducible representations are present in the permutation representation.

An Example with Missing $\mathbf{Z}_6$ Irreducible Representations. These examples relate to the multirhythm states found in Section 3.6. Here the network comprises five cells. Recall Figure 3.14, which we reproduce here as Figure 4.10. Cells 1-3 are identical and arranged in a ring with unidirectional coupling; cells 4-5 are identical (but can differ from 1-3) and are arranged in a 'ring'. Finally, every cell in the first ring is coupled to every cell in the second ring with all such couplings identical. For this example the permutation group is $\mathbf{Z}_6 \equiv \mathbf{Z}_3 \times \mathbf{Z}_2 \subseteq \mathbf{S}_5$ and the permutation representation is generated by the 3-cycle (123) and the 2-cycle (45). The permutation representation of $\mathbf{Z}_6$ on $\mathbf{R}^5$ splits as $2\rho_0 \oplus \rho_2 \oplus \rho_3$. In particular ρ_1 is missing, so there can be no 1/6 twist. Such oscillations can, however, occur as a consequence of resonant Hopf bifurcation, much as we explained in Section 3.6.

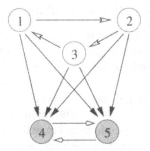

Figure 4.10: Five cells arranged in two rings, coupled together.

Rings of Three Coupled Cells. The second issue is the occurrence of two *different* S^1-actions in certain cases of symmetric Hopf bifurcation — namely, when the action is Γ-simple of *complex* type.

The nature of this problem can be grasped by considering the networks of three cells coupled in a triangle as in Figure 4.11. The global symmetry group here is $\mathbf{Z}_3 \subseteq \mathbf{S}_3$ acting by a 3-cycle. Assume for simplicity that each cell has a 1-dimensional state space. Then the overall state space is 3-dimensional, and it splits into two irreducible components: $\mathbf{R}^3 = W_0 \oplus W_1$ where

$$
\begin{aligned}
W_0 &= \{(x_1, x_1, x_1) : x_1 \in \mathbf{R}\} \\
W_1 &= \{(x_1, x_2, x_3) : x_1 + x_2 + x_3 = 0\}
\end{aligned}
$$

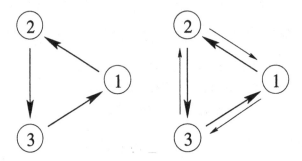

Figure 4.11: Three identical cells with $\mathbf{Z}_3$ symmetry. (Left) a unidirectional ring; (right) an asymmetrically coupled ring.

Say that an invariant subspace of a group Γ *supports* Hopf bifurcation if it can contain the generalized eigenspace for a complex conjugate pair of purely imaginary eigenvalues $\pm i\omega$ of a linear operator that commutes with Γ. Then Hopf bifurcation cannot be supported by W_0, but it can be supported by W_1. The Equivariant Hopf Theorem tells us that under suitable conditions there exists a branch of periodic states corresponding to each **C**-axial subgroup of $\mathbf{Z}_3 \times \mathbf{S}^1$ acting on W_1. Since $\dim W_1 = 2$, the only **C**-axial subgroup is the kernel of the action. This kernel is a twisted $\mathbf{Z}_3$ generated by either $(\rho, -\frac{T}{3})$ or $(\rho, \frac{T}{3})$ where ρ is the 3-cycle (123) and T is the period. Recall (3.4). That is, the periodic state $x(t)$ is a discrete rotating wave in which successive cells are out of phase by $\pm T/3$. In symbols

$$
\begin{aligned}
x(t) &= (X(t), X(t + T/3), X(t + 2T/3)) \\
x(t) &= (X(t), X(t + 2T/3), X(t + T/3))
\end{aligned}
$$

But which of these rotating waves actually occurs. It seems reasonable clear that both rotating waves should occur in the the asymmetrically coupled system. but it also seems reasonable that only one of the discrete rotating waves should occur in the unidirectional system.

In fact, both discrete rotating waves can occur in both cell systems by equivariant Hopf bifurcation, as we now show. The reason is that for this representation there are two alternatives for the circle group action of phase shifts. Both representations of the circle group occur in the loop space of T-periodic maps $\mathbf{S}^1 \to \mathbf{R}^3$, and both can arise through the process of Liapunov-Schmidt reduction that is used to state and prove the Equivariant Hopf Theorem in Section 4.2.

Equations for the asymmetric coupled cell system may be written in the form:

$$dx_1/dt = Ax_1 + Bx_2 + Cx_3 + \phi(x_1, x_2, x_3)$$
$$dx_2/dt = Ax_2 + Bx_3 + Cx_1 + \phi(x_2, x_3, x_1)$$
$$dx_3/dt = Ax_3 + Bx_1 + Cx_2 + \phi(x_3, x_1, x_2)$$

where ϕ and its derivative vanish at the origin. We claim:

Proposition 4.21 *There is a Hopf bifurcation point when*

$$A - \frac{B+C}{2} = 0$$
$$A+B+C \neq 0$$
$$B - C \neq 0$$

At such a point, when $B > C$ the solution given by the Equivariant Hopf Theorem has the form

$$x(t) = (X(t), X(t+T/3), X(t+2T/3))$$

but when $B < C$ this solution has the form

$$x(t) = (X(t), X(t+2T/3), X(t+T/3))$$

In the unidirectional coupled cell system $B = 0$ and $\phi(x_1, x_2, x_3) = \phi(x_1, x_3)$ is independent of x_2. It follows from Proposition 4.21 that which of the two discrete rotating waves occurs depends on the sign of C.

Proof. The linearization is

$$L = \begin{bmatrix} A & B & C \\ C & A & B \\ B & C & A \end{bmatrix}$$

The eigenvalues of L are:

$$A+B+C \qquad \text{with eigenvector } p = (1,1,1)^T$$
$$A + B\omega + C\omega^2 \qquad \text{with eigenvector } q = (1, \omega, \omega^2)^T$$
$$A + B\omega^2 + C\omega \qquad \text{with eigenvector } r = (1, \omega^2, \omega)^T$$

where $\omega = e^{2\pi i/3} = -\frac{1}{2} + i\frac{\sqrt{3}}{2}$. The conditions for Hopf bifurcation are:

$$A + B + C \neq 0$$

$$\text{Re}\left\{ A + B\left(-\frac{1}{2} + i\frac{\sqrt{3}}{2}\right) + C\left(-\frac{1}{2} - i\frac{\sqrt{3}}{2}\right) \right\} = A - \frac{B+C}{2} = 0$$

$$B - C \neq 0$$

with eigenvalues $\pm i(B-C)\frac{\sqrt{3}}{2}$ and period $T = \frac{4\pi}{\sqrt{3}|B-C|}$

To determine the spatiotemporal symmetries of the bifurcating periodic solution, we need to compute the actions of the circle group $e^{\theta L}$ and the permutation ρ on the critical eigenspace W_1. Then we can determine the kernel of the $\mathbf{Z}_3 \times \mathbf{S}^1$ action on W_1. These calculations are made easier by using the eigenvector basis q, r of W_1. In particular, in this basis,

$$L|_{W_1} = \begin{bmatrix} i(B-C)\frac{\sqrt{3}}{2} & 0 \\ 0 & -i(B-C)\frac{\sqrt{3}}{2} \end{bmatrix} \qquad \rho|_{W_1} = \begin{bmatrix} \omega & 0 \\ 0 & \omega^2 \end{bmatrix}$$

To determine the kernel of the action, we need to know whether $e^{TL/3} = \rho$ or $e^{-TL/3} = \rho$ when restricted to W_1. This calculation is straightforward since the two matrices are diagonal in the chosen basis. Moreover, since the second diagonal entry is the complex conjugate of the first diagonal entry in each matrix, we need only ask which of

$$e^{\pm i(B-C)\frac{\sqrt{3}}{2}\frac{T}{3}} = \omega = e^{2\pi i/3}$$

is satisfied. A short calculation shows that

$$i(B-C)\frac{\sqrt{3}}{2}\frac{T}{3} = \operatorname{sgn}(B-C)\frac{2\pi i}{3}$$

We conclude that letting time flow forward by one third of a period acts like ρ when $B > C$, and like ρ^{-1} when $B < C$. Note also that we can interchange the choices of ρ and ρ^{-1} by time reversing the system of ODEs. $\qquad\square$

Note that the transition at $B = C$ corresponds to the period in the Hopf theorem going to infinity, and the condition for purely imaginary eigenvalues to exist is violated there. Note that when $B = C$ the system is $\mathbf{D}_3$ symmetric at linear level and Hopf bifurcation with $\mathbf{D}_3$ symmetry cannot be supported on a 2-dimensional subspace like W_1.

As the above example illustrates, in general the possible distinct circle group actions depend on the representation. The usual way of describing the Equivariant Hopf Theorem involves certain identifications, and when interpreting the abstract theory for coupled cell networks it is important to make these identifications explicit and consider whether they are unique.

Representation Types of $\Gamma \times \mathbf{S}^1$ for Hopf Bifurcation. We begin by recalling:

Lemma 4.22 *Generically, Hopf bifurcation is supported by a finite-dimensional $\Gamma \times \mathbf{S}^1$-irreducible representation.*

Proof. A representation is Γ-simple if and only if it is $\Gamma \times \mathbf{S}^1$-irreducible. See Golubitsky *et al.* [237] Chapter XVI Lemma 3.4(b) p. 273. $\qquad\square$

Thus, to classify Hopf bifurcations, we need to classify the irreducible representations of $\Gamma \times \mathbf{S}^1$. Recall that there are three types of Γ-simple representations that stem from three different types of Γ-irreducible representations:

1) Γ acts absolutely irreducibly on V and Γ-simply on $V \oplus V$.
2) Γ acts nonabsolutely irreducibly on W and is of complex type.
3) Γ acts nonabsolutely irreducibly on W and is of quaternionic type.

Recall also that the *type* of the representation is the isomorphism type of the algebra of commuting linear maps, which is either $\mathbf{R}$, $\mathbf{C}$, or $\mathbf{H}$. Type $\mathbf{R}$ is absolutely irreducible, type $\mathbf{C}$ is nonabsolutely irreducible of complex type, and type $\mathbf{H}$ is nonabsolutely irreducible of quaternionic type.

Lemma 4.23 *There are four types of $\Gamma \times \mathbf{S}^1$-irreducible representations*

Proof. We show that in the real and quaternionic cases there is only one possible $\Gamma \times \mathbf{S}^1$-action, but in the complex case there are two — a complex conjugate pair with respect to the natural complex structure. We treat each of the three cases in turn.

Case 1: *V absolutely irreducible of type* **R**. Here we can write $V \oplus V$ as the set of complex vectors $v + iw$ where $v, w \in V$. The group Γ acts on $V \oplus V \cong V \oplus iV$ by

$$\rho(v + iw) = \rho v + i\rho w$$

and the space of commuting linear maps consists of all block matrix maps

$$M(v + iw) = (av + bw) + i(cv + dw)$$

where

$$M = \begin{bmatrix} a & b \\ c & d \end{bmatrix}$$

Since the S^1-action commutes with that of Γ, it is easy to see that there are two distinct possibilities for the circle group action. Either $\theta \in S^1$ acts by R_θ or by $R_{-\theta}$. We claim these two actions are isomorphic, by a map that commutes with the Γ-action. Specifically, we have two actions of $\Gamma \times S^1$:

$$
\begin{aligned}
(\rho, \theta) \cdot (v + iw) &= e^{i\theta}(\rho v + i\rho w) \\
(\rho, \theta) * (v + iw) &= e^{-i\theta}(\rho v + i\rho w)
\end{aligned}
$$

However, these actions ($\cdot$ and $*$) are isomorphic via the 'complex conjugation' map $\phi(v + iw) = v - iw$ since

$$(\rho, \theta) \cdot \phi(v + iw) = \phi(\rho, \theta) * (v + iw)$$

Case 2: *W non-absolutely irreducible of type* **C**. Here the space of commuting linear maps on W can be written, with respect to a suitable basis, as the 'complex' matrices

$$M = \begin{bmatrix} a & -b \\ b & a \end{bmatrix}$$

which can be identified with the complex scalars $a + ib$. Again there are two possible S^1-actions, by $e^{i\theta}$ and $e^{-i\theta}$. However, these are distinct — not isomorphic by a map that preserves the Γ-action. The reason is that the 'complex conjugation' matrix is not of complex type (in order for it to be so, a must equal both 1 and -1, which is impossible).

Case 3: *W nonabsolutely irreducible of type* **H**. Here the set of commuting matrices is isomorphic to the quaternions **H**, and all circle group actions are quaternionically conjugate by Montaldi *et al.* [406]. (The key point is that $-i$ and i are conjugate: $j^{-1}ij = -i$. So the actions by $e^{\pm i\theta}$ are quaternionically conjugate.) So we have two actions to consider:

$$(\rho, \theta) \cdot (v) = e^{i\theta}(\rho v)$$

and

$$(\rho, \theta) * (v) = q^{-1}e^{i\theta}q(\rho v)$$

Clearly these are isomorphic via the map

$$\phi(v) = qv$$

and since this is left multiplication by a quaternion, it commutes with the Γ-action (as is easily verified directly). $\quad\quad\quad\quad\quad\quad\quad\quad\quad\quad\quad\quad\quad\square$

Representations of $\Gamma \times \mathbf{S}^1$ on Loop Space. Next we show that each type of irreducible representation of $\Gamma \times \mathbf{S}^1$ listed in the proof of Lemma 4.23 actually occurs in kernels of linearized operators and therefore can occur in Hopf bifurcation. Recall from (4.4) that the operator equation in the proof of Hopf bifurcation is

$$N(u, \lambda, \tau) \equiv (1 + \tau)\dot{u} - F(u, \lambda) = 0$$

The linearization of N leads to the linear differential equation

$$Lu = \dot{u} - A_0 u$$

where $A_0 = (dF)_{0,0}$. Let K be the kernel of L inside loop space.

We review the action of $\Gamma \times \mathbf{S}^1$ on K. First, $\gamma \in \Gamma$ acts on $u(s) \in K$ by

$$(\gamma u)(s) = \gamma u(s)$$

Second, $s_0 \in \mathbf{S}^1$ acts on u by

$$(s_0 u)(s) = u(s - s_0)$$

Since Γ acts on the range of u and $\mathbf{S}^1$ acts on the domain of u, the two actions commute and we get an action of $\Gamma \times \mathbf{S}^1$ on K. Moreover, this action factors through Liapunov-Schmidt reduction and leads to the interpretation of symmetries of periodic solutions as spatio-temporal symmetries.

Generically, we can assume that the action of Γ on K is Γ-simple and corresponds to one of the permutation representations associated to the coupled cell system. Indeed, if we permit the coupled cell equations to be the general Γ-equivariant vector field (at least to linear order), then it is possible for anyone of the permutation representations of Γ to generate the Γ-simple representation associated to Hopf bifurcation.

It follows that the action of $\Gamma \times \mathbf{S}^1$ on K is irreducible. In the cases where the action of Γ on K consists of a multiplicity two absolutely irreducible representation or a nonabsolutely irreducible representation of quaternionic type, there is, up to isomorphism, just one irreducible action of $\Gamma \times \mathbf{S}^1$ on K and that irreducible can occur in equivariant Hopf bifurcation. Suppose that the action of $\Gamma \times \mathbf{S}^1$ on K is nonabsolutely irreducible of complex type. Lemma 4.23 shows that, up to isomorphism, there are two possible irreducible representations of $\Gamma \times \mathbf{S}^1$ on K. The question that we are asking is whether both of these representations can occur — or not. The answer is that they both occur and the reason is that the other representation occurs in the linearization of the operator equation

$$\hat{N}(u, \lambda, \tau) = (1 + \tau)\dot{u} + F(u, \lambda) = 0$$

To see this recall that the action of $s_0 \in \mathbf{S}^1$ on K can also be written as multiplication by the matrix $e^{s_0 A_0}$. Changing F to $-F$ changes A_0 to $-A_0$ and that induces the other action. Compare this result with the proof of Proposition 4.21 in the example of three coupled cells with $\mathbf{Z}_3$ symmetry.

4.9 Dynamic Symmetries Associated to Bifurcation

The equivariant approach to Hopf bifurcation introduces an entirely new role for symmetry in nonlinear dynamics: the $\mathbf{S}^1$-symmetry of Hopf bifurcation comes not from the differential equation, but from the type of solution being sought. It represents a symmetry of a model trajectory or attractor.

The simplest example of a dynamic symmetry is a $\mathbf{Z}_2$ symmetry that occur in period-doubling bifurcations, as we now describe.

Period-Doubling for Maps. In certain circumstances, dynamical systems possess symmetries in time instead of (or as well as) in space. Here, we consider an analogous but simpler situation: period-doubling for a mapping. We consider a map that has a fixed point, and analyze possible bifurcation to a period-2 point. This introduces a $\mathbf{Z}_2$ symmetry, and we can apply Liapunov-Schmidt reduction and the Equivariant Branching Lemma.

The basic idea is as follows. Suppose we have a mapping $g : \mathbf{R}^n \times \mathbf{R} \to \mathbf{R}^n$, $g(0,0) = 0$, for which we want to find period-2 points, that is, points x and y such that

$$g(x, \lambda) = y$$
$$g(y, \lambda) = x$$

We form a new system $F : \mathbf{R}^n \times \mathbf{R}^n \times \mathbf{R} \to \mathbf{R}^n \times \mathbf{R}^n$ defined by

$$F(x, y, \lambda) = (g(x, \lambda) - y, g(y, \lambda) - x)$$

so that $F(x, y, \lambda) = 0$ if and only if x and y form a period-2 orbit. We look for zeros by Liapunov-Schmidt reduction. The linearization of F is

$$(dF)_0 = \begin{bmatrix} (dg)_{0,0} & -I \\ -I & (dg)_{0,0} \end{bmatrix}$$

so

$$(dF)_0 \begin{bmatrix} u \\ v \end{bmatrix} = 0$$

if and only if $(dg)_{0,0}u = v$ and $(dg)_{0,0}v = u$. This implies that 1 is an eigenvalue of $(dg)_{0,0}^2$. Therefore 1 or -1 is an eigenvalue of $(dg)_{0,0}$. The eigenvalue 1 corresponds to a bifurcation to a branch of fixed points; the eigenvalue -1 corresponds to a bifurcation to a branch of period-2 points. Therefore, period-doubling is equivalent to $(dg)_{0,0}$ having an eigenvalue -1. Assume that the corresponding eigenspace is V. Then the kernel of dF has the form $\mathcal{K} = \{(u, -u) : u \in V\}$. Liapunov-Schmidt reduction implies the existence of $f : \mathcal{K} \times \mathbf{R} \to \mathcal{K}$ whose zeros are in one-to-one correspondence with period-2 points of g.

Remark 4.24 (1) Suppose g is Γ-equivariant. Then

$$F(\gamma x, \gamma y, \lambda) = \gamma F(x, y, \lambda)$$
$$f(\gamma k, \lambda) = \gamma f(k, \lambda)$$

(2) Note that there exists an extra symmetry $\rho(x, y) = (y, x)$. That is,

$$F(\rho(x, y), \lambda) = \rho F(x, y, \lambda)$$

So there exists an extra $\mathbf{Z}_2$ symmetry on $\mathcal{K}$ which is not a symmetry of the original problem, but is introduced by the bifurcation. Since $\mathcal{K} = \{(u, -u)\}$, the symmetry ρ acts as $-I$ on $\mathcal{K}$. It follows that the reduced equation satisfies $f(-k, \lambda) = -f(k, \lambda)$
(3) For $\Gamma = 1$ the simplest period-doubling bifurcation is the pitchfork, driven by the symmetry ρ.
(4) Suppose that -1 is a double eigenvalue of a $\mathbf{D}_3$-symmetric system. Then $\mathbf{D}_3$ acts on $\mathbf{C}$ in the standard way and $\mathbf{Z}_2(\rho)$ acts as $-I$. Thus the full symmetry group is $\mathbf{D}_3 \oplus \mathbf{Z}_2(\rho) = \mathbf{D}_6$. The bifurcation diagram is shown in Figure 4.12(left) and the axes of symmetry τ_1 and τ_2 are shown in Figure 4.13. Finally, the space symmetries of the period-two solutions are illustrated in Figure 4.12(right).   ◇

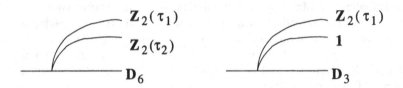

Figure 4.12: (Left) bifurcation diagram for $\mathbf{D}_3$-equivariant period-doubling. (Right) space symmetries of solutions.

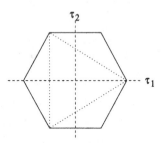

Figure 4.13: Axes of symmetry in $\mathbf{D}_3$ period-doubling bifurcation.

More Sophisticated Dynamic Symmetries. This phenomenon goes considerably deeper, as we now describe. Some of the most interesting issues raised in this section are not yet understood, for example the role of such symmetries in period-doubling cascades, Hao [265, 266].

We have already discussed the 'loop space' approach to Hopf bifurcation in continuous dynamical systems, which leads to the $\mathbf{S}^1$ symmetry. An analogous approach to bifurcation to periodic orbits in discrete systems, generalizing the $\mathbf{Z}_2$-symmetry of period-doubling, has been devised by Vanderbauwhede [509], and independently by Brown [81]. They observe that finding a period-q point of a map f on a space X is equivalent to finding a zero of the map F on X^q defined by

$$F(x_1, \ldots, x_q) = (f(x_q) - x_1, f(x_1) - x_2, \ldots, f(x_{q-1}) - x_q) \qquad (4.32)$$

Such zeros can be found by Liapunov-Schmidt reduction. If the original map f is Γ-equivariant, then the new map F is $\Gamma \times \mathbf{Z}_q$-equivariant, where $\mathbf{Z}_q$ permutes the q coordinates of F cyclicly. With appropriate choice of splittings, the reduced mapping inherits the $\Gamma \times \mathbf{Z}_q$-symmetry, and this can be used to prove existence of period-q points in suitable circumstances. Abstractly, the main difference between this setting and that for Hopf bifurcation is that loop space is replaced by a finite-dimensional space of maps $\mathbf{Z}_q \to X$, and the Liapunov-Schmidt reduced mapping is now $\mathbf{Z}_q$-equivariant rather than $\mathbf{S}^1$-equivariant. A further difference is that the extra period-scaling parameter τ required in the Hopf case does not occur in the discrete case.

Both contexts have the same general structure and employ the same strategy. The idea is to employ a group Ω as a model for the invariant set being sought. The dynamic on this invariant set corresponds to a translation (or in the continuous case to a 1-parameter group of translations) on Ω. The original problem is reinterpreted in terms of zeros of an operator on certain Banach spaces of maps $\Omega \to X$, and Ω-equivariance is used to prove the existence of such zeros.

Period-Doubling Cascades and Adding Machines. An interesting opportunity for introducing a new model group Ω occurs at the Feigenbaum point of a period-doubling cascade. The classic example, of course, is the logistic map, Collet and Eckmann [118]. We can write the map in the form

$$f(x) = \mu x(1 - x)$$

where $x \in [0, 1]$ and $\mu \in [0, 4]$ is a parameter. It is well known that as μ increases towards a value μ_* there is a cascade of periodic solutions with periods 2^k, for $k \to \infty$, with various celebrated universality properties: see Collet and Eckmann [118], Hao [265, 266], and references therein. We call μ_* the *Feigenbaum point*, Figure 4.14.

Figure 4.14: Bifurcation diagram for the quadratic map.

As the figure illustrates, there is also a similar 'reverse cascade' of chaotic intervals for $\mu > \mu_*$. Its main features are attractors that consist of 2^k disjoint intervals, which are cycled by f in the same manner as the corresponding periodic points. (There are also 'windows' of periodic attractors, indeed a fractal hierarchy of periodic and chaotic windows.) These interval attractors merge by way of *crises* in the sense of Grebogi *et al.* [252]. Dellnitz [143] has shown that crises can be studied in terms of iterates of the given mapping. All this suggests that the two 'sides' of the Feigenbaum point are part of the same picture.

The crucial point now is that the dynamics of the quadratic map *at* the Feigenbaum point can be described group-theoretically. Milnor [400] pointed out (in a footnote) that the relevant dynamics is that of a *binary adding machine*. We can define this as the set $\mathcal{A}$ of singly-infinite sequences

$$[a_k] = \ldots a_k a_{k-1} \ldots a_2 a_1 a_0$$

of binary digits $a_k = 0, 1$. There is a map α on $\mathcal{A}$ defined by $\alpha([a_k]) = [b_k]$ where $[b_k] = [a_k] + 1$, where $+$ is interpreted in the sense 'add-and-carry as a binary string'. Specifically, if $a_k = 1$ for all k then $[a_k] + 1 = [b_k]$ where $b_k = 0$ for all k. Otherwise, find the smallest value of k for which $a_k = 0$ and define

$$b_j = \begin{cases} a_j & \text{if } j > k \\ 1 & \text{if } j = k \\ 0 & \text{if } j < k \end{cases}$$

We show later that $\mathcal{A}$ can be interpreted as a (topological) group and, as the notation '$+1$' suggests, α is translation by a fixed element of that group.

The proof that the dynamics of the quadratic map f at the Feigenbaum point is (semiconjugate to) that of α on $\mathcal{A}$ is a consequence of symbolic dynamics. Consider an infinite binary tree $\mathcal{B}$ in which each branch splits into two, infinitely often, as in Figure 4.15.

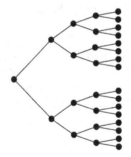

Figure 4.15: Symbolic dynamics for the quadratic map at the Feigenbaum point.

Topologically, the structure of $\mathcal{B}$ corresponds to that part of the bifurcation diagram of f that leads from the trivial equilibrium through the period-doubling cascade and up to (but not including) the Feigenbaum point. The invariant set for the quadratic map at the Feigenbaum point is a Cantor set $\mathcal{C}$, and each point $c \in \mathcal{C}$ is the limit of a sequence of points chosen on successive branches of the bifurcation diagram; that is, each point in $\mathcal{C}$ can be encoded by a singly-infinite binary sequence whose digits, reading from right to left, specify which branch of the binary tree to take in order to limit on c. By considering how f maps points of period 2^k to each other, it follows that the dynamics of f on code symbols is given by α.

Buescu and Stewart [83] prove a generalization of this result. To state it, we must define *multi-base adding machines*. To do this, let $\underline{k} = \{k_n\}_{n\geq 1}$ be a sequence of integers with $k_n > 1$ for all n. Let

$$\Sigma_{\underline{k}} = \prod_{n=1}^{\infty} \{0, \ldots, k_n - 1\}$$

be the space of all one-sided infinite sequences $\underline{i} = (i_n)_{n\geq 1}$ such that $0 \leq i_n \leq k_n - 1$ with the product topology. $\Sigma_{\underline{k}}$ is metrizable; a metric compatible with this topology is $d(\underline{i}, \underline{j}) = \sum_{n=0}^{\infty} |i_n - j_n|/k_n^n$. An elementary check shows that $\Sigma_{\underline{k}}$ is homeomorphic to the Cantor set.

We define the *adding machine* with base $\underline{k} = (k_1, k_2, \ldots)$, which we denote by $\alpha_{\underline{k}} : \Sigma_{\underline{k}} \to \Sigma_{\underline{k}}$, as follows. Let $(i_1, i_2, \ldots) \in \Sigma_{\underline{k}}$. Then

$$\alpha_{\underline{k}}(i_1, i_2, \ldots) =$$
$$\begin{cases} (\underbrace{0, \ldots, 0}_{l-1}, i_l + 1, i_{l+1}, \ldots) & \text{if } i_l < k_l - 1 \text{ and } i_j = k_j - 1 \text{ for } j < l \\ (0, 0, \ldots, 0, \ldots) & \text{if } i_j = k_j - 1 \text{ for all } j \end{cases}$$

Thus the action of $\alpha_{\underline{k}}$ on $x \in \Sigma_{\underline{k}}$ is 'addition of 1 with carry' with respect to a base with an infinite number of digits.

It is well known that $\alpha_{\underline{k}}$ is a topologically minimal homeomorphism of $\Sigma_{\underline{k}}$. The term 'minimal' means that $\omega(x) = \Sigma_{\underline{k}}$ for all $x \in \Sigma_{\underline{k}}$, or equivalently that $\Sigma_{\underline{k}}$ does not contain nontrivial compact invariant subsets. In particular $\Sigma_{\underline{k}}$ admits no periodic orbits under the action of $\alpha_{\underline{k}}$.

The main result of Buescu and Stewart [83] is:

Theorem 4.25 *Suppose that X is a locally connected, locally compact metric space, $f : X \to X$ is a continuous map, and A is a compact transitive set. Assume A is Liapunov stable and has infinitely many components. Then $\tilde{f} : K \to K$ is topologically conjugate to some generalized adding machine.* $\square$

We now return to period-doubling cascades. The above suggests that for 1-dimensional maps, in some sense a binary adding machine is a generic intermediary between a point attractor and an interval attractor, with transitions that necessarily involve period-doubling cascades of periodic points and inverse cascades of crises of interval attractors. As further evidence for this view, the Feigenbaum number 4.669... is associated both with the period-doubling cascade and with the inverse cascade of crises.

Inverse Limits. Associated with the binary adding machine is an intriguing group structure. The binary adding machine is isomorphic to the inverse limit of cyclic groups of order 2^k as $k \to \infty$ (see see Hocking and Young [280] for information about inverse limit systems). Specifically, define homomorphisms

$$\phi_k \quad : \quad \mathbf{Z}_{2^k} \to \mathbf{Z}_{2^{k-1}}$$
$$m \,(\mathrm{mod}\, 2^k) \mapsto m \,(\mathrm{mod}\, 2^{k-1})$$

and let

$$\mathcal{Z}_{2\infty} = \varprojlim_{k \to \infty} \mathcal{Z}_{2^k}$$

Consider the element $\iota = (1, 1, 1, \ldots) \in \mathcal{Z}_{2\infty}$. The dynamics of the quadratic map at the Feigenbaum point is conjugate to that of the map $\alpha(x) = x + \iota$ on $\mathcal{Z}_{2\infty}$. Could some analog of loop space permit a new approach to period-doubling cascades? The analogue of loop space for this group is presumably some topological vector space (preferably a Banach space) of maps

$$\mathcal{C}(\mathcal{Z}_{2\infty}) = \{f : \mathcal{Z}_{2\infty} \to X\}$$

with suitable continuity/smoothness assumptions on f. Finding an invariant Cantor set K for a map g such that $g|_K$ is conjugate to the map α on a binary adding machine is equivalent to solving a $\mathcal{Z}_{2\infty}$-equivariant operator equation on $\mathcal{C}(\mathcal{Z}_{2\infty})$. Specifically, let u belong to loop space and let $u_\gamma = u(\gamma)$ for $\gamma \in \mathcal{Z}_{2\infty}$. Then the operator is Ψ where

$$(\Psi(u))_\gamma = g(x_\gamma) - x_{\gamma + \iota}$$

The analogy with (4.32) is clear.

The group $\mathcal{Z}_{2\infty}$ is a *profinite* group: an inverse limit of finite groups. As such it has a natural topology induced from the product topology on $\prod_k \mathbf{Z}_{2^k}$, making it compact, Hausdorff, and totally disconnected. Moreover, this topology is induced by a group-invariant ultrametric (a metric d that satisfies $d(x, z) \leq \min(d(x, y), d(y, z))$).

A generalized adding machine can also be represented as an inverse limit. Specifically, suppose that we have a sequence of natural numbers $p_1, p_2, \ldots$ and we let $a_j = p_1 \ldots p_j$. Form the inverse limit

$$\mathcal{Z}_{p_1, p_2, \ldots} = \varprojlim_j \mathbf{Z}_{a_j}$$

Again let $\iota = (1, 1, 1, \ldots)$. Then $\mathcal{Z}_{p_1, p_2, \ldots}$ is a generalized adding machine, with dynamics given by $x \mapsto x + \iota$.

By considering generalized adding machines, we may extend the above considerations to general period-multiplying cascades. By this we mean that there exists a series of parameters μ_j and integers a_j such that a_{j+1} is a multiple of a_j and f_{μ_j} has a periodic attractor of period a_j. Such cascades are known to occur in one-humped maps in the case $a_{j+1} = q a_j$, where q is constant (Hao [265]), but note that when $q > 2$, orbits of other periods may occur between consecutive values μ_j. Period-multiplying cascades are also closely associated with attractors of mappings that have infinitely many connected components, Buescu and Stewart [83].

For maximum generality, the group $\mathcal{Z}_{2^\infty}$ should probably be replaced by the *procyclic* group $\mathcal{Z}_\infty$, defined as follows. For all pairs of natural numbers m, n such that m divides n, let

$$\phi_{nm} : \mathbf{Z}_n \;\rightarrow\; \mathbf{Z}_m$$
$$x \,(\mathrm{mod}\, n) \;\mapsto\; x \,(\mathrm{mod}\, m)$$

Then $\mathcal{Z}_\infty$ is the inverse limit of $(\mathbf{Z}_n, \phi_{nm})$. This too is a profinite group and a topological Cantor set. It is associated with general period-multiplying cascades in the same manner that $\mathcal{Z}_{2^\infty}$ is associated with period-doubling cascades. So finding the accumulation point of such a cascade is equivalent to finding a zero of a suitable operator equation posed on $\mathcal{Z}_\infty$-loop space. Connections with renormalization and universality seem equally plausible here, too.

Other features of the mathematics suggest that this approach might prove fruitful. Key items in equivariant bifurcation theory are isotropy subgroups and fixed-point subspaces. For the procyclic group, isotropy subgroups characterize periodic orbits of various periods (and adding machines of various kinds). For example, suppose that we have a sequence of natural numbers $p_1, p_2, \ldots$ as above, with $a_j = p_1 \ldots p_j$, and form the inverse limit

$$\mathcal{Z}_{p_1, p_2, \ldots} = \varprojlim_{j} \mathbf{Z}_{a_j}$$

Consider the subgroup

$$\mathcal{K}_j = \{(1, \ldots, 1, \gamma_{j+1}, \gamma_{j+2}, \ldots\}$$

for arbitrary elements $\gamma_j \in \mathbf{Z}_{a_j}$. Then in an obvious notation

$$\mathrm{Fix}(\mathcal{K}_j) = \{(x_1, \ldots, x_j, 0, 0, \ldots)\}$$

and the normalizer action cycles the x_j. So here the isotropy subgroup picks out periodic cycles. For more general adding machines, the isotropy subgroup can also pick out 'quotient' adding machines. The occurrence of these related types of dynamics suggests that the group-theoretic setting is natural for such problems, and is a strong hint that an explicitly equivariant treatment is worth seeking. We leave this as an open question for future research.

Chapter 5

Steady-State Bifurcations in Euclidean Equivariant Systems

Many pattern-forming phenomena are modeled by partial differential equations that are invariant under translations, rotations, and reflections on an infinite plane. Lateral boundaries are deemed unimportant when describing these phenomena. Examples include certain kinds of reaction-diffusion models, convection, Navier-Stokes flow, geometric hallucination patterns in the visual cortex, the appearance of stripes and spots on animal skins, and nematic liquid crystals.

Many types of solutions can be found in these models. In this chapter we focus on those solutions that are spatially doubly periodic and can be obtained by bifurcation from a trivial spatially homogeneous state, since these are among the most striking that appear in experiments. See Figures 5.1 and 5.2. In Section 6.7 we also consider uniformly rotating spiral waves in the infinite plane and meandering states that can be obtained from spiral waves by Hopf bifurcation.

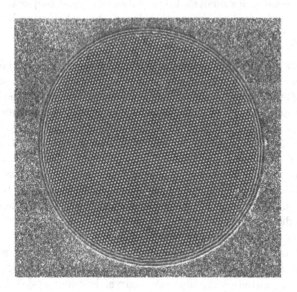

Figure 5.1: Hexagonal state in non-Boussinesq Raleigh-Bénard convection (courtesy of Eberhard Bodenschatz [69]).

Spatially periodic solutions to Euclidean equivariant systems can be found by a direct method. The equations are restricted to the space of functions that are doubly periodic with respect to a planar lattice (square, hexagonal, or rhombic). The resulting system has two types of symmetry: the 2-torus $\mathbf{T}_{\mathcal{L}}^2$ of translations modulo the lattice and the *holohedry* $H_{\mathcal{L}}$ of rotation and reflection symmetries that

Figure 5.2: (Left) Striped pattern in a reaction-diffusion in a chemical reactor (courtesy of Harry L. Swinney [424]). (Right) Squares pattern in vibrated granular layers (courtesy of Paul Umbanhowar [507]).

preserve the lattice. In function spaces the irreducible representations of the compact group $\Gamma_{\mathcal{L}} = H_{\mathcal{L}} \dot{+} \mathbf{T}_{\mathcal{L}}^2$ can be found (using Fourier series) and axial subgroups for each of these representations can be computed. This process results in a series of *planforms* — patterns associated to solutions — whose existence follows from the Equivariant Branching Lemma.

When performing this analysis there is an additional complication due to the infinite plane — we need to choose the size of the lattice as well as the lattice type. In applications this size is determined using dispersion curves in a way that is perhaps nonintuitive.

For example, in Rayleigh-Bénard convection, where a planar fluid layer is contained between two horizontal plates and is uniformly heated from below, the Rayleigh number is proportional to the temperature difference between the top and bottom plates. Models of convection usually consist of a system of PDE that model the velocity of the fluid coupled with a PDE describing the evolution of temperature. In such models there is a trivial Euclidean invariant solution — the *pure conduction solution* — in which the fluid velocity is zero. Heat conducts from the bottom plate to the top, but the fluid does not move. In experiments, the Rayleigh number is increased until the pure conduction state loses stability to a convective state (the fluid moves). A model can be tested by how well it predicts the transition value. The prediction process proceeds as follows. The pure conduction solution is tested for instability to a perturbation that has a plane wave factor $e^{i\mathbf{k}\cdot x}$ for some wave vector $\mathbf{k}$ with wave number $k = |\mathbf{k}|$. The Rayleigh number is increased to the first value R_k where an instability to such a plane wave occurs and a graph of k versus R_k is plotted — the dispersion curve. In many models that dispersion curve has a unique minimum value R_{k_*} at the critical wave number k_*, and R_{k_*} is the candidate for the Rayleigh number where instability occurs. Dispersion curves were discussed for a single reaction-diffusion equation on the line in Section 2.8 and will be discussed again for reaction-diffusion systems in Section 5.4.

In our analysis we choose the size of a planar lattice $\mathcal{L}$ so that some plane wave with critical wave number $k = k_*$ is doubly periodic with respect to that lattice. There are a countable number of lattice sizes for which that statement is valid. In most studies the size of the lattice is chosen to be minimal, since the planforms usually seen in experiments are the ones that are obtained in this way. There is evidence, however, of bifurcation to solutions obtained when the lattice size is next to smallest, see Figure 5.3. Planforms associated with second smallest states were originally studied in Kirchgässner [311] and planforms on all possible lattice sizes that can be obtained using the Equivariant Branching Lemma are classified in Dionne and Golubitsky [160]. The planforms involving lattices of the second smallest size are revisited by Silber *et al.* [303, 439] where the relationship between theory and experiment in the Faraday experiments of Kudrolli *et al.* [333] is discussed.

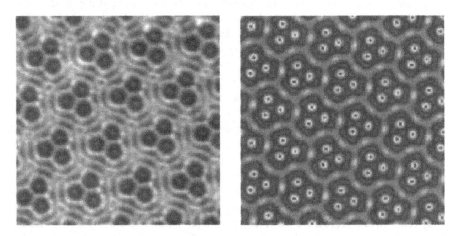

Figure 5.3: (Left) Hexagonally periodic state in the Faraday experiment (courtesy of Kudrolli, Pier, and J.P. Gollub [333]). (Right) Theoretically derived planform (courtesy of Judd and Silber [303]).

In this chapter we show that the kinds of planform that appear depend on the precise form of the action of the Euclidean group on the domain of the differential operator, and on additional symmetries. We discuss steady-state bifurcation in six models: reaction-diffusion systems, two-dimensional Navier-Stokes equations, the visual cortex, two variants of Rayleigh-Bénard convection, and liquid crystals. These examples all have different symmetry structures and all produce different planforms, with the simplest case being steady-state bifurcation in reaction-diffusion systems — sometimes called the *scalar* equation. The action of the Euclidean group in *pseudoscalar* equations involves a simultaneous action on the domain and range of functions. An example is the equation for the stream function derived from the two-dimensional Navier-Stokes equations. The possibility of a pseudoscalar action was first discussed by Bosch Vivancos, Chossat, and Melbourne [71]. Pseudoscalar actions also play a role in visual cortex models and, most likely, in nematic liquid crystal models. The results are complicated further when extra symmetries are present: the midplane reflection, which can be present in convection and liquid crystals, is an example.

Planform selection (the question of stability of solutions near bifurcation) is discussed in Section 5.9.

We end our introductory remarks with a summary of the steps needed to study bifurcations to steady patterned solutions in Euclidean equivariant systems of PDEs. These steps will be described in more detail in this chapter.

1) We assume that there is a trivial Euclidean invariant equilibrium valid for each λ. Translation symmetry implies that eigenfunctions of the linearized equations have *plane wave* factors $e^{i\mathbf{k}\cdot x}$ where $\mathbf{k}, x \in \mathbf{R}^2$.

2) Linear analysis leads to a dispersion curve as follows: for each *wave number* $|\mathbf{k}|$ there is a smallest λ at which the trivial solution loses stability to a disturbance with this wave number. Dispersion curves often have a unique absolute minimum at k_* — the *critical wave number*. The first instability is assumed to occur with wave number equal to k_*.

3) Because of rotational symmetry, the kernel of the linearization is infinite-dimensional. To circumvent this difficulty we look for spatially doubly periodic solutions on a fixed planar lattice. Typically, the kernel of the linearization restricted to doubly periodic functions on a fixed planar lattice is finite-dimensional. Therefore, we can apply Liapunov-Schmidt reduction to the problem restricted to a fixed lattice.

4) The symmetries of the restricted problem are related to Euclidean symmetry — but are not exactly Euclidean symmetry. There are two types of symmetry: the 2-torus of translations modulo the lattice $\mathbf{T}_{\mathcal{L}}^2 = \mathbf{R}^2/\mathcal{L}$ and the *holohedry* $H_{\mathcal{L}}$ of the lattice consisting of rotations and reflections that preserve the lattice. The symmetry group of the restricted problem is the compact group $\Gamma_{\mathcal{L}} = H_{\mathcal{L}}\dot{+}\mathbf{T}_{\mathcal{L}}^2$.

5) To perform this analysis we need to choose a planar lattice type (hexagonal, square, or rhombic) and the size of the lattice. Having made these choices, we perform a Liapunov-Schmidt reduction — at least in principle.

6) The *wave vectors* $\mathbf{k}$ for which the associated plane waves are doubly periodic with respect to a lattice $\mathcal{L}$ form the dual lattice $\mathcal{L}^*$. Once the size of the lattice is chosen, the kernel of the linearization is spanned by a finite number of plane waves. The action of $\Gamma_{\mathcal{L}}$ depends on the choice of critical dual wave vectors.

7) Finally, we find solutions to the reduced equations using the Equivariant Branching Lemma and then compute the stability of these solutions — at least with respect to $\mathcal{L}$-periodic disturbances.

5.1 Translation Symmetry, Rotation Symmetry, and Dispersion Curves

Suppose that a system of partial differential equations is defined on the domain $\mathbf{D} = \mathbf{R}^2 \times \Omega$. We will see that the set Ω is a point for planar reaction-diffusion systems, an interval for Rayleigh-Bénard convection, and a circle in some models of the primary visual cortex.

We think of this system in operator form. More precisely, let $\mathcal{F}$ be a space of vector-valued functions defined on $\mathbf{D}$, that is, $\mathcal{F}$ consists of $f : \mathbf{D} \to \mathbf{R}^m$ where m is the number of unknown functions in the system of differential equations. Let $\mathcal{P}$ be the operator on $\mathcal{F}$ associated to this system of differential equations. We assume that u_0 is a Euclidean invariant equilibrium of $\mathcal{P}$ (that is, a spatially homogeneous state), and L is the linearization of $\mathcal{P}$ about u_0. In this chapter we consider only those $\mathcal{P}$, and hence L, that commute with an action of the Euclidean group $\mathbf{E}(2)$ and only those steady-state solutions that bifurcate from u_0 as a parameter λ in $\mathcal{P}$ is varied.

We assume that the action of the Euclidean group $\mathbf{E}(2)$ on $\mathbf{R}^2$ is as rigid motions and is generated by

$$
\begin{aligned}
T_y(x) &= x + y \\
R_\theta(x) &= \begin{bmatrix} \cos\theta & -\sin\theta \\ \sin\theta & \cos\theta \end{bmatrix} \begin{bmatrix} x_1 \\ x_2 \end{bmatrix} \\
\kappa(x) &= (x_1, -x_2)
\end{aligned}
\tag{5.1}
$$

where $x = (x_1, x_2) \in \mathbf{R}^2$, T_y is translation by y, R_θ is rotation counterclockwise through angle θ, and κ is reflection across the x_1-axis. We assume that the action of $\mathbf{E}(2)$ on $\mathbf{D}$ is the rigid motion action (5.1) on $\mathbf{R}^2$. Finally, we assume that translations act trivially on Ω, whereas the actions of rotations and reflections are presently unspecified — but are denoted as follows:

$$
\begin{aligned}
T_y(x, \omega) &= (x + y, \omega) \\
R_\theta(x, \omega) &= (R_\theta(x), \theta\omega) \\
\kappa(x, \omega) &= (\kappa(x), \kappa\omega)
\end{aligned}
\tag{5.2}
$$

Translations are assumed to act on $f \in \mathcal{F}$ in the standard way

$$(yf)(x, \omega) = f(x - y, \omega)$$

A symmetry $\gamma \in \mathbf{O}(2)$ acts on $f \in \mathcal{F}$ by acting on the domain of f in the standard way and perhaps on the range of f; that is,

$$\gamma f(x, \omega) = \rho_\gamma f(\gamma^{-1}x, \gamma^{-1}\omega) \tag{5.3}$$

where ρ is a representation of $\mathbf{O}(2)$ on $\mathbf{R}^m$.

The purpose of the first two sections is to explore in turn the restrictions placed on the eigenfunctions of a Euclidean-equivariant linear operator L by translations, rotations, and reflections. The determination of the kernel of L is made clearer by understanding these general representation theoretic principles.

The Effect of Translational Symmetry. Translation symmetries $\mathbf{T}$ simplify the quest for solutions in two ways. First, the eigenfunctions of the linearized operator must have plane wave factors. Second, translation symmetries allow us to look specifically for solutions that are doubly periodic with respect to a planar lattice. We discuss the first point now; our discussion here is formal but can be made rigorous once we reduce to lattices and doubly periodic functions in Section 5.2.

Lemma 5.1 *The subspaces*

$$W_{\mathbf{k}} = \{a(\omega)e^{2\pi i \mathbf{k} \cdot x} + \text{c.c.} : a(\omega) \in \mathbf{C}\} \tag{5.4}$$

of $\mathcal{F}$ (with $W_{\mathbf{k}} = W_{-\mathbf{k}}$) are the isotypic components of the action of translations on $\mathcal{F}$. Thus, eigenfunctions of L have the form $a(\omega)e^{2\pi i \mathbf{k} \cdot x} + \text{c.c.}$ for some dual wave vector $\mathbf{k}$.

Proof. Note that translation by y acts on $v(x, \omega) = a(\omega)e^{2\pi i \mathbf{k} \cdot x}$ by

$$T_y v(x, \omega) = v(x - y, \omega) = \left(e^{-2\pi i \mathbf{k} \cdot y} a(\omega)\right) e^{2\pi i \mathbf{k} \cdot x} + \text{c.c.}$$

It follows that $W_{\mathbf{k}}$ is the sum of $\mathbf{T}$-invariant subspaces

$$W_{\mathbf{k}}^a = \{za(\omega)e^{2\pi i \mathbf{k} \cdot x} + \text{c.c.} : z \in \mathbf{C}\} \tag{5.5}$$

each of which is $\mathbf{T}$-isomorphic to the two-dimensional space

$$V_{\mathbf{k}} = \{ze^{2\pi i \mathbf{k} \cdot x} + \text{c.c.} : z \in \mathbf{C}\} \tag{5.6}$$

Thus each space $W_{\mathbf{k}}$ is the sum of $\mathbf{T}$-isomorphic irreducible representations.

Note that $V_{-\mathbf{k}} = V_{\mathbf{k}}$. Moreover, except for $\mathbf{k}$ and $-\mathbf{k}$, these representations are all distinct (just verify that the kernels of the actions of $\mathbf{T}$ on the spaces $V_{\mathbf{k}}$ are all different). There are analytic difficulties in proving that the subspaces $W_{\mathbf{k}}$ are isotypic components of the action of $\mathbf{T}$ on $\mathcal{F}$. This point can be proved easily using Fourier series once we restrict attention to spaces of doubly periodic functions, as we do in Section 5.2. Assuming that $W_{\mathbf{k}}$ is an isotypic component, the equivariance of the linear operator L implies that L must map $W_{\mathbf{k}}$ into itself for each dual wave vector $\mathbf{k}$ and that eigenfunctions of L must lie in $W_{\mathbf{k}}$ for some $\mathbf{k}$. □

Effect of Rotational Symmetry. Let $\gamma \in \mathbf{O}(2)$. Then γ acts on a function $f \in \mathcal{F}$ as in (5.3), and

$$\gamma W_{\mathbf{k}} = W_{\gamma \mathbf{k}} \tag{5.7}$$

To verify (5.7), observe that

$$\gamma\left(a(\omega)e^{2\pi i \mathbf{k}\cdot x}\right) = \gamma(a(\omega))e^{2\pi i \mathbf{k}\cdot\gamma^{-1}x} = \gamma(a(\omega))e^{2\pi i \gamma\mathbf{k}\cdot x} \in W_{\gamma\mathbf{k}}$$

with the last equality proved using the orthogonality of γ.

It follows from the rotation equivariance of L that when we look for eigenfunctions having a zero eigenvalue, we can perform our search with $\mathbf{k} = k(1,0)$ where k is the dual wave number. It also follows from (5.7) that $\ker \mathsf{L}$ is infinite-dimensional (there is an independent eigenfunction in $W_{\mathbf{k}}$ for each $|\mathbf{k}| = k$). This analytic difficulty will be circumvented once we restrict attention to doubly periodic solutions in Section 5.2.

Dispersion Curves. Dispersion curves are computed as follows. We assume that $\mathcal{P}$ and hence L depends on a parameter λ and that the trivial solution u_0 is linearly stable when $\lambda = 0$. Fix the wave number k and find the smallest value λ_k of λ for which L has a zero eigenvalue. This calculation is performed by calculating just with $\mathsf{L}_k = \mathsf{L}|W_{\mathbf{k}} : W_{\mathbf{k}} \to W_{\mathbf{k}}$, where $\mathbf{k} = k(1,0)$. The *dispersion curve* is the graph of λ_k as a function of k.

We will see later that in certain cases the action of reflections (and other symmetries) can divide the space $W_{\mathbf{k}}$ into proper L_k invariant subspaces, see (5.13). In these instances, we must compute a dispersion curve for each such invariant subspace.

5.2 Lattices, Dual Lattices, and Fourier Series

Definition 5.2 A *planar lattice* is the set

$$\mathcal{L} = \{m_1\ell_1 + m_2\ell_2 : m_1, m_2 \in \mathbf{Z}\}$$

where $\ell_1, \ell_2 \in \mathbf{R}^2$ are linearly independent vectors. The space $\mathcal{F}_{\mathcal{L}}$ of *doubly periodic* functions with respect to $\mathcal{L}$ is

$$\mathcal{F}_{\mathcal{L}} = \{f \in \mathcal{F} : f(x + \ell, \omega) = f(x, \omega) \quad \forall \ell \in \mathcal{L}\}$$

◇

There are two ways to understand why it is reasonable to look specifically for doubly periodic solutions to systems of Euclidean equivariant differential equations.

(a) The lattice $\mathcal{L}$ is a subgroup of the group of translations, and $\mathcal{F}_{\mathcal{L}}$ is the fixed-point subspace of the action of $\mathcal{L}$ on $\mathcal{F}$. Hence $\mathcal{F}_{\mathcal{L}}$ is an invariant subspace for $\mathcal{P}$ and solutions of $\mathcal{P}|\mathcal{F}_{\mathcal{L}}$ are solutions of the original system.

(b) Equivalently, let $\widehat{D}$ be the fundamental domain of the lattice $\mathcal{L}$, that is, $\widehat{D} = D \times \Omega$, where

$$D = \{d_1\ell_1 + d_2\ell_2 : 0 \leq d_j < 1, j = 1, 2\}$$

Then solutions in $\mathcal{F}_\mathcal{L}$ may be thought of as solutions of $\mathcal{P}$ that satisfy periodic boundary conditions on the boundary of $\widehat{D}$.   ◇

Lattice Symmetries. There are five kinds of planar lattice: rhombic, square, hexagonal, rectangular, and oblique. The lattices that are most relevant for bifurcation theory are those generated by two vectors of the same length: rhombic, square, and hexagonal. These three lattices types are distinguished by their symmetries in $\mathbf{O}(2)$.

Definition 5.3 The *holohedry* $H_\mathcal{L}$ of a planar lattice $\mathcal{L}$ is the subgroup of $\mathbf{O}(2)$ that preserves $\mathcal{L}$.   ◇

The holohedries and lattice generators of the three relevant lattices are listed in Table 5.1. For example, up to scaling, rotation, and reflection, the square lattice is generated by $\ell_1 = (1, 0)$ and $\ell_2 = (0, 1)$, and the holohedry of a square lattice is the symmetry group $\mathbf{D}_4$ of a square.

Lemma 5.4 *Euclidean symmetry induces an action of*

$$\Gamma_\mathcal{L} \equiv H_\mathcal{L} \dotplus \mathbf{T}_\mathcal{L}^2 \tag{5.8}$$

on $\mathcal{F}_\mathcal{L}$ where $H_\mathcal{L}$ is the holohedry of $\mathcal{L}$ and $\mathbf{T}_\mathcal{L}^2 = \mathbf{R}^2/\mathcal{L}$ consists of translations modulo the lattice $\mathcal{L}$.

Proof. The translation group leaves the space $\mathcal{F}_\mathcal{L}$ invariant. If $f(x, \omega)$ is $\mathcal{L}$-periodic then $g(x, \omega) = f(x - y, \omega)$ is also $\mathcal{L}$-periodic. Similarly, suppose that $f \in \mathcal{F}_\mathcal{L}$ and $h \in H_\mathcal{L}$. Then

$$(hf)(x + \ell) = f(h^{-1}(x + \ell)) = f(h^{-1}x + h^{-1}\ell) = f(h^{-1}x)$$

since $h^{-1}\ell \in \mathcal{L}$ and f is $\mathcal{L}$-periodic. Therefore

$$(hf)(x + \ell) = (hf)(x)$$

and $hf \in \mathcal{F}_\mathcal{L}$. As previously noted, translation by vectors in $\mathcal{L}$ act trivially on $\mathcal{F}_\mathcal{L}$, so translations act on $\mathcal{F}_\mathcal{L}$ as the group $\mathbf{R}^2/\mathcal{L}$.   □

The classification of planforms found in bifurcation problems with Euclidean symmetry restricted to a lattice $\mathcal{L}$ depends on the precise nature of the symmetry group $\Gamma_\mathcal{L}$ and its action on $\mathcal{F}_\mathcal{L}$.

Dual Lattices. Define the dual lattice $\mathcal{L}^*$ of $\mathcal{L}$ as follows.

Definition 5.5 Let $\mathbf{k}, x$ be vectors in $\mathbf{R}^2$. Then $\mathbf{k}$ is a *dual lattice vector* if the function

$$x \mapsto e^{2\pi i \mathbf{k} \cdot x} \tag{5.9}$$

is $\mathcal{L}$-periodic. The *dual lattice* $\mathcal{L}^*$ is the set of all dual lattice vectors. The functions in (5.9) are called *plane waves*.   ◇

We make two remarks about dual lattices.

(a) The set $\mathcal{L}^*$ is a lattice. More precisely, the identity

$$e^{2\pi i \mathbf{k} \cdot (x + \ell)} = e^{2\pi i \mathbf{k} \cdot \ell} e^{2\pi i \mathbf{k} \cdot x} \tag{5.10}$$

implies that $\mathbf{k} \in \mathcal{L}^*$ if and only if $\mathbf{k} \cdot \ell$ is an integer for every $\ell \in \mathcal{L}$.

(b) The generators of the dual lattice can be chosen to satisfy

$$\mathbf{k}_i \cdot \boldsymbol{\ell}_j = \delta_{ij} \tag{5.11}$$

(where δ_{ij} is the Kronecker delta), which implies that the dual of a rhombic lattice is rhombic, the dual of a square lattice is square, and the dual of a hexagonal lattice is hexagonal. The generators for these lattices are listed in Table 5.1.   ◇

Lattice	$H_\mathcal{L}$	$\boldsymbol{\ell}_1$	$\boldsymbol{\ell}_2$	$\mathbf{k}_1$	$\mathbf{k}_2$
Square	$\mathbf{D}_4$	$(1,0)$	$(0,1)$	$(1,0)$	$(0,1)$
Hexagonal	$\mathbf{D}_6$	$(1,\frac{1}{\sqrt{3}})$	$(0,\frac{2}{\sqrt{3}})$	$(1,0)$	$\frac{1}{2}(-1,\sqrt{3})$
Rhombic	$\mathbf{D}_2$	$(1,-\cot\eta)$	$(0,\csc\eta)$	$(1,0)$	$(\cos\eta,\sin\eta)$

Table 5.1 Holohedry and generators for the planar lattices and their dual lattices. For rhombic lattices: $0 < \eta < \frac{\pi}{2}$; $\eta \neq \frac{\pi}{3}$. The lattice sizes have been chosen so that dual lattice generators have length 1.

Fourier Series and Irreducible Representations. We assume throughout that $\mathcal{F}_\mathcal{L}$ consists of functions that can be expanded in convergent Fourier series Therefore, $f \in \mathcal{F}_\mathcal{L}$ has the form

$$f(x,\omega) = \sum_{\mathbf{k}\in\mathcal{L}^*} \left(a_\mathbf{k}(\omega)e^{2\pi i \mathbf{k}\cdot x} + \text{c.c.}\right) \tag{5.12}$$

where c.c. indicates the complex conjugate, and $a_\mathbf{k}$ is a complex-valued vector. We say that a term of the form $a_\mathbf{k}(\omega)e^{2\pi i \mathbf{k}\cdot x}$ in (5.12) has the *plane wave factor* $e^{2\pi i \mathbf{k}\cdot x}$.

We use Fourier series to decompose $\mathcal{F}_\mathcal{L}$ into isotypic components of irreducible representations of the action of the translations $\mathbf{T}_\mathcal{L}^2$. It follows from (5.12) that

$$\mathcal{F}_\mathcal{L} = \bigoplus_{\mathbf{k}\in\mathcal{L}^*} W_\mathbf{k}$$

where $W_\mathbf{k}$ is the subspace of $\mathcal{F}_\mathcal{L}$ defined in (5.4). It also follows from (5.12) that each $W_\mathbf{k}$ is an isotypic component of the action of $\mathbf{T}_\mathcal{L}^2$, thus completing the discussion begun in the proof of Lemma 5.1.

Lattices provide an elegant way around the difficulties associated with infinite-dimensional kernels in Euclidean equivariant bifurcation problems. Suppose that we look only for solutions of $\mathcal{P}$ that are doubly periodic with respect to a fixed lattice $\mathcal{L}$. Then the relevant kernel of the linearized equations is $\ker(\mathsf{L}|\mathcal{F}_\mathcal{L})$, rather than $\ker \mathsf{L}$. The lattice restricted bifurcation problem has only a finite number of rotations (those in the holohedry); therefore symmetry does not force infinite-dimensional kernels when the bifurcation problem is restricted to a lattice. Nevertheless, the lattice-restricted problem is still complicated in a variety of ways.

The Connection between Lattice Size and Dispersion Curves. Suppose that the dispersion curve for a bifurcation problem has a unique minimum at wave number k_* and that we look only for doubly periodic solutions with that wave number. In this situation, let us be more precise about the constraints placed on $\ker(\mathsf{L}|\mathcal{F}_\mathcal{L})$. Suppose that this kernel has an eigenvector with plane wave factor $v_\mathbf{k} = e^{2\pi i \mathbf{k}\cdot x}$ where $|\mathbf{k}| = k_*$. Then, it follows from (5.7) that there is an eigenfunction with plane wave factor $v_\mathbf{k}$ for every $\mathbf{k} \in \mathcal{L}^*$ for which $|\mathbf{k}| = k_*$.

For example, suppose $\mathcal{L}$, and hence $\mathcal{L}^*$, is a hexagonal lattice, as pictured in Figure 5.4. The number of such dual wave vectors with $|\mathbf{k}| = k_*$ is given geometrically by the intersection of the circle of radius k_* about the origin with the dual lattice $\mathcal{L}^*$. See Figure 5.4. This number is at least six, but it can be 12 or higher (depending on certain diophantine conditions). Most pattern formation analyses are performed on lattices where the critical dual wave vectors lie on one of the circles of small radius — leading to the cases of six or sometimes 12 critical dual wave vectors. We discuss why below.

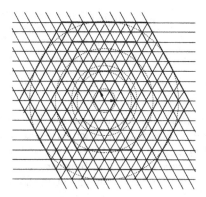

Figure 5.4: Rings of dual lattice vectors for the hexagonal lattice.

What may seem surprising at first is the fact that in planar systems of PDEs all of these circles of dual wave vectors are simultaneously critical — but with respect to different lattices. Here is the reason. Let $\mathcal{L}$ be a lattice whose dual lattice is generated by wave vectors of length 1, such as the examples given in Table 5.1. Let $s\mathcal{L}$ denote the lattice consisting of all vectors $s\boldsymbol{\ell}$ where $\boldsymbol{\ell} \in \mathcal{L}$. It follows from (5.11) that the dual lattice $(s\mathcal{L})^*$ is generated by vectors of length $\frac{1}{s}$. Thus, by making the lattice $s\mathcal{L}$ very large, we can arrange for any circle of dual wave vectors to have the critical wave length k_*, and hence for the lattice-restricted problem to have a bifurcation whose kernel is generated by dual wave vectors on the given circle. As long as the original system of differential equations is posed on the infinite plane, we can in principle scale the lattice to be as large as we like. We have thus proved:

Theorem 5.6 *Let $\mathcal{L}$ be a fixed lattice. Then the possible lengths of dual wave vectors in $\mathcal{L}^*$ form a sequence*

$$0 < p_1 < p_2 < \cdots$$

By scaling the length of $\mathcal{L}$ it is possible to have $p_j = k_*$ for any j.

Note, however, that in order to get instabilities corresponding to p_j for j large, we need to choose a lattice $\mathcal{L}$ which is large. That is, we need to assume double periodicity of solutions with periodicity on a large scale. For the remainder of this chapter we consider only those lattices $\mathcal{L}$ for which the critical dual wave vectors are dual wave vectors of shortest length. That is, we choose to study doubly periodic patterns with the smallest possible spatial period. After normalization, we may assume that the shortest dual wave vectors are of length 1, as is done in Table 5.1.

Effects of Reflectional Symmetry. Bosch Vivancos, Chossat, and Melbourne [71] observed that reflection symmetries can further decompose the L-invariant subspace $W_{\mathbf{k}}$ into two L-invariant subspaces. To see why this is so, let $\rho_{\mathbf{k}}$ be the planar reflection that fixes the dual wave vector $\mathbf{k}$ and compute the action of $\rho_{\mathbf{k}}$ on $W_{\mathbf{k}}$ (dropping the +c.c. and using the fact that $\rho_{\mathbf{k}}^{-1} = \rho_{\mathbf{k}}$)

$$\rho_{\mathbf{k}}\left(a(\omega)e^{2\pi i \mathbf{k}\cdot x}\right) = \rho_{\mathbf{k}}(a(\omega))e^{2\pi i \mathbf{k}\cdot \rho_{\mathbf{k}}(x)} = \rho_{\mathbf{k}}(a(\omega))e^{2\pi i \rho_{\mathbf{k}}(\mathbf{k})\cdot x} = \rho_{\mathbf{k}}(a(\omega))e^{2\pi i \mathbf{k}\cdot x}$$

where $\rho_{\mathbf{k}}(a(\omega))$ indicates the action of $\rho_{\mathbf{k}}$ on the function $a(\omega)$. Since $\rho_{\mathbf{k}}^2 = 1$, the subspace $W_{\mathbf{k}}$ itself decomposes into two subspaces according to whether $\rho_{\mathbf{k}}$ acts as ± 1. They are:

$$W_{\mathbf{k}} = W_{\mathbf{k}}^+ \oplus W_{\mathbf{k}}^- \tag{5.13}$$

where $\rho_{\mathbf{k}}$ acts trivially on $W_{\mathbf{k}}^+$ and as minus the identity on $W_{\mathbf{k}}^-$. We call functions in $W_{\mathbf{k}}^+$ *even* and functions in $W_{\mathbf{k}}^-$ *odd*; these functions satisfy

$$\rho_{\mathbf{k}}(a(\omega)) = a(\omega) \qquad \rho_{\mathbf{k}}(a(\omega)) = -a(\omega)$$

respectively.

The decomposition in (5.13) has implications for the eigenvector structure of L. More precisely, since L commutes with $\rho_{\mathbf{k}}$, each subspace in (5.13) is $\rho_{\mathbf{k}}$-invariant and eigenfunctions of $\mathsf{L}|W_{\mathbf{k}}$ appear in two types: even and odd. Bifurcations based on even eigenfunctions are called *scalar* and bifurcations based on odd eigenfunctions are called *pseudoscalar*. We discuss this point in the next section.

5.3 Actions on Kernels and Axial Subgroups

Suppose that $\ker \mathsf{L}$ has an eigenfunction with critical wave number k_*. We then choose a lattice $\mathcal{L}$ so that there is a dual lattice vector $\mathbf{k} \in \mathcal{L}^*$ such that $|\mathbf{k}| = k_*$. Moreover, as just discussed in the previous section, we focus only on those bifurcations where $\mathbf{k}$ is a dual wave vector of shortest length in $\mathcal{L}^*$.

Generically, $\ker(\mathsf{L}|\mathcal{F}_{\mathcal{L}})$ will be generated by one eigenfunction and symmetry. That is, we expect to be able to choose

$$v(x,\omega) = a(\omega)e^{2\pi i \mathbf{k}\cdot x} + c.c. \tag{5.14}$$

for some function $a(\omega)$. It follows from translation symmetry that the two-dimensional subspace $W_{\mathbf{k}}^a$ is contained in $\ker(\mathsf{L}|W_{\mathbf{k}})$, where $W_{\mathbf{k}}^a$ is defined in (5.5). The genericity assumption states precisely that

$$W_{\mathbf{k}}^a = \ker \mathsf{L} \cap W_{\mathbf{k}}$$

Moreover, we can be more explicit about our choice of function $a(\omega)$, as we now show. We say that a is R_{π}-*invariant* if

$$R_{\pi}(a(\omega)) = \overline{a(\omega)} \tag{5.15}$$

Lemma 5.7 *Generically, there is one eigenfunction in $W_{\mathbf{k}}$ of the form (5.14) where a is either even or odd and R_{π}-invariant. Moreover, $\ker(\mathsf{L}|\mathcal{F}_{\mathcal{L}})$ is generated by this eigenfunction and symmetry.*

Proof. Let $v(x,\omega) = a(\omega)e^{2\pi i \mathbf{k}\cdot x} + c.c. \in W_{\mathbf{k}}$ be an eigenfunction of $\mathsf{L}|\mathcal{F}_{\mathcal{L}}$. The decomposition (5.13) implies, along with genericity, that we may assume that $a(\omega)$ is either odd or even.

It follows from (5.7) that $R_{\pi}(W_{\mathbf{k}}) = W_{-\mathbf{k}} = W_{\mathbf{k}}$. Since $R_{\pi}^2 = I$, we can write

$$W_{\mathbf{k}} = U^+ \oplus U^-$$

where R_π acts trivially on U^+ and by $-I$ on U^-. Since L commutes with R_π we can assume that eigenfunctions lie either in U^+ or U^-. Suppose that $v \in U^-$ is an eigenfunction for L. Then, translation invariance implies $v\left(x - \frac{1}{4}\mathbf{k}, \omega\right) = iv(x, \omega)$ is also an eigenfunction of L. We claim that $iv \in U^+$; hence, we can assume that the eigenfunction is R_π invariant.

If we show that $R_\pi(iv) = -iR_\pi(v)$, then $R_\pi(iv) = -i(-v) = iv$, as claimed. Compute the action of R_π on $v(x, \omega)$ (dropping the $+c.c.$)

$$\begin{aligned} R_\pi\left(ia(\omega)e^{2\pi i \mathbf{k} \cdot x}\right) &= R_\pi(ia(\omega))e^{2\pi i \mathbf{k} \cdot R_\pi(x)} &= R_\pi(ia(\omega))e^{-2\pi i \mathbf{k} \cdot x} \\ &= \overline{R_\pi(ia(\omega))}e^{2\pi i \mathbf{k} \cdot x} &= -i\overline{R_\pi(a(\omega))}e^{2\pi i \mathbf{k} \cdot x} \end{aligned} \quad (5.16)$$

Thus $R_\pi(iv) = -iR_\pi(v)$.

Note that if $R_\pi(v) = v$, then it follows from the calculation in (5.16) that $a(\omega)$ is R_π-invariant. □

Coordinates on Kernels. We can now construct the generic kernel of a Euclidean equivariant linear operator on doubly periodic functions corresponding to critical dual wave vectors of shortest length. If $k' = R_\theta k$, then $V_{k'} = R_\theta(V_k)$ is also in $\ker \mathsf{L}$. Therefore

$$\ker(\mathsf{L}|\mathcal{F}_\mathcal{L}) = \sum_{R_\theta(k)\in\mathcal{L}^*} R_\theta(V_k)$$

There is a subtlety. In general, $\mathbf{k}' = R_\theta \mathbf{k}$ and $\mathbf{k}, \mathbf{k}' \in \mathcal{L}^*$ do not imply that $R_\theta \in H_\mathcal{L}$. (Consider the example of a rhombic lattice.) Nevertheless, we can determine coordinates on the kernel in this way. The eigenspaces are listed in Table 5.2 and the eigenfunctions are listed in Table 5.3.

Lattice	$\ker \mathsf{L}$	k_1	k_2	k_3
Square	$V_{k_1} \oplus V_{k_2}$	$(1,0)$	$(0,1) = R_{\frac{\pi}{2}}k_1$	$-$
Hexagonal	$V_{k_1} \oplus V_{k_2} \oplus V_{k_3}$	$(1,0)$	$\frac{1}{2}(-1,\sqrt{3}) = R_{\frac{2\pi}{3}}k_1$	$\frac{1}{2}(-1,-\sqrt{3}) = R_{\frac{4\pi}{3}}k_1$
Rhombic	$V_{k_1} \oplus V_{k_2}$	$(1,0)$	$(\cos\eta, \sin\eta) = R_\eta k_1$	$-$

Table 5.2 Kernels corresponding to shortest dual wave vectors.

Lattice	$\ker \mathsf{L}$
Square	$z_1 a(\omega)e^{2\pi i \mathbf{k}_1 \cdot x} + z_2 a(\frac{\pi}{2} \cdot \omega)e^{2\pi i \mathbf{k}_2 \cdot x} + c.c.$
Hexagonal	$z_1 a(\omega)e^{2\pi i \mathbf{k}_1 \cdot x} + z_2 a(\frac{2\pi}{3} \cdot \omega)e^{2\pi i \mathbf{k}_2 \cdot x} + z_3 a(\frac{4\pi}{3} \cdot \omega)e^{2\pi i \mathbf{k}_3 \cdot x} + c.c.$
Rhombic	$z_1 a(\omega)e^{2\pi i \mathbf{k}_1 \cdot x} + z_2 a(\eta \cdot \omega)e^{2\pi i \mathbf{k}_2 \cdot x} + c.c.$

Table 5.3 Eigenfunctions corresponding to $\Gamma_\mathcal{L}$ irreducible representation. The function $a(\omega)$ is nonzero, R_π-invariant, and either odd or even.

Action of $\Gamma_\mathcal{L}$ in Coordinates. The complex numbers z_j in Table 5.3 provide an explicit system of coordinates on $\ker(\mathsf{L}|\mathcal{F}_\mathcal{L})$. Moreover, given the properties of $a(\omega)$, the action of $\Gamma_\mathcal{L}$ can be computed explicitly in these coordinates.

The action of the torus $\mathbf{T}_\mathcal{L}^2$ is derived as follows. Write $\theta \in \mathbf{T}_\mathcal{L}^2$ as

$$\theta = \theta_1 \ell_1 + \theta_2 \ell_2 \equiv [\theta_1, \theta_2] \quad (5.17)$$

where $0 \le \theta_j < 1$. The $\mathbf{T}_\mathcal{L}^2$-action can be computed using the identity $\mathbf{k}_i \cdot \boldsymbol{\ell}_j = \delta_{ij}$. In particular,

$$\theta\left(e^{2\pi i \mathbf{k} \cdot x}\right) = e^{2\pi i \mathbf{k} \cdot (x-\theta)} = e^{-2\pi i \mathbf{k} \cdot \theta} e^{2\pi i \mathbf{k} \cdot x}$$

In coordinates on the square and rhombic lattices the torus action is

$$\theta(z_1, z_2) = \left(e^{-2\pi i \theta_1} z_1, e^{-2\pi i \theta_2} z_2\right)$$

On the hexagonal lattice $\mathbf{k}_3 = -(\mathbf{k}_1 + \mathbf{k}_2)$, so the torus action is

$$\theta(z_1, z_2, z_3) = \left(e^{-2\pi i \theta_1} z_1, e^{-2\pi i \theta_2} z_2, e^{2\pi i (\theta_1 + \theta_2)} z_3\right)$$

The holohedries $H_{\mathcal{L}}$ are $\mathbf{D}_4$, $\mathbf{D}_6$, and $\mathbf{D}_2$ for the square, hexagonal, and rhombic lattices respectively. In each case the generators for these groups are a reflection and a rotation. For the square and hexagonal lattices, the reflection is κ. For the rhombic lattice, the reflection is κ_η, the reflection that interchanges $\mathbf{k}_1$ and $\mathbf{k}_2$. The counterclockwise rotation ξ (through angles $\frac{\pi}{2}$, $\frac{\pi}{3}$, and π) is the rotation generator for the three lattices. The actions of $H_{\mathcal{L}}$ are given in Table 5.4.

$\mathbf{D}_2$	Action	$\mathbf{D}_4$	Action	$\mathbf{D}_6$	Action
1	(z_1, z_2)	1	(z_1, z_2)	1	(z_1, z_2, z_3)
ξ	$(\overline{z_1}, \overline{z_2})$	ξ	$(\overline{z_2}, z_1)$	ξ	$(\overline{z_2}, \overline{z_3}, \overline{z_1})$
κ_η	$\varepsilon(z_2, z_1)$	ξ^2	$(\overline{z_1}, \overline{z_2})$	ξ^2	(z_3, z_1, z_2)
$\kappa_\eta \xi$	$\varepsilon(\overline{z_2}, \overline{z_1})$	ξ^3	$(z_2, \overline{z_1})$	ξ^3	$(\overline{z_1}, \overline{z_2}, \overline{z_3})$
		κ	$\varepsilon(z_1, \overline{z_2})$	ξ^4	(z_2, z_3, z_1)
		$\kappa\xi$	$\varepsilon(\overline{z_2}, \overline{z_1})$	ξ^5	$(\overline{z_3}, \overline{z_1}, \overline{z_2})$
		$\kappa\xi^2$	$\varepsilon(\overline{z_1}, z_2)$	κ	$\varepsilon(z_1, z_3, z_2)$
		$\kappa\xi^3$	$\varepsilon(z_2, z_1)$	$\kappa\xi$	$\varepsilon(\overline{z_2}, \overline{z_1}, \overline{z_3})$
				$\kappa\xi^2$	$\varepsilon(z_3, z_2, z_1)$
				$\kappa\xi^3$	$\varepsilon(\overline{z_1}, \overline{z_3}, \overline{z_2})$
				$\kappa\xi^4$	$\varepsilon(z_2, z_1, z_3)$
				$\kappa\xi^5$	$\varepsilon(\overline{z_3}, \overline{z_2}, \overline{z_1})$
$[\theta_1, \theta_2]$	$\left(e^{-2\pi i\theta_1} z_1, e^{-2\pi i\theta_2} z_2\right)$		$\left(e^{-2\pi i\theta_1} z_1, e^{-2\pi i\theta_2} z_2\right)$		$\left(e^{-2\pi i\theta_1} z_1, e^{-2\pi i\theta_2} z_2, e^{2\pi i(\theta_1+\theta_2)} z_3\right)$

Table 5.4 (Left) $\mathbf{D}_2 \dotplus \mathbf{T}_{\mathcal{L}}^2$ action on rhombic lattice. (Center) $\mathbf{D}_4 \dotplus \mathbf{T}_{\mathcal{L}}^2$ action on square lattice. (Right) $\mathbf{D}_6 \dotplus \mathbf{T}_{\mathcal{L}}^2$ action on hexagonal lattice. For scalar actions $\varepsilon = +1$; for pseudoscalar actions $\varepsilon = -1$.

Axial Subgroups. Using Table 5.4, it is straightforward to verify that the subgroups listed in Table 5.5 are axial subgroups for scalar bifurcations. Moreover, it is not difficult to show that up to conjugacy there are precisely two axial subgroups on the rhombic and square lattices. Verifying this assertion on the hexagonal lattice is more involved, see [93].

As noted in [71], the axial subgroups for the pseudoscalar action are different from those in the scalar case — even though the fixed-point subspaces are mostly the same. It is only when the interpretation of the corresponding eigenfunctions is made in physical space that the potential confusion between say scalar and pseudoscalar hexagons is resolved.

Bressloff *et al.* [74] show that there are exactly four axial subgroups in the hexagonal pseudoscalar case, while there are only two in the scalar case: see Table 5.6. As in the scalar case, it is straightforward to check that the subgroups listed in Table 5.6 are pseudoscalar axial subgroups.

In the next two sections we consider two examples where Ω is a point. In these cases the function $a(\omega)$ is a constant, and by (5.15) $a(\omega)$ may be taken to be 1 in the list of eigenfunctions in Table 5.3. We will see that in systems of planar

Lattice	Axial	Isotropy Subgroup	Fixed Vector
Rhombic	Rhombs	$\mathbf{D}_2(\kappa_\eta, \xi)$	$(1,1)$
	Rolls	$\mathbf{O}(2)$	$(1,0)$
Square	Squares	$\mathbf{D}_4$	$(1,1)$
	Rolls	$\mathbf{O}(2) \oplus \mathbf{Z}_2(\kappa)$	$(1,0)$
Hexagonal	Hexagons$^+$	$\mathbf{D}_6$	$(1,1,1)$
	Hexagons$^-$	$\mathbf{D}_6$	$(-1,-1,-1)$
	Rolls	$\mathbf{O}(2) \oplus \mathbf{Z}_2(\kappa)$	$(1,0,0)$

Table 5.5 Summary of scalar axial subgroups. $\mathbf{O}(2)$ is generated by $[0, \theta_2] \in \mathbf{T}^2$ and rotation by π (ξ on rhombic lattice, ξ^2 on square lattice, and ξ^3 on hexagonal lattice). On the hexagonal lattice the points $(1,1,1)$ and $(-1,-1,-1)$ have the same isotropy subgroup ($\mathbf{D}_6$) — but are not conjugate by a group element. Therefore, the associated eigenfunctions generate different planforms.

Lattice	Axial	Isotropy Subgroup	Fixed Vector
Rhombic	Anti-rhombs	$\mathbf{D}_2\left(\kappa_\eta\left[\frac{1}{2},\frac{1}{2}\right], \xi\right)$	$(1,1)$
	Anti-rolls	$\mathbf{O}(2)$	$(1,0)$
Square	Anti-squares	$\mathbf{D}_4\left(\kappa\left[\frac{1}{2},\frac{1}{2}\right], \xi\right)$	$(1,1)$
	Anti-rolls	$\mathbf{O}(2) \oplus \mathbf{Z}_2\left(\xi^2\kappa\left[\frac{1}{2},0\right]\right)$	$(1,0)$
Hexagonal	Anti-hexagons	$\mathbf{Z}_6$	$(1,1,1)$
	Anti-triangles	$\mathbf{D}_3(\kappa\xi, \xi^2)$	(i,i,i)
	Anti-rectangles	$\mathbf{D}_2(\kappa, \xi^3)$	$(0,1,-1)$
	Anti-rolls	$\mathbf{O}(2) \oplus \mathbf{Z}_4\left(\xi^3\kappa\left[\frac{1}{2},0\right]\right)$	$(1,0,0)$

Table 5.6 Summary of pseudoscalar axial subgroups. $\mathbf{O}(2)$ is generated by $[0, \theta_2] \in \mathbf{T}^2$ and rotation by π (ξ on rhombic lattice, ξ^2 on square lattice, and ξ^3 on hexagonal lattice).

reaction-diffusion equations $W_\mathbf{k} = W_\mathbf{k}^+$, and in two-dimensional Navier-Stokes flow $W_\mathbf{k} = W_\mathbf{k}^-$. Using these two examples, we illustrate the difference in axial subgroup symmetry between scalar and pseudoscalar bifurcations. We also consider examples where both scalar and pseudoscalar bifurcations can occur: the visual cortex, where Ω is $\mathbf{S}^1$, and liquid crystals, where Ω is again a point.

5.4 Reaction-Diffusion Systems

A planar reaction-diffusion system is a system of PDEs of the form

$$u_t = D\Delta u + F(u) \tag{5.18}$$

where $u = u(x,t) = (u_1, \ldots, u_m)^\mathrm{T}$ is an m-vector of functions, D is an $m \times m$ matrix of diffusion coefficients, F is an m-dimensional smooth mapping (the reaction term), and $x \in \mathbf{R}^2$. Note that $\Delta u = (\Delta u_1, \ldots, \Delta u_m)^\mathrm{T}$. Moreover, both reaction terms and diffusion coefficients may depend on parameters. For planar reaction-diffusion systems, the domain $\mathbf{D}$ is $\mathbf{R}^2$ and Ω is a point. The group of symmetries for (most) reaction-diffusion systems is $\Gamma = \mathbf{E}(2)$ with $\mathbf{E}(2)$ acting on $\mathbf{R}^2$ by its standard action (5.1).

Since Ω is a point, every eigenfunction in $W_\mathbf{k}$ (see (5.4)) is independent of ω. By (5.15) we may assume that $a \in \mathbf{R}^m$. Since reflections act trivially on a, all bifurcations in reaction-diffusion systems are scalar bifurcations.

Equilibria of (5.18) are defined by the elliptic equation

$$D\Delta u + F(u) = 0 \tag{5.19}$$

For simplicity. assume that $F(0) = 0$ so that $u = 0$ is an equilibrium for all parameter values.

The system (5.19) is invariant under the standard action of the Euclidean group $\mathbf{E}(2) = \mathbf{O}(2) \dotplus \mathbf{R}^2$ on the plane (5.1). The linearization of (5.19) at $u = 0$ is

$$Lu = D\Delta u + Au \tag{5.20}$$

where $A = (\mathrm{d}F)_0$ is an $m \times m$ reaction matrix.

Dispersion Curves. Translation invariance implies that eigenfunctions of L have the form

$$e^{2\pi i \mathbf{k}\cdot x} v \tag{5.21}$$

for a vector $v \in \mathbf{R}^m$. To determine which v, compute

$$\Delta(e^{2\pi i \mathbf{k}\cdot x} v) = -4\pi^2 |\mathbf{k}|^2 e^{2\pi i \mathbf{k}\cdot x} v$$

Therefore, the kernel of L is spanned by functions (5.21) where

$$L_{\mathbf{k}} v \equiv (A - 4\pi^2 |\mathbf{k}|^2 D) v = 0$$

We assume that L depends on a parameter λ. The *dispersion curve*, where λ is a function of $|\mathbf{k}|^2$, is defined by the equation

$$\det(L_{\mathbf{k}}) \equiv \det(A - 4\pi^2 |\mathbf{k}|^2 D) = 0 \tag{5.22}$$

Typically, this relation is quadratic-like in $|\mathbf{k}|^2$ as illustrated in Figure 5.5. There is then a smallest value of λ, denoted by λ_*, at which the trivial solution $u = 0$ goes unstable to linear perturbations. In this situation the value λ_* occurs for a unique critical wave number $k_* \geq 0$. When $k_* > 0$, pattern formation occurs since the linearized eigenfunctions have nontrivial spatial dependence. We assume that this discussion is valid for a given reaction-diffusion system (with $k_* > 0$) and ask what kinds of planform we should expect as solutions to the nonlinear system.

For example, consider the (linearized) Brusselator [226] consisting of two equations (normalized so that $u = 0$ is an equilibrium):

$$D = \begin{bmatrix} D_1 & 0 \\ 0 & D_2 \end{bmatrix} \qquad A = \begin{bmatrix} b-1 & a^2 \\ -b & -a^2 \end{bmatrix}$$

For definiteness, let $a^2 = 3$, $D_1 = 1$, and $D_2 = 2$, and let $b = \lambda$ be the bifurcation parameter. Let $\mu = 4\pi^2 |\mathbf{k}|^2$. Then the dispersion relation (5.22) is

$$\det(A - \mu D) = \det \begin{bmatrix} \lambda - 1 - \mu & 3 \\ -\lambda & -3 - 2\mu \end{bmatrix} = 2\mu^2 + 5\mu + 3 - 2\mu\lambda = 0$$

Therefore

$$\lambda = 4\pi^2 |\mathbf{k}|^2 + \frac{3}{2} \frac{1}{4\pi^2 |\mathbf{k}|^2} + \frac{5}{2}$$

whose graph is given in Figure 5.5. This graph has a minimum $\lambda_* \approx 10$ at $k_*^2 \approx 0.05$.

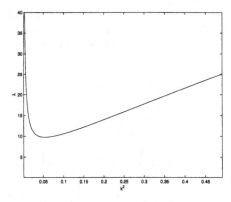

Figure 5.5: Sample dispersion relation for the Brusselator.

Reaction-Diffusion Planforms. In many reaction-diffusion system models, the unknown functions represent chemical concentrations. It is often the case that some observed feature in the experiment (color for instance) depends on whether a particular chemical concentration is greater or less than some threshold. Thus, the pattern that is observed in physical space is delineated by a level contour of that concentration. For that reason we will use level contours of one component of the eigenfunctions to illustrate the patterns that are associated with axial subgroups.

In this section we determine only those solutions that can be found by applying the Equivariant Branching Lemma to kernels corresponding to k_* scaled to 1 (that is, the wave vectors on the innermost ring of the dual lattice). The form of one component of the eigenfunctions in these kernels is given in Table 5.2. The corresponding eigenfunctions are listed in Table 5.7. The formulas for the axial eigenfunctions are presented in Table 5.8.

Lattice	ker L
Square	$z_1 e^{2\pi i k_1 \cdot x} + z_2 e^{2\pi i k_2 \cdot x} + c.c.$
Hexagonal	$z_1 e^{2\pi i k_1 \cdot x} + z_2 e^{2\pi i k_2 \cdot x} + z_3 e^{2\pi i k_3 \cdot x} + c.c.$
Rhombic	$z_1 e^{2\pi i k_1 \cdot x} + z_2 e^{2\pi i k_2 \cdot x} + c.c.$

Table 5.7 Eigenfunctions corresponding to $\Gamma_{\mathcal{L}}$ irreducible representation.

Lattice	Axial	Planform Eigenfunction
Square	Squares	$\cos(2\pi x_1) + \cos(2\pi x_2)$
	Rolls	$\cos(2\pi x_1)$
Hexagonal	Hexagons	$\cos(2\pi k_1 \cdot x) + \cos(2\pi k_2 \cdot x) + \cos(2\pi k_3 \cdot x)$
	Rolls	$\cos(2\pi x_1)$
Rhombic	Rhombs	$\cos(2\pi x_1) + \cos(2\pi k_2 \cdot x)$
	Rolls	$\cos(2\pi x_1)$

Table 5.8 Axial eigenfunctions for scalar action.

The threshold level contour need not be 0 — and that point is important for the correct symmetries to be visible in the contours. Level contours (asymmetrically placed around 0) of the axial eigenfunctions listed in Table 5.8 are plotted in Figures 5.6 and 5.7.

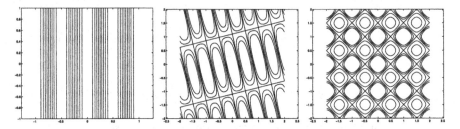

Figure 5.6: Contours of axial eigenfunctions: (Left) Rolls. (Center) Rhombs ($\eta = 0.47$). (Right) Squares.

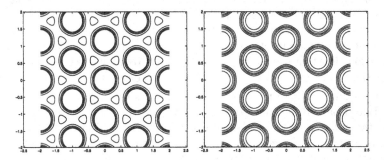

Figure 5.7: Contours of axial eigenfunctions: (Left) Hexagons$^+$. (Right) Hexagons$^-$.

5.5 Pseudoscalar Equations

Bosch Vivancos, Chossat, and Melbourne [71] observed that there is a class of planar partial differential equations, called *pseudoscalar* equations, where the action of $\mathbf{E}(2)$ on functions is different from that of reaction-diffusion equations, the *scalar* case. The scalar action of $\mathbf{E}(2)$ on functions for reaction-diffusion equations is:

$$\gamma u(x) = u(\gamma^{-1}x)$$

for any $\gamma \in \mathbf{E}(2)$. For pseudoscalar equations the action becomes

$$\gamma u(x) = \det(\gamma)u(\gamma^{-1}x) \tag{5.23}$$

where $\det(\gamma) = 1$ for translations and rotations and $\det(\gamma) = -1$ for reflections. As pointed out in [71], this small change in group action changes the types of planform that are expected in steady-state bifurcations on lattices. Just compare the lists of axial subgroups in Tables 5.5 and 5.6. In fact, the scalar and pseudoscalar actions are the only possible actions: see Melbourne [388]. Note that for a planar system of PDE (Ω is a point) that has pseudoscalar symmetry, $W_{\mathbf{k}} = W_{\mathbf{k}}^-$ in (5.13) and all bifurcations are pseudoscalar bifurcations.

Two Dimensional Navier-Stokes. The planar Navier-Stokes equations lead to a pseudoscalar action; we follow [71]. It is well known that the two-dimensional Navier-Stokes equations reduce to a single PDE for a quantity called the *stream function*. This PDE is pseudoscalar.

The Navier-Stokes equations (in the plane) have the form

$$\partial_t \mathbf{V} + (\mathbf{V} \cdot \nabla)\mathbf{V} = \nu \Delta \mathbf{V} + \nabla P + \mathbf{F} \tag{5.24}$$
$$\nabla \cdot \mathbf{V} = 0$$

where $\mathbf{V} : \mathbf{R}^2 \to \mathbf{R}^2$ is the velocity of an incompressible fluid, $P : \mathbf{R}^2 \to \mathbf{R}$ is the pressure, $\mathbf{F} : \mathbf{R}^2 \to \mathbf{R}^2$ is a force which may depend on $\mathbf{V}$, and ν is the kinematic viscosity of the fluid. With the possible exception of $\mathbf{F}$, this system is equivariant under the standard action of $\mathbf{E}(2)$ on *vector fields*

$$\phi \cdot \mathbf{V}(x) = \begin{cases} \phi \mathbf{V}(\phi^{-1} x) & \phi \in \mathbf{E}(2) \text{ a rotation or a reflection.} \\ \mathbf{V}(\phi^{-1} x) & \phi \in \mathbf{E}(2) \text{ a translation.} \end{cases} \tag{5.25}$$

We shall assume that $\mathbf{F}$, and hence the system (5.24), is equivariant with respect to this action of $\mathbf{E}(2)$. (Strictly speaking, the PDE (5.24) is an equation in $\mathbf{V}$ and P. However this is of no consequence, since P is eliminated shortly.)

The stream function ψ is defined by the equation

$$\Delta \psi = \text{curl}(\mathbf{V}) = \frac{\partial V_2}{\partial x_1} - \frac{\partial V_1}{\partial x_2}$$

Conversely, given ψ we can recover a divergence-free vector field $\mathbf{V}$ with stream function ψ from the equations

$$V_1 = -\frac{\partial \psi}{\partial x_2} \qquad V_2 = \frac{\partial \psi}{\partial x_1} \tag{5.26}$$

A calculation shows that if $\mathbf{V}$ transforms under $\mathbf{E}(2)$ as a vector field, then $\text{curl}(\mathbf{V})$ transforms as a pseudoscalar. More precisely,

$$\phi \cdot \text{curl}(\mathbf{V}) = \text{curl}(\phi \cdot \mathbf{V}) \qquad \phi \in \mathbf{E}(2)$$

where the actions of $\mathbf{E}(2)$ on the left-hand and right-hand sides are those in (5.23) and (5.25) respectively. Hence by elementary vector calculus the system (5.24) reduces to the single pseudoscalar PDE

$$\partial_t \Delta \psi + \text{curl}[\Delta \psi \nabla \psi] = \nu \Delta^2 \psi + \text{curl}(\mathbf{F}) \tag{5.27}$$

It is traditional to use ψ rather than u for the stream function $\text{curl}(\mathbf{V})$.

Pseudoscalar Planforms. As noted previously, the axial subgroups in scalar and pseudoscalar bifurcations differ even though the associated eigenfunctions are identical. To visualize this difference, we view the symmetries in physical space coordinates, that is, in terms of the velocity field of the two-dimensional Navier-Stokes equations rather than in terms of the stream function. We could plot the *streamlines*, that is, the level contours of the stream function (as we plotted the level contours of the eigenfunctions for scalar actions) but then we would lose the vector field direction along the streamlines — which is important when considering symmetry.

Consider the eigenfunction $\psi(x) = \cos(2\pi x_1)$ and use (5.26) to convert this eigenfunction to the vector field

$$V_{ar}(x) = 2\pi(0, -\sin(2\pi x_1))$$

See Figure 5.8(left) for a graph of this vector field. As with the rolls planform in the scalar reaction-diffusion case, this vector field (anti-rolls) is invariant under translations in the x_2 direction. However, in this case, if we use κ and reflect the image across the x_1 axis, then we must translate the image one-half a period in the x_1 direction (a glide reflection) in order to recover the same vector field. In the scalar case reflecting by κ alone is a symmetry of the solution. Similarly, the vector field associated to anti-squares is

$$V_{as}(x) = 2\pi(\sin(2\pi x_2), -\sin(2\pi x_1))$$

and is pictured in Figure 5.8(right). Again note the glide reflection symmetry.

The vector fields associated with the planforms on the hexagonal lattice are as follows. Let

$$s_1 = \sin(2\pi \mathbf{k}_1 \cdot x) \qquad s_2 = \sin(2\pi \mathbf{k}_2 \cdot x) \qquad s_3 = \sin(2\pi \mathbf{k}_3 \cdot x)$$

Then

$$V_h(x) = 2\pi(-\mathbf{k}_1(2)s_1 - \mathbf{k}_2(2)s_2 - \mathbf{k}_3(2)s_3, \mathbf{k}_1(1)s_1 + \mathbf{k}_2(1)s_2 + \mathbf{k}_3(1)s_3)$$

and

$$V_r(x) = 2\pi(\mathbf{k}_2(1)s_2 - \mathbf{k}_3(1)s_3, -\mathbf{k}_2(2)s_2 + \mathbf{k}_3(2)s_3)$$

The graphs of these vector fields are shown in Figures 5.9 and 5.10. Note the $\mathbf{Z}_6$ symmetry of the pseudoscalar anti-hexagons in contrast to the $\mathbf{D}_6$ symmetry of the scalar hexagons.

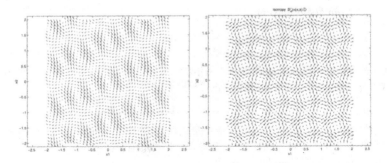

Figure 5.8: Axial planforms: (Left) Anti-rhombs. (Right) Anti-squares.

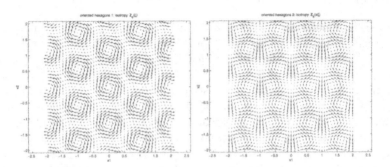

Figure 5.9: Axial planforms: (Left) Anti-hexagons. (Right) Anti-triangles.

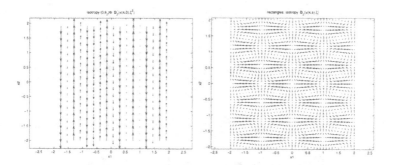

Figure 5.10: Axial planforms: (Left) Anti-rolls. (Right) Anti-rectangles.

5.6 The Primary Visual Cortex

In mammalian vision, the primary visual cortex (V1) is known to be sensitive to the orientation of contours in the visual field. As discussed in [75], the pattern of neuronal connections in V1 leads to an action of the Euclidean group on $D = R^2 \times S^1$ and to both scalar and pseudoscalar actions at bifurcation. Ermentrout and Cowan [176] and Bressloff et al. [75, 76] argue that it is possible that common geometric patterns seen in visual hallucinations may be the result of spontaneous symmetry-breaking activity patterns on V1. Using microelectrodes, voltage sensitive dyes, and optical imaging, scientists have accumulated much information about the distribution of orientation-selective cells in V1, and about their pattern of interconnection. In this discussion we present an abbreviated version of Bressloff et al. [75].

Each cortical cell in the V1 layer corresponds to one position in the retina, and has a propensity to fire when a contour or contrast bar with a *fixed orientation* occurs at that point in the retinal image. The V1 cells themselves are coupled in a fascinating pattern — which is the source of the Euclidean group action mentioned above.

The V1 layer is approximately a square, $40mm$ on a side. Hubel and Wiesel [286, 287, 288] noted that V1 is divided into small areas of about $1mm$ diameter, called *hypercolumns*, and the cells in each hypercolumn receive signals from one small area in the retina. A hypercolumn contains all cortical cells that correspond to such an area: its architecture allows it to determine whether a contour occurs at that point in the retinal image, and if so, what its orientation is. This task is accomplished by having all pairs of cells in a hypercolumn connected by inhibitory coupling — so if a contour is detected by one neuron, it tends to suppress the other neurons in that hypercolumn, a local *winner-take-all* strategy. Experimental confirmation of the existence of hypercolumns is found in Blasdel [67], see the iso-orientation patches in Figure 5.11.

What is curious — and crucial from the symmetry point of view — is how hypercolumns themselves are coupled. In recent years information has been obtained about connections using, for example, optical imaging with voltage-sensitive dyes [208, 72, 141, 178]. These studies show that cells that selectively fire for one orientation make contact only every millimeter or so along their axons with cells that fire selectively in the same orientation. See Figure 5.12, which illustrates the inhomogeneity in lateral coupling.

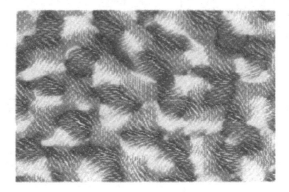

Figure 5.11: Distribution of orientation preferences in V1 obtained via optical imaging. Redrawn from Blasdel [67].

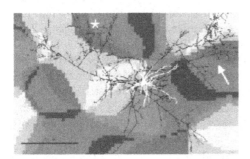 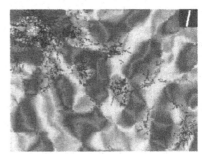

Figure 5.12: (Left) Connections made by an inhibitory interneuron in V1. Redrawn from Eysel [178]. (Right) Lateral Connections made by a cell in V1. Redrawn from Bosking *et al.* [72].

In addition, it appears that the long axons that support such connections, known as *intrinsic lateral* or horizontal connections, tend to be oriented more or less along the direction of their cells' preference. See the schematic diagram in Figure 5.13. Note that the strength of the lateral connection between hypercolumns is small when compared to the strength of the local connections within hypercolumns.

Ermentrout and Cowan [176] observed that drug-induced visual hallucination patterns can be thought of as patterns formed on V1 by spontaneous symmetry-breaking. When an individual is under the influence of a drug, the entire visual cortex is stimulated uniformly by the drug and not by the retina. When this forced stimulus is sufficiently large, patterns of activation are formed on V1 and interpreted by the brain as visual images — often with a distinctly geometric flavor.

The work in [176] was completed before the pattern of coupling in V1 was understood. Thus [176] assumed that models of V1 are Euclidean-invariant as in reaction-diffusion systems. The pattern of coupling in V1, as illustrated in Figure 5.13, shows that the standard action of rotations does not preserve the structure. More recently, Bressloff *et al.* [75] have incorporated the coupling pattern

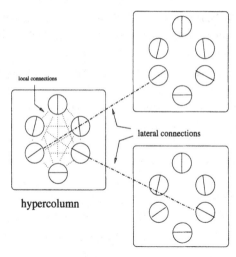

Figure 5.13: Short and long range connections in the visual cortex. $\cdots$ inhibitory; $\cdot - \cdot - \cdot -$ excitatory.

illustrated in Figure 5.13 into a model, the Wiener-Cowan model, based on the Wilson-Cowan equation [518].

Symmetries of the Model. In the Wiener-Cowan model [75] hypercolumns are made infinitesimal and modeled by circles, with each point on the circle corresponding to a boundary orientation. Moreover, points ϕ and $\phi+\pi$ both correspond to boundaries with the same orientation. Thus, the physical space of V1 (ignoring the boundary of V1) is $\mathbf{D} = \mathbf{R}^2 \times \mathbf{S}^1$. The action of the Euclidean group on $\mathbf{D}$ is dictated by the couplings and is generated by:

$$
\begin{aligned}
y(x, \phi) &= (x + y, \phi) & (y \in \mathbf{R}^2) \\
\theta(x, \phi) &= (R_\theta x, \phi + \theta) & (\theta \in \mathbf{S}^1) \\
\kappa(x, \phi) &= (\kappa x, -\phi)
\end{aligned}
\tag{5.28}
$$

Note that when the cortex is rotated the long range connections connect cells that have the wrong firing selectivity — but that can be compensated for by rotating the cells within each hypercolumn as indicated in (5.28). This representation of $\mathbf{E}(2)$ is called the *shift-twist* representation. In this model the state of the cortex is given by an *activity variable* $u(x, \phi)$ that represents the likelihood that there is a contour at the place in the visual field corresponding to the hypercolumn x in the visual cortex that is oriented in the direction ϕ. Such an activity variable could be measured by the voltage potential of the group of neurons in the hypercolumn at x that are sensistive to orientations in the direction ϕ.

The Wiener-Cowan model has been studied in Bressloff *et al.* [75] from the point of view of symmetry. We summarize their pattern formation results here. It is assumed that the patterns seen in visual hallucinations are the retinal images that would produce activity patterns on the visual cortex that are identical to those created by symmetry-breaking bifurcations. Methods for translating V1 patterns to retinal images are discussed in [176, 75], along with comparisons of pictures drawn by individuals who have experienced visual hallucinations with patterns obtained by bifurcation. We assume that the base state is the constant function $u(x, \phi)$ (all orientations are equally likely), which we can take to be $u = 0$.

Eigenfunctions and Planforms. Let L be $E(2)$-equivariant. Recall that the isotypic components of the action of translations on $\mathcal{F}_\mathcal{L}$ are parametrized by pairs $\pm \mathbf{k}$ and defined by

$$W_\mathbf{k} = \{a(\phi)e^{2\pi i\mathbf{k}\cdot x} + \text{c.c.} : a(\phi + \pi) = a(\phi) \text{ and } a(\phi) \in \mathbf{C}\} \qquad (5.29)$$

Contours in the visual field are assumed to be unoriented; hence the angle ϕ is π-periodic and $a(\phi)$ is π-periodic. Lemma 5.7 (and genericity) imply that we may assume that eigenfunctions are R_π-invariant as in (5.15), and have the form

$$a(\phi)e^{2\pi i\mathbf{k}\cdot x} + \text{c.c.} \qquad (5.30)$$

where $a(\phi)$ is real-valued.

Moreover, suppose we fix $\mathbf{k} = (1,0)$. Then

$$\kappa a(\phi) = a(-\phi)$$

Commutativity of L with κ implies that

$$\begin{aligned}
W_\mathbf{k}^+ &= \{a(\phi)e^{2\pi i\mathbf{k}\cdot x} + \text{c.c.} : a(-\phi) = a(\phi)\} \\
W_\mathbf{k}^- &= \{a(\phi)e^{2\pi i\mathbf{k}\cdot x} + \text{c.c.} : a(-\phi) = -a(\phi)\}
\end{aligned}$$

are invariant subspaces for L and $W_\mathbf{k} = W_\mathbf{k}^+ \oplus W_\mathbf{k}^-$. Hence eigenfunctions of L have the form (5.30) where $a(\phi)$ is either even or odd. The even case leads to the scalar action, while the odd case leads to the pseudoscalar action. Thus, in principle, both scalar and pseudoscalar bifurcations can coexist in this example. In fact, there are two dispersion curves; one for each of $W_\mathbf{k}^+$ and $W_\mathbf{k}^-$ and Bressloff *et al.* [75] show that reasonable forms of the Wilson-Cowan equation [518] can have either dispersion curve as the first bifurcation. So both bifurcations are possible in these models.

Axial Planforms. The eigenfunctions corresponding to the scalar and pseudoscalar actions are given in Tables 5.9 and 5.10. We form pictures of the planform $f(x, \phi)$ as follows. For each x choose the direction ϕ that maximizes the function $f(x, \cdot)$ on the circle $\mathbf{S}^1$, the continuos version of the winner-take-all strategy mentioned previously. Then graph a small line element at x oriented in the direction ϕ. Note that this process can and does lead to discontinuities at points x where f has more than one absolute maximum. Indeed, we plot all maximal directions in these figures.

Planform	Eigenfunction
Rolls	$a(\phi)\cos(2\pi x_1)$
Rhombs	$a(\phi)\cos(2\pi \mathbf{k}_1 \cdot x) + a(\phi - \eta)\cos(2\pi \mathbf{k}_2 \cdot x)$
Squares	$a(\phi)\cos(2\pi x_1) + a\left(\phi - \frac{\pi}{2}\right)\cos(2\pi x_2)$
Hexagons	$a(\phi)\cos(2\pi \mathbf{k}_1 \cdot x) + a\left(\phi - \frac{2\pi}{3}\right)\cos(2\pi \mathbf{k}_2 \cdot x) +$ $a\left(\phi - \frac{4\pi}{3}\right)\cos(2\pi \mathbf{k}_3 \cdot x)$

Table 5.9 Scalar axial subgroups: $a(-\phi) = a(\phi)$ is even.

Planform	Eigenfunction
Anti-rolls	$a(\phi)\cos(2\pi x_1)$
Anti-rhombs	$a(\phi)\cos(2\pi\mathbf{k}_1\cdot x) + a(\phi-\eta)\cos(2\pi\mathbf{k}_2\cdot x)$
Anti-squares	$a(\phi)\cos(2\pi x_1) - a\left(\phi-\frac{\pi}{2}\right)\cos(2\pi x_2)$
Anti-hexagons	$a(\phi)\cos(2\pi\mathbf{k}_1\cdot x) + a\left(\phi-\frac{2\pi}{3}\right)\cos(2\pi\mathbf{k}_2\cdot x) +$ $a\left(\phi-\frac{4\pi}{3}\right)\cos(2\pi\mathbf{k}_3\cdot x)$
Anti-triangles	$a(\phi)\sin(2\pi\mathbf{k}_1\cdot x) + a\left(\phi-\frac{2\pi}{3}\right)\sin(2\pi\mathbf{k}_2\cdot x) +$ $a\left(\phi-\frac{4\pi}{3}\right)\sin(2\pi\mathbf{k}_3\cdot x)$
Anti-rectangles	$a\left(\phi-\frac{2\pi}{3}\right)\cos(2\pi\mathbf{k}_2\cdot x) -$ $a\left(\phi-\frac{4\pi}{3}\right)\cos(2\pi\mathbf{k}_3\cdot x)$

Table 5.10 Pseudoscalar axial subgroups: $a(-\phi) = -a(\phi)$ is odd.

The Choice of the Amplitude Function $a(\phi)$. Symmetry can, in a perhaps unanticipated way, aid in determining the form of the amplitude $a(\phi)$. As we mentioned previously, the strength of the lateral coupling β is small when compared to the strength of the local coupling. It therefore makes sense, as discussed in Bressloff *et al.* [75], to consider the bifurcation problem as a small perturbation of the $\beta = 0$ case. Wolf and Geisel [523] note that when $\beta = 0$ the bifurcation problem has symmetry group $\mathbf{E}(2) \oplus \mathbf{SO}(2)$ rather than just $\mathbf{E}(2)$, since local coupling is isotropic. That is,

$$\psi(x,\phi) = (x, \phi+\psi) \qquad (\psi \in \mathbf{S}^1)$$

is a model symmetry. It follows that when $\beta = 0$, the subspace $W_{\mathbf{k}}$ decomposes by Fourier series into a direct sum of $\mathbf{SO}(2)$-invariant subspaces $W_{\mathbf{k},p}$ where

$$W_{\mathbf{k},p} = \{ze^{p\phi i}e^{2\pi i\mathbf{k}\cdot x} + \text{c.c.} : z \in \mathbf{C}\}$$

and that eigenfunctions of L generically lie in $W_{\mathbf{k},p}$. Moreover, the one-dimensional space $W_{\mathbf{k},p}^{+}$ is generated by $\cos(p\phi)e^{2\pi i\mathbf{k}\cdot x}$ and the one-dimensional space $W_{\mathbf{k},p}^{-}$ is generated by $\sin(p\phi)e^{2\pi i\mathbf{k}\cdot x}$. Bressloff *et al.* [75] show that reasonable Wilson-Cowan models lead to primary $p = 0$ or $p = 1$ bifurcations in the scalar case and $p = 1$ bifurcations in the pseudoscalar case.

Thus, when the lateral coupling strength $\beta \approx 0$, we can assume that the π-periodic function $a(\phi)$ is, in the pseudoscalar case, near $a(\phi) = \sin(2\phi)$. This choice, however, is nongeneric for $\mathbf{E}(2)$ symmetry, since equal maxima occur only when the eigenfunction is identically zero. To verify this assertion observe, for example, that the eigenfunction for odd hexagons

$$e^{2\pi i\mathbf{k}_1\cdot x}\sin(2\phi) + e^{2\pi i\mathbf{k}_2\cdot x}\sin\left(2\left(\phi-\frac{2\pi}{3}\right)\right) + e^{2\pi i\mathbf{k}_3\cdot x}\sin\left(2\left(\phi+\frac{2\pi}{3}\right)\right)$$

can be transformed to

$$A(x)\cos(2(\phi-\phi_0(x))$$

In this form maxima can occur only when $A(x) = 0$ or $\phi = \phi_0(x)$. For a given x, the latter value of ϕ is unique. So multiple maxima occur only when $A(x) = 0$, that is, when the eigenfunction is identically zero. To avoid this source of nongenericity, we perform our calculations using $a(\phi) = \sin(2\phi) - 0.05\sin(4\phi)$.

In the even case, the first nonzero term is the constant term, which has no effect when choosing maxima. So, in the even case, we choose the function $a(\phi) = \cos(2\phi) - 0.05\cos(4\phi)$. Line field drawings of the various axial planforms are given in Figures 5.14, 5.15, 5.16, and 5.17.

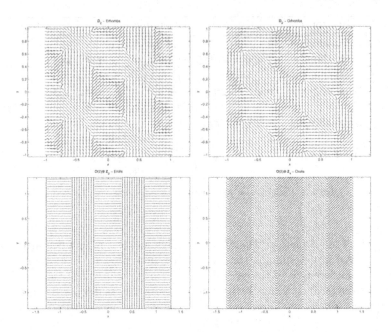

Figure 5.14: Rhombic lattice ($\eta = \frac{\pi}{2.8}$): even Rhombs (upper left); odd Rhombs (upper right); even Rolls (lower left); odd Rolls (lower right).

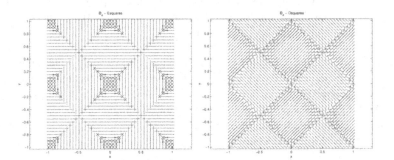

Figure 5.15: Square lattice: (Left) Even squares. (Right) Odd squares.

Structure of Discontinuities in Planform Line Field. Let $f(x, \phi)$ be an eigenfunction and let $\mathcal{X}_f$ be the set of points x at which $f(x, \cdot)$ has multiple absolute maxima. The set $\mathcal{X}_f$ is the set of points where the line field planform has discontinuities. Note that $\mathcal{X}_f$ is a codimension one algebraic variety; that is, $\mathcal{X}_f$ is a collection of (perhaps intersecting) curves that divides the plane into cells. On each cell, the line field varies smoothly, and in the examples we consider it is more or less constant. Moreover, suppose that $\sigma \in \Gamma_{\mathcal{L}}$. Then σ acts both on x and ϕ. We assert that if σ is a symmetry of an eigenfunction $f(x, \phi)$ then $\sigma \mathcal{X}_f = \mathcal{X}_f$, that is, the set $\mathcal{X}_f$ is invariant under the isotropy subgroup of the eigenfunction. For example, this remark allows us to see a difference in planforms between $\mathbf{D}_6$ hexagons on the even hexagonal lattice and $\mathbf{Z}_6$ hexagons on the odd hexagonal lattice. Note also the similarities between the regions where the line field varies smoothly and the

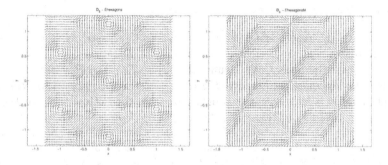

Figure 5.16: Hexagonal lattice: (Left) Even hexagons$^+$. (Right) Even hexagons$^-$.

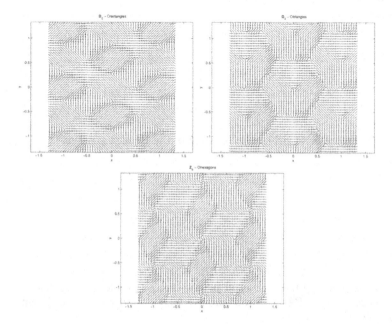

Figure 5.17: Hexagonal lattice: (Upper Left) Odd rectangles. (Upper Right) Odd triangles. (Lower) Odd hexagons.

iso-orientation patches shown in Figure 5.11. That figure was obtained by stimulating the retina with a bar image, rather than by symmetry-breaking bifurcations induced on V1 by the uniform action of a stimulus. Thus, we do not expect exact correspondence.

The symmetries of individual solutions give some hint about discontinuities in the associated line fields. For example, suppose that $f(x, \phi)$ is invariant under the reflection κ. Then $f(x_1, -x_2, -\phi) = f(x_1, x_2, \phi)$, so $f(x_1, 0, -\phi) = f(x_1, 0, \phi)$. Therefore the maximum value of $f(x_1, 0, \cdot)$ can be unique only when that maximum occurs at $\phi = 0$ or $\phi = \frac{\pi}{2}$. If f is an even rolls solution, then multiple maxima can be forced on open regions. This effect is barely visible in Figure 5.14, since we are near the $\beta = 0$ case.

See Bressloff $et\ al.$ [75, 76] for a discussion of how these planforms can be viewed as geometric visual hallucinations.

5.7 The Planar Bénard Experiment

The planar Bénard experiment consists of a viscous fluid layer between two parallel infinite plates held at different (but constant) temperatures: T_u for the upper plate and T_ℓ for the lower. We assume that $T_l < T_u$. The *Rayleigh number R* is a dimensionless quantity proportional to $T_u - T_\ell$. See Figure 5.18. When the Rayleigh number is small, heat is transmitted from the bottom plate to the top plate by pure conduction (there is no fluid motion). Convection (heat conduction coupled with fluid motion) occurs for $R > R_c$, where R_c is a critical Rayleigh number. When R is slightly greater than R_c, experiments often lead to rolls: see Figure 5.19(left). Under certain circumstances it is also possible to obtain hexagonal convection cells. In this case the fluid usually rises at the center and falls at the edges: see Figure 5.19(right).

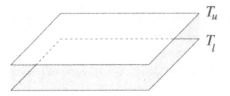

Figure 5.18: Schematic drawing of the Bénard experiment.

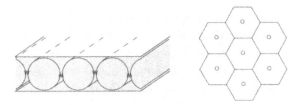

Figure 5.19: (Left) Rolls solution in planar Bénard problem (side view). (Right) Hexagons solution in Bénard problem (top view): fluid rising in cell centers and falling near cell boundary.

We assume that convection is modeled by the Navier-Stokes equations in the Boussinesq approximation. That is, the Navier-Stokes equations for incompressible flow plus buoyancy, and the heat equation plus convection. We will not write the equations here: they may be found in [237], p. 149. The variables in this model are the velocity field $v(x)$ in $\mathbf{R}^3$ and the deviation $\theta(x)$ of the temperature from the pure conduction state. The parameters in the model are the Rayleigh number, the Prandtl number (a dimensionless ratio of viscosity to trivial conductivity), pressure, and gravity.

Two classes of boundary conditions are relevant for this experiment: those where the boundary conditions on the top and bottom plates are the same, and those where they are different. When they are different, as was the case for Bénard's original experiment, the model symmetries are the planar Euclidean symmetries. When the boundary conditions are the same, an additional symmetry (reflection about the midplane) is introduced. The Euclidean group $\mathbf{E}(2)$ acts on v and θ as follows. The action of $\gamma \in \mathbf{O}(2)$ is

$$\gamma(v(x), \theta(x)) = (\gamma v(\gamma^{-1}x), \theta(\gamma^{-1}x))$$

Axial	Isotropy	Fixed Vector	Eigenfunction
I_R	$\mathbf{O}(2) \oplus \mathbf{D}_2\left(\kappa, \mu\left[\frac{1}{2}, 0\right]\right)$	$(1,0,0)$	$\cos(\mathbf{k}_1 \cdot x)$
I_H	$\mathbf{D}_6$	$(1,1,1)$	$\cos(\mathbf{k}_1 \cdot x) + \cos(\mathbf{k}_2 \cdot x) + \cos(\mathbf{k}_3 \cdot x)$
I_T	$\mathbf{D}_3 \oplus \mathbf{Z}_2(\mu \xi^3)$	(i,i,i)	$\sin(\mathbf{k}_1 \cdot x) + \sin(\mathbf{k}_2 \cdot x) + \sin(\mathbf{k}_3 \cdot x)$
I_P	$\mathbf{Z}_2^3\left(\xi^3, \mu\kappa, \mu\left[0, \frac{1}{2}\right]\right)$	$(0,1,-1)$	$\cos(\mathbf{k}_2 \cdot x) - \cos(\mathbf{k}_3 \cdot x)$

Table 5.11 Axial subgroups of action on hexagonal lattice with midplane reflection corresponding to smallest dual wave vector. See Table 5.4.

where

$$\gamma x = \left[\gamma \begin{bmatrix} x_1 \\ x_2 \end{bmatrix} \right]$$
$$\quad x_3$$

Planar translations $y \in \mathbf{R}^2$ act by

$$y(v, \theta)(x) = (v, \theta)(x - (y, 0)).$$

The midplane reflection μ acts by

$$\mu(v, \theta)(x_1, x_2, x_3) = (v, -\theta)(x_1, x_2, -x_3) \tag{5.31}$$

Solution Branches. We now describe how symmetries act on the kernel of the linearized Boussinesq equations at a steady-state bifurcation from the trivial pure conduction solution. The eigenfunctions for these equations all have a nonzero component in the θ direction. It follows that the action of Euclidean symmetries on this kernel is given by the action on the θ variable — which is the same action as the one that occurs in reaction-diffusion equations. Thus, in the case when no midplane reflection is present, the planforms (visualized as level contours of temperature) are identical to those in Figure 5.6 for the square lattice and Figure 5.7 for the hexagonal lattice.

When the midplane reflection is present, the situation is quite different — at least on the hexagonal lattice. The θ component of the eigenfunctions are invariant under $x_3 \mapsto -x_3$; hence, according to (5.31), the midplane reflection acts on the kernel as multiplication by $-I$. This small change in the group (substitute $(\mathbf{D}_6 \dotplus \mathbf{T}_\mathcal{L}^2) \oplus \mathbf{Z}_2(\mu)$ for $\mathbf{D}_6 \dotplus \mathbf{T}_\mathcal{L}^2$) and its action on the six-dimensional kernel changes the number of axial subgroups. When the mid-plane reflection symmetry is included, there are four planforms: stripes, hexagons, regular triangles, and patchwork quilt. To verify the existence of these planforms see Table 5.7. Pictures of regular triangles and patchwork quilt are given in Figure 5.20.

5.8 Liquid Crystals

Liquid crystals exhibit phases that are intermediate between liquids and crystals. See Sluckin [478] for an introduction. The modeling of liquid crystalline flow is based on a variable that represents molecular orientation. Having said that, there are several different kinds of model; see Roy and Tsuji [453]. The classical models are attributed to Leslie and Eriksen [355] and Doi [166]. The Leslie-Eriksen model is described in terms of a vector quantity called the *director*. The Doi model uses a probability density function for molecular orientation, and the director at each point is the direction of maximum probability — very much like the maximization process employed in the visual cortex model. A third model that is often used is the Landau - de Gennes model [142, 59, 453, 198]. This model assumes that the

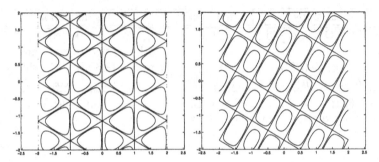

Figure 5.20: Contours of axial eigenfunctions on the hexagonal lattice with a mid-plane reflection: (Left) Regular triangles. (Right) Patchwork quilt.

density function in the Doi model is Gaussian at each point and is therefore posed in terms of a 3×3 symmetric trace zero matrix — the matrix of second moments of the probability density function normalized to have trace zero. In this model the director is the eigendirection corresponding to the largest eigenvalue.

In a homeotropic nematic liquid crystal the directors line up in a (more or less) fixed direction. Following Chillingworth and Golubitsky [102], we consider a planar layer of nematic liquid crystal, whose molecules align vertically. We determine patterns that might form as the director field bifurcates to a nonconstant line field by spontaneous symmetry-breaking. In the spirit of this chapter, we assume that the defining equations for the 3×3 symmetric trace zero matrix are Euclidean-invariant and we look for states that are spatially doubly periodic. Experiments of Peacock *et al.* [428] show interesting periodic states.

Symmetries. In the Landau - de Gennes model the state of a nematic liquid crystal is determined by $Q(x)$ where, at each point x, Q is a 3×3 symmetric trace zero matrix. We assume that the liquid crystal is in a planar layer (that is, $x \in \mathbf{R}^2$), though the director is a line in $\mathbf{R}^3$. The Euclidean group $\mathbf{E}(2)$ acts on Q by

$$
\begin{aligned}
(yQ)(x) &= Q(x - y) & y \in \mathbf{R}^2 \\
(\gamma Q)(x) &= \gamma Q(\gamma^{-1}x)\gamma^{-1} & \gamma \in \mathbf{O}(2)
\end{aligned}
$$

We consider bifurcation (and spatial pattern formation) from a Euclidean invariant state Q. Translation invariance implies that $Q(x) = Q_0$ is a constant matrix and rotation invariance implies that $Q_0 = \gamma Q_0 \gamma^{-1}$ for all $\gamma \in \mathbf{O}(2)$. It follows that

$$
Q_0 = \alpha \begin{bmatrix} -1 & 0 & 0 \\ 0 & -1 & 0 \\ 0 & 0 & 2 \end{bmatrix}
$$

for some $\alpha \in \mathbf{R}$. Note that when $\alpha > 0$ the director field points in the vertical direction at every point and the nematic liquid crystal is homeotropic. When $\alpha < 0$, the liquid crystal is in an *isotropic* state — the directors are horizontal but are equally likely to point in any horizontal direction.

Note that Q_0 is also invariant under the up-down reflection

$$
\tau = \begin{bmatrix} 1 & 0 & 0 \\ 0 & 1 & 0 \\ 0 & 0 & -1 \end{bmatrix}
$$

Thus, the full symmetry group for this state is $\Gamma = \mathbf{E}(2) \oplus \mathbf{Z}_2(\tau)$ — just like states in Bénard convection when the midplane reflection is present.

Linear Theory. Denote the symmetric matrix by

$$Q = \begin{bmatrix} Q_1 & Q_3 & Q_4 \\ Q_3 & Q_2 & Q_5 \\ Q_4 & Q_5 & * \end{bmatrix}$$

where $*$ is chosen so that $tr(Q) = 0$. Because of translation symmetry, eigenfunctions have the form

$$e^{2\pi i \mathbf{k} \cdot x} Q + \text{c.c.} \tag{5.32}$$

where Q is complex-valued, so

$$W_{\mathbf{k}} = \{e^{2\pi i \mathbf{k} \cdot x} Q + \text{c.c.} : Q \text{ is complex-valued}\} \tag{5.33}$$

Here $W_{\mathbf{k}}$ is a ten-dimensional real vector space. Rotations and reflections act on eigenfunctions by

$$\gamma(e^{2\pi i \mathbf{k} \cdot x} Q) = e^{2\pi i \gamma \mathbf{k} \cdot x} \gamma Q \gamma^{-1} \tag{5.34}$$

When performing the linear analysis we can assume, after rotation, that $\mathbf{k} = (1, 0)$ is a critical dual wave vector.

We will show that there are four types of bifurcation, based on the scalar/pseudoscalar dichotomy and on whether the up-down symmetry τ acts trivially or nontrivially on Q. We begin with a discussion of scalar versus pseudoscalar.

To determine the form of the scalar and pseudoscalar matrices (that is, to determine matrices $Q^+ \in W_{\mathbf{k}}^+$ and $Q^- \in W_{\mathbf{k}}^-$), we compute the action of κ on eigenfunctions. Since κ fixes $\mathbf{k}$ and

$$\kappa = \begin{bmatrix} 1 & 0 & 0 \\ 0 & -1 & 0 \\ 0 & 0 & 1 \end{bmatrix}$$

it follows from (5.34) that

$$\kappa(e^{2\pi i \mathbf{k} \cdot x} Q) = e^{2\pi i \mathbf{k} \cdot x} \begin{bmatrix} Q_1 & -Q_3 & Q_4 \\ -Q_3 & Q_2 & -Q_5 \\ Q_4 & -Q_5 & * \end{bmatrix} \tag{5.35}$$

Hence (5.35) implies that those Q on which κ acts as the identity are

$$Q^+ = \begin{bmatrix} Q_1 & 0 & Q_4 \\ 0 & Q_2 & 0 \\ Q_4 & 0 & * \end{bmatrix}$$

and those Q on which κ acts as minus the identity are

$$Q^- = \begin{bmatrix} 0 & Q_3 & 0 \\ Q_3 & 0 & Q_5 \\ 0 & Q_5 & 0 \end{bmatrix}$$

Lemma 5.7 implies that typically eigenspaces are two-dimensional subspaces of $W_{\mathbf{k}}^+$ or $W_{\mathbf{k}}^-$, and we can choose Q^+ and Q^- to be R_π-invariant. The formula that corresponds to (5.15) is $R_\pi(e^{2\pi i \mathbf{k} \cdot x} Q) = e^{2\pi i \mathbf{k} \cdot x} Q$. Since

$$R_\pi = \begin{bmatrix} -1 & 0 & 0 \\ 0 & -1 & 0 \\ 0 & 0 & 1 \end{bmatrix}$$

it follows that

$$\mathbf{R}_\pi(e^{2\pi i \mathbf{k} \cdot x} Q) = e^{2\pi i \mathbf{k} \cdot x} \begin{bmatrix} Q_1 & Q_3 & -Q_4 \\ Q_3 & Q_2 & -Q_5 \\ -Q_4 & -Q_5 & * \end{bmatrix} \qquad (5.36)$$

Thus by R_π-invariance we may assume that

$$Q^+ = \begin{bmatrix} a & 0 & ci \\ 0 & b & 0 \\ ci & 0 & * \end{bmatrix} \qquad Q^- = \begin{bmatrix} 0 & a & 0 \\ a & 0 & bi \\ 0 & bi & 0 \end{bmatrix} \qquad (5.37)$$

where $a, b, c \in \mathbf{R}$.

Moreover, since L commutes with τ we can further subdivide

$$W_{\mathbf{k}}^+ = W_{\mathbf{k}}^{++} \oplus W_{\mathbf{k}}^{+-} \quad \text{and} \quad W_{\mathbf{k}}^- = W_{\mathbf{k}}^{-+} \oplus W_{\mathbf{k}}^{--}$$

into subspaces on which τ acts trivially and on which τ acts by $-I$, and each of these subspaces is L-invariant. Since

$$\tau Q \tau = \begin{bmatrix} Q_1 & Q_3 & -Q_4 \\ Q_3 & Q_2 & -Q_5 \\ -Q_4 & -Q_5 & * \end{bmatrix}$$

we can see that

$$Q^{++} = \begin{bmatrix} a & 0 & 0 \\ 0 & b & 0 \\ 0 & 0 & -a-b \end{bmatrix} \qquad Q^{+-} = \begin{bmatrix} 0 & 0 & i \\ 0 & 0 & 0 \\ i & 0 & 0 \end{bmatrix}$$

and

$$Q^{-+} = \begin{bmatrix} 0 & 1 & 0 \\ 1 & 0 & 0 \\ 0 & 0 & 0 \end{bmatrix} \qquad Q^{--} = \begin{bmatrix} 0 & 0 & 0 \\ 0 & 0 & i \\ 0 & i & 0 \end{bmatrix}$$

Each $Q^{\pm\pm}$ generates a two-dimensional space $V_{\mathbf{k}}^{\pm\pm}$ that in turn generates ker L, as we now discuss.

Square Lattice Planforms. The holohedry of the square lattice is $\mathbf{D}_4$, which is generated by κ and ξ, where ξ is counterclockwise rotation of the plane by $\frac{\pi}{2}$. The action of ξ on Q is

$$\xi(Q) = \xi Q \xi^{-1} \qquad (5.38)$$

On the square lattice the kernel of L is (typically) four-dimensional:

$$V_{\mathbf{k}}^{\pm\pm} \oplus \xi\left(V_{\mathbf{k}}^{\pm\pm}\right)$$

Therefore we can write the general eigenfunction as

$$z_1 e^{2\pi i \mathbf{k}_1 \cdot x} Q^{\pm\pm} + z_2 e^{2\pi i \mathbf{k}_2 \cdot x} \xi Q^{\pm\pm} \xi^{-1} + \text{c.c.}$$

There are two axial subgroups in each case — one corresponding to rolls and one to squares. For rolls $z_1 = 1$ and $z_2 = 0$ and for squares $z_1 = z_2 = 1$ (up to conjugacy by a group element). Write $x = (x_1, x_2)$. Then $e^{2\pi i \mathbf{k}_1 \cdot x} = e^{2\pi i x_1}$ and $e^{2\pi i \mathbf{k}_2 \cdot x} = e^{2\pi i x_2}$.

We assume that the trivial solution corresponds to all molecules oriented in the x_3 direction, and that a symmetry-breaking bifurcation occurs. Therefore, we picture the bifurcating solution as follows. At each point (x_1, x_2) we choose the eigendirection corresponding to the largest eigenvalue of the symmetric 3×3 matrix at (x_1, x_2). Then we plot only the x_1, x_2 components of that line field at (x_1, x_2). In this picture, a line element that degenerates to a point corresponds to a vertical eigendirection. So the trivial solution looks like at array of points. Indeed, when τ acts trivially, every solution is invariant under up-down reflection and hence the line field remains vertical. The eigenvalues corresponding to horizontal directions change — but this change will not appear in our figures — so we do not picture the cases where τ acts trivially. In Figures 5.21 and 5.22 we plot solutions corresponding to scalar and pseudoscalar rolls and squares, when τ acts nontrivially as $-I$.

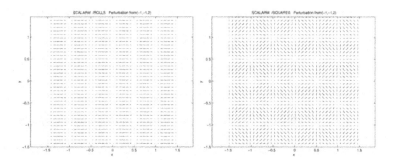

Figure 5.21: Square lattice with scalar representation with τ acting as $-I$: (Left) Rolls. (Right) Squares.

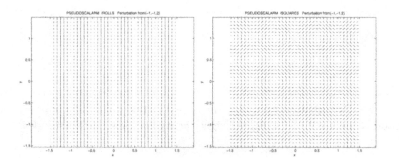

Figure 5.22: Square lattice with pseudoscalar representation with τ acting as $-I$: (Left) Anti-rolls. (Right) Anti-squares.

Hexagonal Lattice Planforms. The holohedry of the hexagonal lattice is $\mathbf{D}_6$, which is generated by κ and ξ, where ξ is counterclockwise rotation of the plane by $\frac{\pi}{3}$. The action of ξ on Q is the same as in (5.38).

On the hexagonal lattice, we first study the case when the kernel of L is six-dimensional. In this case the dual wave vectors can be chosen to be

$$k_1 = (1, 0) \qquad k_2 = \frac{1}{2}(-1, \sqrt{3}) \qquad k_3 = \frac{1}{2}(-1, -\sqrt{3})$$

and the eigenspaces are

$$V_{\mathbf{k}}^{\pm\pm} \oplus \xi^2 \left(V_{\mathbf{k}}^{\pm\pm} \right) \oplus \xi^4 \left(V_{\mathbf{k}}^{\pm\pm} \right)$$

Therefore, we can write the general eigenfunction in each of the four cases as

$$z_1 e^{2\pi i k_1 \cdot x} Q^{\pm\pm} + z_2 e^{2\pi i k_2 \cdot x} \xi^2 Q^{\pm\pm} \xi^4 + z_3 e^{2\pi i k_3 \cdot x} \xi^4 Q^{\pm\pm} \xi^2 + \text{c.c.}$$

On the scalar hexagonal lattice there are two branches of axial solutions — hexagons and rolls — and hexagons come in two types: hexagons$^+$ and hexagons$^-$.

Let $z = (z_1, z_2, z_3)$. For rolls we may take $z = (1, 0, 0)$ and for hexagons$^+$ and hexagons$^-$ we may take $z = \pm(1, 1, 1)$. Again we do not picture the planforms on which τ acts trivially.

The scalar representation $\tau = -1$ behaves just like bifurcation in Bénard convection with the midplane reflection. There are four axial subgroups [237] and the associated planforms are presented in Figure 5.23.

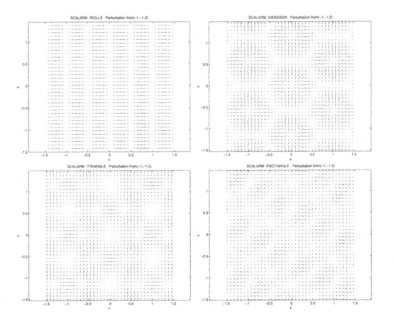

Figure 5.23: Perturbation from Q_0. Hexagonal lattice with scalar $\tau = -1$ representation: (upper left) rolls; (upper right) hexagons; (lower left) triangles; (lower right) rectangles.

Bosch Vivancos *et al.* [71] and Bressloff *et al.* [74] show that in the pseudoscalar representation hexagons are given by $(z_1, z_2, z_3) = (1, 1, 1)$, triangles by $(z_1, z_2, z_3) = (i, i, i)$, and rectangles by $(z_1, z_2, z_3) = (1, -1, 0)$. Here τ acts trivially.

Bifurcation in the fourth representation, the pseudoscalar with $\tau = -1$ representation, again behaves as Bénard with the midplane reflection symmetry. The planforms are shown in Figure 5.24.

Pattern Formation from the Isotropic State. We end this discussion by indicating the patterns that form in Rolls solutions on bifurcation from a isotropic nematic state $\alpha < 0$. See Figure 5.25. A more complete discussion of pattern formation resulting from this bifurcation may be found in Chillingworth and Golubitsky [102].

5.9 Pattern Selection: Stability of Planforms

In Section 5.3 we listed all axial subgroups on the rhombic, square, and hexagonal lattices in both the scalar and pseudoscalar cases corresponding to critical wave vectors of shortest length in the dual lattice. We also illustrated our findings using line field pictures associated to eigenfunctions of the visual cortex model. In this section we consider the stability of these planforms and, where relevant, secondary

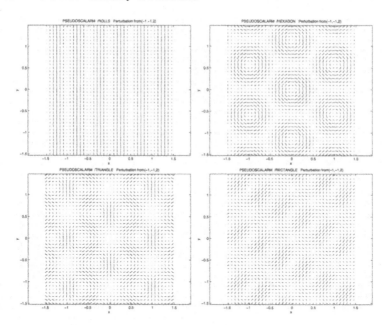

Figure 5.24: Perturbation from Q_0. Hexagonal lattice with pseudoscalar $\tau = -1$ representation: (upper left) rolls; (upper right) hexagons; (lower left) triangles; (lower right) rectangles.

bifurcations. Stability is a difficult issue and we address only one part of this complex problem.

Fix a planar lattice $\mathcal{L}$. We answer the following question: What is the stability of axial solutions to perturbations that are doubly periodic with respect to $\mathcal{L}$? We can find solutions near bifurcation by reducing the original differential equations model to a system of ordinary differential equations on a center manifold. Moreover, the reduction to center manifold can be done in a way that preserves the symmetries of the system. Existence of equilibria for the center manifold equations is found using the Equivariant Branching Lemma. To determine stability, we discuss the structure of $\Gamma_{\mathcal{L}}$-equivariant vector fields and isotypic decompositions of the associated Jacobian matrices at equilibrium points.

Except for the pseudoscalar representation on the hexagonal lattice, the information has been computed in a number of different places. For completeness, we begin by sketching the previously known material. In each case we assume that $f(z)$ is a $\Gamma_{\mathcal{L}}$-equivariant mapping.

Rhombic Lattice. The scalar and pseudoscalar actions of $\Gamma_{\mathcal{L}}$ on $\mathbf{C}^2$ are identical in the sense that that the two actions generate precisely the same set of linear mappings on $\mathbf{C}^2$. This assertion is verified by noting that the torus element $\theta = \left[\frac{1}{2}, \frac{1}{2}\right]$ acts by $(z_1, z_2) \mapsto (-z_1, -z_2)$. It follows that $\theta\kappa$ acts in the pseudoscalar representation exactly as κ acts in the scalar representation. Thus the two representations are generated by the same set of matrices on $\mathbf{C}^2$, and the two actions produce the same set of equivariant mappings and the same abstract bifurcation theory. The planforms for the two actions are different since the symmetry groups of the planforms are different (though isomorphic as groups). The corresponding statements

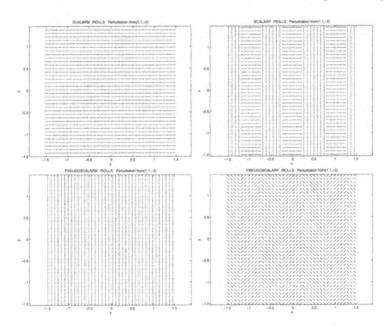

Figure 5.25: Rolls perturbation from $-Q_0$: (upper left) scalar $\tau = +1$; (upper right) scalar $\tau = -1$; (lower left) pseudoscalar $\tau = +1$; (lower right) pseudoscalar $\tau = -1$.

are false for the hexagonal lattice, since the scalar and pseudoscalar actions consist of different matrices.

Lemma 5.8 *For the rhombic lattice, the general equivariant mapping on the shortest wave vector representation is*

$$f(z_1, z_2) = (p(|z_1|^2, |z_2|^2)z_1, p(|z_2|^2, |z_1|^2)z_2) \tag{5.39}$$

where p is real-valued and odd.

Proof. Let $f : \mathbf{C}^2 \to \mathbf{C}^2$ be $\Gamma_{\mathcal{L}}$-equivariant. Commutativity with respect to the torus action implies that

$$f(z_1, z_2) = (P(|z_1|^2, |z_2|^2)z_1, Q(|z_1|^2, |z_2|^2)z_2)$$

for some complex-valued functions P and Q. Commutativity with respect to $\xi z = \bar{z}$ implies that P and Q are real-valued. Indeed, f is determined by a mapping $g : \mathbf{R}^2 \to \mathbf{R}^2$ of the form

$$g(x_1, x_2) = (p(x_1^2, x_2^2)x_1, q(x_1^2, x_2^2)x_2)$$

Moreover, g has $\mathbf{D}_2$ symmetry, since it commutes with $(x_1, x_2) \mapsto (x_2, x_1)$ and $x \mapsto -x$. Therefore

$$g(x_1, x_2) = (p(x_1^2, x_2^2)x_1, p(x_2^2, x_1^2)x_2)$$

where p is an odd function. $\qquad\square$

Suppose that f depends on a bifurcation parameter λ. Equilibria appear in two-dimensional group orbits created by the action of $\mathbf{T}_{\mathcal{L}}^2$. These orbits of solutions have unique representatives of the form $(x_1, x_2) \in \mathbf{R}^2$ with $x_1, x_2 \geq 0$. Therefore we find all equilibria by solving the restricted system of equations $f : \mathbf{R}^2 \times \mathbf{R} \to \mathbf{R}^2$.

Up to conjugacy these equilibria are written in terms of equations in $p(x_1^2, x_2^2, \lambda)$ in Table 5.12.

Solution Type	Bifurcation Equations
trivial	$x_1 = x_2 = 0$
rolls	$p(x_1^2, 0, \lambda) = 0;\ x_1 > 0;\ x_2 = 0$
rhombs	$p(x_1^2, x_1^2, \lambda) = 0;\ x_1 = x_2 > 0$
submaximal	$p(x_1^2, x_2^2, \lambda) = 0;\ p(x_2^2, x_1^2, \lambda) = 0;\ 0 < x_1 < x_2$

Table 5.12 Classification up to conjugacy of center manifold equilibria on rhombic lattice.

We can expand f to third order by writing

$$p(x_1^2, x_2^2, \lambda) = \lambda + Ax_1^2 + Bx_2^2 + \cdots$$

where we have scaled the coefficient of λ to be 1. Note that when $A \neq B$, there are no submaximal solutions (that is, solutions whose isotropy subgroup is contained in some larger isotropy subgroup) near the origin.

The stability of solutions is determined by the signs of eigenvalues of the 4×4 Jacobian matrix of $f : \mathbf{C}^2 \times \mathbf{R} \to \mathbf{C}^2$. The torus action forces two of these eigenvalues to be 0; the other two eigenvalues are obtained from the 2×2 Jacobian matrix of $f : \mathbf{R}^2 \times \mathbf{R} \to \mathbf{R}^2$. That Jacobian is (supressing explicit dependence on λ)

$$\begin{bmatrix} p(x_1^2, x_2^2) + 2x_1^2 p_1(x_1^2, x_2^2) & 2x_1 x_2 p_2(x_1^2, x_2^2) \\ 2x_1 x_2 p_2(x_2^2, x_1^2) & p(x_2^2, x_1^2) + 2x_2^2 p_1(x_2^2, x_1^2) \end{bmatrix}$$

where p_j indicates differentiation of p with respect to its jth variable. Using the entries in Table 5.12 we can compute these matrices in the case of trivial, rolls, and rhombs solutions. The results are:

$$p(0, 0, \lambda)I_2 \qquad \begin{bmatrix} 2x_1^2 p_1(x_1^2, 0, \lambda) & 0 \\ 0 & p(0, x_1^2, \lambda) \end{bmatrix} \qquad 2x_1^2 \begin{bmatrix} p_1(x_1^2, x_1^2, \lambda) & p_2(x_1^2, x_1^2, \lambda) \\ p_2(x_1^2, x_1^2, \lambda) & p_1(x_1^2, x_1^2, \lambda) \end{bmatrix}$$

Theorem 5.9 *Assume that $A \neq 0$ and $A \neq \pm B$. Then near the origin the direction of branching of solutions and their stability are as stated in Table 5.13.*

Solution Type	Direction of Branching	Asymptotic Stability
trivial	$x_1 = x_2 = 0$	$\lambda < 0$
rolls	$\lambda = -Ax_1^2$	$A < 0;\ B < A$
rhombs	$\lambda = -(A + B)x_1^2$	$A + B < 0;\ A < B$

Table 5.13 Stability of equilibria in generic bifurcations on rhombic lattice.

The theorem is proved by elementary applications of the Implicit Function Theorem and direct calculations. Note that if the center manifold equation depended on a second parameter ρ, then for special values of ρ these nondegeneracy conditions could fail, forcing changes in stability of the axial solutions and the existence of submaximal equilibria.

Square Lattice. As with rhombic lattices, the scalar and pseudoscalar actions of $\Gamma_{\mathcal{L}}$ on $\mathbf{C}^2$ on the square lattice yield identical equivariants and bifurcation theories.

Lemma 5.10 *For the rhombic lattice, the general equivariant mapping on the shortest wave vector representation has the form (5.39).*

Proof. Let $f : \mathbf{C}^2 \to \mathbf{C}^2$ be a square lattice equivariant. As in the proof of Lemma 5.8, the torus symmetry and $\xi^2(z) = \bar{z}$ symmetry imply that f is determined by a mapping $g : \mathbf{R}^2 \to \mathbf{R}^2$ of the form

$$g(x_1, x_2) = (p(x_1^2, x_2^2)x_1, q(x_1^2, x_2^2)x_2)$$

where p and q are real-valued. The effective action of $\Gamma_{\mathcal{L}}$ on the real subspace $(x_1, x_2) \in \mathbf{R}^2$ is identical with the action in the rhombic case, since κ acts trivially on $\mathbf{R}^2$. $\square$

It follows from Lemma 5.10, and its proof, that the bifurcation theory for the square lattice is identical to the bifurcation theory for the rhombic lattice (in the shortest wave vector case) with the single exception that rhombs solutions are identified in the analysis with squares solutions. Indeed, the solutions on the square lattice have different symmetries, and the discussion in physical space (rather than center manifold space) reflects these differences.

Hexagonal Lattice: Scalar Case. Buzano and Golubitsky [93] made a detailed analysis of bifurcations corresponding to shortest wave vectors on the hexagonal lattice. By examining the general $\Gamma_{\mathcal{L}}$-equivariant mapping they proved that generically the only solutions that bifurcate from the trivial solution are rolls and hexagons. Recall that Theorem 2.14 states that whenever there is a nonzero quadratic equivariant map, then generically all solutions obtained using the Equivariant Branching Lemma are unstable. In the case of the scalar representation on the hexagonal lattice there is precisely one nonzero equivariant quadratic mapping:

$$Q(z_1, z_2, z_3) = (\bar{z}_2 \bar{z}_3, \bar{z}_1 \bar{z}_3, \bar{z}_1 \bar{z}_2)$$

As observed much earlier by Busse [90] when studying transitions in convection systems, there is a way around this pattern selection conundrum. By varying a second parameter ρ, it is possible to arrive at certain critical values of ρ where the equivariant quadratic vanishes at transition. Then, for small perturbations of ρ away from these critical values, the branches of hexagons and rolls can undergo secondary bifurcations and become stable. For Rayleigh-Bénard convection, the determination of such a critical value of ρ is straightforward, since the quadratic term vanishes for an ideal Boussinesq fluid. In that case, ρ may be interpreted as a deviation of a fluid from ideal Boussinesq. In general, it may be quite difficult to find such critical values of ρ. Depending on cubic and higher order terms, many bifurcation problems are possible at critical ρ values. Buzano and Golubitsky [93] used singularity theory methods to classify all of the generic bifurcations at such critical values and their universal unfoldings. They also showed precisely which cubic, fourth, and fifth order terms need to be calculated to prove that higher order terms in the expansion of the center manifold vector field f do not matter — in the sense that the bifurcation diagrams and stability of equilibria are qualitatively the same regardless of the exact values of coefficients in higher order terms.

The secondary bifurcations from axial subgroups produce solutions with submaximal symmetry. In this case, triangles and rectangles are produced. The triangles solutions can be stable in certain cases — depending only on the sign of a fifth order term. Triangle solution eigenfunctions have the form (z_1, z_1, z_1) with $z_1 \notin \mathbf{R}$. Rectangles solutions have the form (x, y, y) with $x, y \in \mathbf{R}, x \neq y, y \neq 0$. Planforms associated to these solution types are shown in Figure 5.20 with $z_1 = 1 - i$ (triangles) and $x = 2; y = 1$ (rectangles). The planforms will change with different choices of z_1 and x, y.

Hexagonal Lattice: Pseudoscalar Case. We have shown that there are four axial subgroups: hexagons, rolls, rectangles, and triangles. The complete discussion of stability and secondary bifurcations of axial solutions differs drastically from any of the other cases. We can show, for example, that variation of one additional parameter ρ can lead to a variety of secondary time-periodic solutions. Here, we just consider the consequences of the fact that there are no nonzero equivariant quadratics in the pseudoscalar hexagonal lattice (though there is an equivariant quartic).

Recall that there are two bifurcation variants in Bénard convection, depending on boundary conditions. When the boundary conditions on the top and bottom plates are identical, then an extra midplane reflection symmetry is introduced into the model. The midplane reflection happens to act on center subspaces as $z \mapsto -z$, so the midplane reflection forces equivariant quadratics to be zero — but in a way that is different from the assumption of an ideal Boussinesq fluid.

Golubitsky *et al.* [238] studied the scalar bifurcation on the hexagonal lattice with the midplane reflection in the shortest wave vector case. They found that four solutions exist (hexagons, triangles, rectangles, and rolls). Moreover, rectangles are never stable near bifurcation; rolls can be determined to be stable at third order; and at third order either triangles or hexagons can be stable — exactly which solution is stable depends on the sign of a fifth order coefficient.

Since there are no nonzero quadratics in the hexagonal lattice odd case, the cubic truncation has $z \mapsto -z$ as a symmetry; that is, the cubic truncation is identical in form to the cubic truncation in the even hexagonal lattice with the midplane reflection. Therefore, it can be determined at third order whether rolls are stable and whether either hexagons or triangles are stable. For reasons that differ slightly from the midplane relection case, rectangles are generically not stable near bifurcation, and hexagons and triangles may be stable, but that determination requires the computation of fifth order terms. As mentioned previously, an important difference between the even and odd hexagonal lattice calculations concerns the type of secondary states obtained by varying ρ.

Chapter 6

Bifurcation From Group Orbits

In this chapter we address two issues. The first concerns the relation between bifurcating states in phase space and experimental observations in physical space. The second issue is that of bifurcation from group orbits. In equivariant dynamics, all states occur as entire group orbits, and especially when the group has a nontrivial continuous part — that is, when it is not finite — the fact that a group orbit is a manifold has a substantial effect on possible bifurcations. The situation becomes even more complicated — and interesting — when the symmetry group is the Euclidean group, which is not compact.

The Couette-Taylor Experiment. We motivate the above ideas in terms of a classic pattern-forming experiment, the Couette-Taylor system. The standard model for the Couette-Taylor system is well equipped with continuous symmetries, but its group-theoretic aspects are relatively simple and easy to analyze. The experimental apparatus consists of two coaxial cylinders, with fluid between them. Each cylinder can rotate. As the rotation speeds of the cylinders change, many different patterned flow states appear. Figure 6.1 shows an experimental classification of the observed states in terms of the angular velocities of the two cylinders: here R_o is the Reynolds number of the outer cylinder and R_i is the Reynolds number of the inner cylinder. Figure 6.2 illustrates four common flow patterns. Reading from right to left, these are Couette flow, Taylor vortices, spiral vortices, and wavy vortices.

Couette flow is observed when R_i is small: it is a homogeneous (that is, fully symmetric) state of the system. In the classic experiments of Taylor, which we describe in Section 6.1, the inner cylinder is rotated at increasing angular velocities while the outer one is held fixed . The first bifurcation is to Taylor vortices, which resemble a stack of donuts. In more recent experiments, it was found that if instead the inner cylinder is held fixed and the outer one is rotated in the opposite direction, then the primary bifurcation is to spiral vortices, which form helical traveling waves. At other combinations of speeds, many different patterns occur — for example, the first bifurcation from Taylor vortices in the classical experiment leads to wavy vortices, which are rotating waves in which the boundaries between vortices become wavy, but rotate as if the pattern is rigid. There are also time-periodic states, turbulent (chaotic) states, and even patterened turbulent states.

Early work on the Couette-Taylor system concentrated on solving the Navier-Stokes equations in cyclindrical geometries, either by perturbation theory or by numerical simulation. It turns out, however, that much of the phenomenology of the Couette-Taylor system can be understood as a natural consequence of the symmetries of the standard model, at least if a few crucial pieces of information from the Navier-Stokes equations are taken into account. In Section 6.1 we show that the primary bifurcations of this system — the branches that emanate from Couette flow — can be organized by symmetry considerations. Very little information is

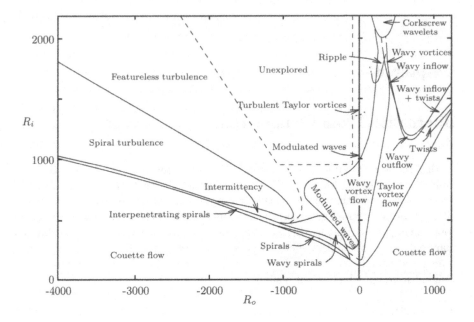

Figure 6.1: Flow states in the Couette-Taylor experiment. Taken from Andereck *et al.* [5]

Figure 6.2: Couette-Taylor Experiment. From left to right: Couette flow, Taylor vortices, spiral vortices, and wavy vortices. Pictures courtesy of H.L. Swinney.

needed from the Navier-Stokes equations to achieve this: the range of patterns expected in primary bifurcations is model-independent. (The main role of the Navier-Stokes equations, in this approach, is to determine the appropriate 'modes', that is, representations of the symmetry group.)

However, secondary bifurcations from these primary branches are much less straightforward. It turns out that they too can be analyzed using symmetry, but only when we have developed some new techniques. The extra ingredient involved is the presence of continuous symmetries, creating continuous group orbits of states.

Suppose that Γ is the overall symmetry group of the Couette-Taylor system, and that a primary branch occurs with isotropy subgroup Σ. Then the vector field along this branch is $N(\Sigma)/\Sigma$-equivariant. Can we conclude that generic secondary bifurcations from this branch behave like generic $N(\Sigma)/\Sigma$-equivariant bifurcations? The answer is 'no', in general. One reason is that, associated to any solution $x(t)$ with isotropy subgroup Σ, there are other solutions $\gamma x(t)$ for any $\gamma \in \Gamma$. If $\dim \Sigma < \dim \Gamma$, then the group orbit $\Gamma x(t)$ is a manifold of dimension ≥ 1. This means that $x(t)$ is not isolated — which is nongeneric in $N(\Sigma)/\Sigma$-equivariant bifurcation problems.

Phase Space and Physical Space. The Couette-Taylor system raises a number of other important issues related to the modeling of symmetric systems, and we discuss these in general terms before examining the Couette-Taylor system in detail. We focus on three main questions:

- What are the correct symmetries for models of a particular application?
- How do symmetries of solutions manifest themselves in physical space?
- What kinds of symmetric states do we expect to observe near transitions?

We now discuss each question in turn.

What are the correct symmetries? In models arising in the physical sciences, exact symmetry is often a justifiable assumption, whereas in biological models, symmetries are often approximate. For example in the speciation model of Section 1.5, organisms in the same species are not, in fact, identical. Nonetheless they are 'nearly' the same, and this approximate resemblance can be idealized as a symmetry. Generally speaking, an approximately symmetric system will behave in a very similar manner to an exactly symmetric system, and it will *not* behave like a typical asymmetric system — an issue we discussed briefly in Sections 1.5 and 2.7. In our discussion of the Couette-Taylor experiment in Section 6.1 we shall see that even in the physical sciences appropriate model symmetries are sometimes only approximate symmetries. Thus, each symmetry that is found in a mathematical model must be thought of as a *modeling assumption*.

How do symmetries manifest themselves in physical space? Answers to this question involve at least two separate issues. First, the answer depends on the dynamics of the solution: Is the solution steady, time-periodic, quasiperiodic, or chaotic? For equilibria the answer is relatively straightforward — any symmetry that fixes an equilibrium in phase space fixes the corresponding state in physical space. As we have seen, the answer is somewhat subtler for time-periodic states. Even so, symmetry of a state in physical space is often seen through special observable features. For example, in the phase space of speciation models, the division of a species into two subspecies is detected by a change in symmetry ($\mathbf{S}_N$ to $\mathbf{S}_k \times \mathbf{S}_{N-k}$) of an equilibrium, whereas in physical space we observe this change of symmetry because the population divides into two distinguishable clumps. Individuals in the same clump are phenotypically roughly the same, whereas individuals in different clumps have easily distinguishable phenotypes.

In phase space, a symmetry that fixes the trajectory of a periodic solution corresponds, because of uniqueness of solutions to initial value problems, to a spatio-temporal symmetry in physical space. We saw in our discussion of Hopf bifurcation with $\mathbf{O}(2)$ symmetry (Example 4.12 in Section 4.2) that there are states whose trajectories are fixed by all rotations (the trajectories are actual circles in phase space) and these states correspond to rotating waves in physical space — such as spiral wave solutions to reaction-diffusion equations, Section 4.6.

We can give a more detailed answer for symmetric PDE systems, where shape is interpreted as pattern formation. For equilibria and time-periodic states of PDEs obtained by bifurcation, the shape of the bifurcating solutions is largely determined by the geometry of the linearized eigenfunctions. The Liapunov-Schmidt procedure implies that near bifurcation the actual solutions of the nonlinear equations must be close (in the sense of the appropriate Banach space of functions) to the corresponding linearized eigenfunctions. We find the linearized eigenfunctions by solving linear PDEs, and we use the symmetry of the bifurcating solution to select the relevant combination of linearized eigenfunctions.

What kinds of symmetric states do we expect near transitions? This question is the most studied of these three questions, as much of equivariant bifurcation theory and the theory of spontaneous symmetry-breaking have been developed to answer this question — in specific cases. For example, $O(2)$ symmetry-breaking Hopf bifurcation (Example 4.12) generically leads to two types of periodic states — rotating waves and standing waves — which are easily distinguished by their symmetries. More generally, suppose that a particular model has a compact Lie group Γ as its group of symmetries. Then Theorems 1.27 and 4.5 state that there is

(a) a steady-state bifurcation corresponding to each absolutely irreducible representation of Γ, and

(b) a Hopf bifurcation corresponding to each irreducible representation of Γ.

These bifurcations are candidates for typical transitions from Γ-invariant equilibria. Taken together, answers to our third question lead to model-independent results: once we know the appropriate symmetry group Γ of a model (but not necessarily the model itself) and how Γ acts on physical space, we can guess (predict) a list of states that are likely to be observed in the class of models with Γ symmetry. This list does not depend on the specific model: just on its symmetry structure. This idea was the basis of our study of animal gaits in Section 3.1, which led to the prediction of the jump gait in Table 3.4.

There are several caveats to be made when using model-independent arguments to study a particular application: we mention two here. First, we must know the symmetry group of the model exactly. In Chapter 5 we discussed an example where the addition of just one extra symmetry, the midplane reflection in Bénard convection, changes the expected transition structure quite substantially. Second, we must assume that generically there are no restrictions on transitions in the model other than those imposed by the known symmetries. This assumption can be tricky to verify because of the previously discussed issue of approximate symmetries and the issue of hidden symmetries, which will be discussed in Chapter 7.

Bifurcation from Relative Equilibria. An especially important aspect of the third question above arises when we consider typical transitions from equilibria whose symmetry group is smaller than Γ, and what kinds of transition can be expected when Γ is noncompact? Because the symmetry group of the equilibrium is smaller that Γ, its group orbit is nontrivial. If Γ has a continuous subgroup that is not contained in the isotropy subgroup of the equilibrium, then the group orbit of the equilibrium is a manifold of nonzero dimension. Thus the equilibrium is not isolated. This issue is much simpler to understand in phase space than it is in physical space, because in physical space it is related to *potential* states of the system near to those that actually occur, not to the observed states alone.

More generally, we can study bifurcations from 'relative equilibria': trajectories that lie inside a single group orbit (and hence become equilibria when projected into orbit space). New phenomena here include the possibility of continuous 'drifting'.

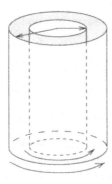

Figure 6.3: Couette-Taylor apparatus.

Many of the most interesting examples of bifurcation from a relative equilibrium arise in PDEs with Euclidean symmmetry, where noncompactness of the symmetry group rasies new technical issues, and also — intriguingly — leads to genuinely different phenomena compared to the compact case.

The general study of bifurcations from group orbits of equilibria (or *relative equilibria*) starts with the work of Field [184] and Krupa [329]. (Relative equilibria have been studied much more extensively in Hamiltonian systems, see Chapter 10.) Bifurcations from the quintessential noncompact group — the Euclidean group $\mathbf{E}(2)$ — have been studied for many years. Many of these studies consider only bifurcation to states that are doubly periodic with respect to some planar lattice, as discussed in Chapter 5. However, states like planar spiral waves are not doubly periodic, so must be derived by different techniques, and these states have also been studied for many years. However, a general bifurcation theorem for the transition from a homogeneous steady-state to a spiral wave in planar reaction-diffusion systems has been accomplished only recently, see Scheel [466]. Descriptions of Hopf bifurcations from spiral waves to quasiperiodic motions rely on using the full Euclidean group, as we discuss below.

In this chapter we explore the above issues in several physical contexts:

- Pattern formation in the Couette-Taylor system: fluid flow between coaxial rotating cylinders.

- The effects of symmetry on pattern in quasiperiodic motion obtained by Hopf bifurcation from a rotating wave.

- Meandering of spiral waves.

6.1 The Couette-Taylor Experiment

As stated in the introduction to this chapter, the Couette-Taylor apparatus consists of a fluid contained between two coaxial independently rotating cylinders, as in Figure 6.1. In 1986 Andereck, Liu, and Swinney [5] published a tabulation of states observed in this experiment. Their results, shown above in Figure 6.1, are plotted in the parameter plane given by the speeds R_i, R_o of the inner and outer cylinders respectively. Some of the observed states are reproduced in Figure 6.2 above.

Studies of the Couette-Taylor experiment show that some of the transitions observed in this experiment are best understood using symmetry. In this section we survey four of these transitions:

(1) Couette flow to Taylor vortices
(2) Couette flow to spiral vortices [103]
(3) Couette flow simultaneously to spiral vortices and Taylor vortices [228, 352, 217]
(4) Transitions from Taylor vortices to time-periodic states [298]

A Short History. Donnelly [167] has written a history of the Couette-Taylor experiment — from which we excerpt one or two points. The Navier-Stokes equations, which are now thought to be an excellent model of fluid flow, were derived by Navier in 1821 and then rederived in more rigorous fashion by Stokes [496] in 1856. However, it was not until 1923 that G.I. Taylor [501] gave the first experimental confirmation of these equations, over a hundred years after they were first published by Navier.

What makes this confirmation particularly interesting is the method Taylor used to compare theory and experiment. He observed that when the outer cylinder is held fixed ($R_o = 0$) and the speed of the inner cylinder R_i is increased, there is a critical value of R_i at which featureless Couette flow loses stability. A new state, now called Taylor vortices, appears, which is very nearly spatially periodic in the axial direction. Taylor computed the instability of Couette flow to Taylor vortices using the Navier-Stokes equations, finding that theory and experiment agreed on the transition value to within experimental error.

Thus the Navier-Stokes equations were confirmed by determining both experimentally and theoretically the value at which a symmetry-breaking bifurcation occurs. In his theory, Taylor assumed periodic boundary conditions in the axial direction — presumably motivated by the experimental results. From a modeling perspective this assumption is equivalent to presuming $SO(2)$ translation symmetry along the common axes of the cylinders, that is, translations modulo the period. This symmetry is only approximate — but it is a sufficiently good approximation to be assumed exact in many studies of the Couette-Taylor system, including the ones discussed here.

Symmetries in Model Equations. The physical symmetries in this experiment are the $SO(2)$ azimuthal symmetry and the up-down reflection symmetry κ. In addition, as just discussed, there is an approximate $SO(2)$ axial translation symmetry, which we assume is exact. Thus the complete symmetry group of the model equations is $\Gamma = O(2) \times SO(2)$, where $O(2)$ is generated by axial translations $SO(2)$ and κ. Given this group and its action on physical space, can we anticipate features of an equilibrium state or a time-periodic state that bifurcates from Couette flow? The answer is 'yes', and the discussion yields another example of model-independent reasoning. We will assume that the new states break axial symmetry.

Transition from Couette flow to Taylor Vortices. We can assume by genericity, motivated by Theorem 1.27, that $O(2) \times SO(2)$ acts absolutely irreducibly on the kernel of the linearization at a point of bifurcation from a Γ-invariant equilibrium. A theorem from representation theorem states that any irreducible representation of a direct product of two groups is the tensor product of irreducible representations of each group individually (see for example Dionne et al. [163]). Moreover, the representation of the cross product is absolutely irreducible if and only if the two component representations are each absolutely irreducible (Dionne *et al.* [163]). The irreducible representations of the axial symmetry group $O(2)$ are one- or two-dimensional. The subgroup $SO(2) \subseteq O(2)$ is in the kernel of the one-dimensional actions, which implies that in these cases all solutions bifurcating from Couette flow will be invariant under axial translations. But we have assumed that bifurcating solutions break axial symmetry; therefore the action of $O(2)$ must be on $C = R^2$. The irreducible representations of the azimuthal $SO(2)$ symmetry group are one- or two-dimensional — but the action on $C = R^2$ is nonabsolutely irreducible. Therefore, the action of Γ on the kernel of the linearization, at the point of transition from Couette flow, must be $C \otimes R \cong C$.

In this representation the azimuthal subgroup $SO(2)$ acts trivially — so the new bifurcating state must be constant on circles around the apparatus — which is the case for Taylor vortices. Dividing out by the kernel of the representation of Γ on C yields the standard action of $O(2)$ on C. Steady-state bifurcations lead to a group orbit of equilibria, each having isotropy subgroup Z_2. Up to conjugacy we may assume that the new solution has κ symmetry (in addition to the azimuthal $SO(2)$ symmetry). The Taylor vortex state does have these symmetries. Indeed, κ symmetry forces the Taylor vortex flow to be horizontal along the reflection plane of κ and hence to look like stacks of doughnuts.

The comments made to this point are model-independent — they depend only on the symmetries of the system and not on the exact description of the Navier-Stokes equations. However, there are two points concerning the transition to Taylor vortices that do depend specifically on the Navier-Stokes equations. First, as discussed previously, the exact values where transitions occur depend on the model equations. Second, the exact form of Taylor vortices is found by inspecting the eigenfunctions of the linearized Navier-Stokes equations at bifurcation: see Figure 6.4.

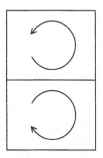

Figure 6.4: Schematic picture of pair of vortices in Taylor vortex flow.

Transition from Couette Flow to Spiral Vortices. A similar analysis can be performed for Hopf bifurcation in the presence of $O(2) \times SO(2)$ symmetry. Since we are looking for states that break axial translation symmetry, we can assume that $O(2)$ acts irreducibly on $R^2 = C$. Suppose that we also look for time-periodic states that break azimuthal rotational symmetry; then we can assume that $SO(2)$ acts irreducibly on $R^2 = C$ and that $O(2) \times SO(2)$ acts irreducibly on $C^2 = C \oplus C$. This action is not absolutely irreducible, Therefore generically we can assume that the center subspace is C^2 by Theorem 4.5. To simplify the discussion, we assume that no element in Γ acts trivially on C^2. Finally, the phase shift S^1 action on C^2

commutes with the action of Γ, as does the action of the azimuthal rotations $\mathbf{SO}(2)$. It is not hard to show that the only linear maps that commute with Γ on $\mathbf{C}^2$ are real multiples of $\mathbf{SO}(2)$, and it follows that $\mathbf{SO}(2)$ and $\mathbf{S}^1$ act either identically on $\mathbf{C}^2$ or as inverses of each other. Thus the analysis of this bifurcation is the same as that of the symmetry-breaking Hopf bifurcation of $\mathbf{O}(2)$. In addition, however, all states produced by this bifurcation will have $(0, \theta, \theta)$ symmetry inside $\mathbf{O}(2) \times \mathbf{SO}(2) \times \mathbf{S}^1$; that is, all states are rotating waves with respect to azimuthal rotation.

Thus Hopf bifurcation with $\mathbf{O}(2) \times \mathbf{SO}(2)$ symmetry acting on $\mathbf{C}^2$ produces the same states that Hopf bifurcation with $\mathbf{O}(2)$ symmetry acting on $\mathbf{C}^2$ produces: a rotating wave, with $\mathbf{SO}(2)$ axial symmetry as the spatial part of the spatio-temporal symmetries of the rotating wave, and a standing wave with $\mathbf{Z}_2(\kappa)$ symmetry (see Chossat and Iooss [105]). A simple group-theoretic argument shows that the rotating wave must be fixed by $(\theta, \theta, 0) \in \mathbf{O}(2) \times \mathbf{SO}(2) \times \mathbf{S}^1$ for all time, that is, it lies on a cylinder. The rotating wave can be identified geometrically with spirals. See Figure 6.5. The standing wave states are called *ribbons* and have $\mathbf{Z}_2(\kappa)$ symmetry (Chossat and Iooss [105]); they will also rotate in the azimuthal direction. Thus, using model-independent reasoning, we are led to expect that time-periodic states found by a smooth transition from Couette flow will have the features of either spiral vortices or ribbons.

As noted previously (see Theorem 4.19), generically spirals and ribbons cannot both be asymptotically stable at bifurcation. Chossat, Demay, and Iooss [103] showed that ribbons can be stable at certain parameter values. Tagg and Swinney [499] confirmed the existence of ribbons experimentally.

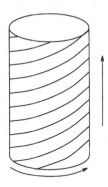

Figure 6.5: Spiral vortices in Couette-Taylor system.

Simultaneous Transition from Couette Flow to Taylor Vortices and Spiral Vortices: A Mode Interaction. The experiments of Andereck *et al.* [5] indicate the existence of a steady-state/Hopf mode interaction when the cylinders are counterrotating. This point was established numerically for the Navier-Stokes equations by Kreuger *et al.* [327], and is reproduced in [217, 352]. Based on these observations, it is possible to give model-independent descriptions of all secondary states that appear in the unfolding of this mode interaction. The results of [228, 237] are sketched below.

The discussion of the transitions from Couette flow to Taylor vortices and from Couette flow to spiral vortices imply that the action of $O(2) \times S^1$ on C^3 is:

$$\theta(z_0, z_1, z_2) = (z_0, e^{i\theta}z_1, e^{i\theta}z_2) \quad \theta \in S^1$$
$$\kappa(z_0, z_1, z_2) = (\bar{z}_0, z_2, z_1)$$
$$\phi(z_0, z_1, z_2) = (e^{i\phi}z_0, e^{i\phi}z_1, e^{-i\phi}z_2) \quad \phi \in SO(2) \subseteq O(2)$$

and the action of the azimuthal $SO(2)$ symmetry can be identified with the action of the phase shift S^1 symmetry. The lattice of isotropy subgroups for this Hopf/steady-state mode interaction is given in Figure 6.6, and the names of the corresponding flows are given in Figure 6.7.

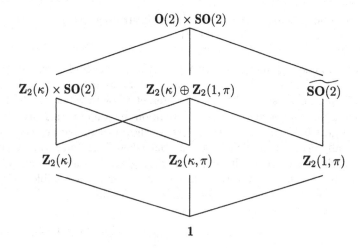

Figure 6.6: Lattice of isotropy subgroups in 6-dimensional $O(2)$ symmetry-breaking Hopf/steady-state mode interaction.

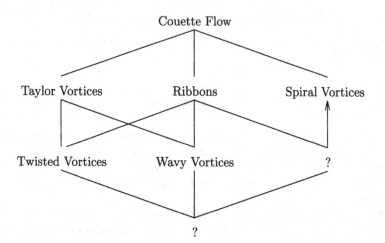

Figure 6.7: Flows in the Couette-Taylor system that correspond to isotropy subgroups in Figure 6.6.

If we flatten the cylinder, the flow patterns can be represented as in Figure 6.8. Taylor vortices are invariant under flips and rotations, that is, $\mathbf{Z}_2(\kappa) \times \mathbf{SO}(2)$. Wavy vortices are invariant under a flip followed by a half-period spatial axial rotation π, that is, $\mathbf{Z}_2(\kappa, \pi)$. The twisted vortices are symmetric only with respect to the flip $\mathbf{Z}_2(\kappa)$.

Figure 6.8: Schematic pictures of isoflow lines illustrating the symmetries of Taylor vortices, wavy vortices, and twisted vortices.

Using Liapunov-Schmidt reduction from the Navier-Stokes equations, Golubitsky and Langford [217] produced a circulant bifurcation diagram in the two-dimensional parameter space of this mode interaction by traveling in a circle around the codimension two degeneracy. The bifurcation diagram for this circular path is shown in Figure 6.9. The stable wavy vortices shown there had not been observed experimentally near this mode interaction, but when Tagg and Swinney [499] performed experiments in the predicted parameter region, the wavy vortices were seen. The existence of stable wavy vortices in an experimentally accessible region of parameter space is an example of a model-dependent result.

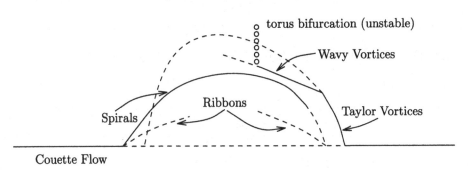

Figure 6.9: Circulant bifurcation diagram indicating stable wavy vortices [217].

When wavy vortices lose stability in the experiment, the system goes to a state that oscillates between two symmetrically related states in a heteroclinic-like way. See Figure 6.10. Knobloch and Dangelmayr [139] have shown that by slightly breaking the $\mathbf{O}(2)$ axial symmetry in $\mathbf{O}(2)$ Hopf bifurcation, states resembling this can occur.

Similar analyses were carried out by Chossat, Demay, and Iooss [103] for Hopf-Hopf mode interactions, where it was shown that interpenetrating spiral vortices result. The book by Chossat and Iooss [106] describes these and many other uses of symmetry in the analysis of the Couette-Taylor experiment. We discuss transitions from Taylor vortices to time-periodic states at the end of the next section.

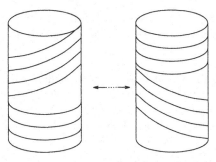

Figure 6.10: Schematic diagram of states occurring in the experiment when wavy vortices lose stability.

6.2 Bifurcations From Group Orbits of Equilibria

There are two possible ways to study secondary bifurcations from a group-invariant equilibrium. The first, as just mentioned in the context of the Couette-Taylor experiment, uses unfoldings of mode interactions. This approach has the advantage that it uses local methods near an (often) explicitly computable group-invariant equilibrium. The second approach is to consider bifurcations from equilibria (or time-periodic solutions) with less than full symmetry. That second approach is the subject of this section; references include Field [184], Krupa [329] and Vander-bauwhede *et al.* [510].

Relative Equilibria. Let $f : \mathbf{R}^n \times \mathbf{R} \to \mathbf{R}^n$ be a Γ-equivariant vector field, where $\Gamma \subseteq \mathbf{O}(n)$, and let Γx_0 denote the group orbit through x_0. By equivariance, $f(x_0) = 0$ implies that $\Gamma x_0 = 0$: that is, equilibria always come in group orbits. We now study bifurcations from a continuous group orbit of equilibria.

Assume that $\dim \Gamma x_0 \geq 1$ and let Σ_{x_0} be the isotropy subgroup of x_0. The group orbit $\Gamma x_0 \subseteq \mathbf{R}^n$ is always a smooth manifold, as pictured in Figure 6.11, of dimension $\dim \Gamma - \dim \Sigma_{x_0}$. More generally, a group orbit can be flow-invariant rather than just consisting of equilibria; in that case the group orbit is called a *relative equilibrium*.

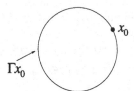

Figure 6.11: Group orbit of Γ through x_0.

In this subsection we show that flows on relative equilibria generically fill out k-dimensional tori, where the number k is determined uniquely by the isotropy subgroup Σ_{x_0}: see Theorem 6.4. It follows that steady-state bifurcation from group orbits of equilibria can lead to relative equilibria rather than just new equilibria. In order to motivate the general theory of relative equilibria, we consider two examples of continuous group orbits in which the phase space is $\mathbf{R}^2$ and the groups are $\mathbf{SO}(2)$ and $\mathbf{O}(2)$ respectively. As we will see, these two actions lead to very different phenomena.

Example 6.1 Let $\Gamma = \mathbf{O}(2)$ with its standard action on $\mathbf{R}^2 \equiv \mathbf{C}$. Every equivariant vector field is of the form $g(|z|^2)z$. In polar coordinates (r, θ) we have

$$\dot{r} = rg(r^2) \qquad \dot{\theta} = 0$$

reflecting the fact that the vector field is purely radial. This is a consequence of equivariance under the reflections in $\mathbf{O}(2)$. Since θ is constant, we can restrict attention to the r-equation only. Relative equilibria are given by the zeros of g. In this case, a relative equilibrium is an $\mathbf{O}(2)$ group orbit of equilibria and equilibria other than the origin occur in circles. $\qquad \diamond$

Example 6.2 The $\mathbf{SO}(2)$ case is different, and in an interesting way. Now we let $\Gamma = \mathbf{SO}(2)$ with its standard action on $\mathbf{R}^2 \equiv \mathbf{C}$. Every equivariant vector field is of the form $g(|z|^2)z + h(|z|^2)iz$. In polar coordinates (r, θ) we have

$$\dot{r} = rg(r^2) \qquad \dot{\theta} = h(r^2)$$

and the vector field has tangential as well as radial components. Again, relative equilibria other than the origin are given by the zeros of g, and relative equilibria other than the origin occur in circles ($\mathbf{SO}(2)$ group orbits).

Generically, h is nonzero at a zero r_0^2 of g. It follows that the flow around the circle $r = r_0$ is given by

$$\dot{\theta} = h(r_0^2)$$

and the solution is a time-periodic rotating wave (time evolution is the same as spatial rotation). In this case, relative equilibria are generically time-periodic — not a group orbit of equilibria. $\qquad \diamond$

The following observation will prove useful, here and elsewhere:

Lemma 6.3 *Suppose that $Y \subseteq \mathbf{R}^n$, and all elements $y \in Y$ have the same isotropy subgroup, that is, $\Sigma_y = \Sigma$ for some $\Sigma \subseteq \Gamma$. Let $x(t)$ be a solution of a Γ-equivariant ODE, and assume that $x(t) \in \Gamma Y$ for all t and $x(0) \in Y$. Then $x(t) \in N(\Sigma)Y$ for all t.*

Proof. Since $x(t) \in \Gamma Y$ for all t we have

$$x(t) = \gamma_t y_t$$

for all t, where $\gamma_t \in \Gamma$ and $y_t \in Y$. We must show that $\gamma_t \in N(\Sigma)$. Since $x(0) \in \text{Fix}(\Sigma)$, Theorem 1.17 implies that $x(t) \in \text{Fix}(\Sigma)$ for all t. Therefore, for all $\sigma \in \Sigma$,

$$\gamma_t y_t = x(t) = \sigma x(t) = \sigma \gamma_t y_t$$

So $(\gamma_t^{-1}\sigma\gamma_t)y_t = y_t$, implying that $\gamma_t^{-1}\sigma\gamma_t \in \Sigma_{y_t} = \Sigma$. Therefore $\gamma_t \in N(\Sigma)$ and $x(t) \in N(\Sigma)Y$, as required.

$\qquad\qquad\qquad\qquad\qquad\qquad\qquad\qquad\qquad\qquad\qquad\qquad\qquad\qquad\qquad \square$

We next discuss the dynamics on a relative equilibrium Γx_0 of the Γ-equivariant ODE

$$\frac{dx}{dt} = f(x)$$

where $x(0) = x_0$ and the isotropy subgroup of x_0 is Σ_{x_0}. This solution lies in Γx_0 for all time t, whence $x(t) = \gamma_t x_0$ for a curve $\gamma_t \in \Gamma$. In fact, γ_t must lie in the normalizer $N(\Sigma_{x_0})$ of Σ_{x_0} by Lemma 6.3.

The choice of γ_t is not unique, since

$$\gamma_t x_0 = \gamma_t \sigma x_0$$

for any choice of $\sigma \in \Sigma_{x_0}$. However, γ_t does project onto a unique curve in $N(\Sigma_{x_0})/\Sigma_{x_0}$. Let $\hat{\gamma}$ denote the projection of $\gamma \in N(\Sigma_{x_0})$ in $N(\Sigma_{x_0})/\Sigma_{x_0}$.

Finally, uniqueness of solutions implies that

$$\hat{\gamma}_{t_1+t_2} = \hat{\gamma}_{t_2}\hat{\gamma}_{t_1}$$

Thus, $\hat{\gamma}_t$ is a one-parameter group; the closed subgroup $G = \overline{\{\hat{\gamma}_t\}}$ is abelian and connected; and, since $G \subseteq N(\Sigma_{x_0})/\Sigma_{x_0}$, G is also compact. Hence, G must be a torus of some dimension — say k. It follows that $\overline{\{x(t)\}} \subseteq \mathbf{R}^n$ must be $\mathbf{T}^k\{x_0\}$ for some k-torus $\mathbf{T}^k$. Hence solutions that are relative equilibria are quasiperiodic with k frequencies. Moreover, symmetry insures that every trajectory on a given relative equilibrium leads to a k-torus with the same k.

The next theorem shows that the number k has a typical value in the world of all possible Γ-invariant vector fields. Recall that the *rank* of a Lie group is the maximal dimension of any torus subgroup contained in that group.

Theorem 6.4 *Let Γx_0 be a relative equilibrium and let Σ_{x_0} be the isotropy subgroup of x_0. Then relative equilibria are quasiperiodic motions with k frequencies where generically $k = \text{rank}(N(\Sigma_{x_0})/\Sigma_{x_0})$.*

Proof. The argument above shows that k cannot exceed $\text{rank}(N(\Sigma_{x_0})/\Sigma_{x_0})$. Generically, we expect there to be no further restrictions on the size of k (because generic tori in compact Lie groups have the largest possible dimension). Detailed proofs may be found in Field [184] or Krupa [329].   □

Example 6.5 Consider $\Gamma = \mathbf{SO}(2), \Sigma_0 = 1$. Then

$$N(\Sigma_{x_0})/\Sigma_{x_0} = N(1)/1 = \mathbf{SO}(2)$$

and $\text{rank}(\mathbf{SO}(2)) = 1$. Therefore, generically, relative equilibria are time-periodic solutions, as in Example 6.2.

Example 6.6 Consider $\Gamma = \mathbf{O}(2), \Sigma_0 = \mathbf{Z}_2$. Then

$$N(\Sigma_{x_0})/\Sigma_{x_0} = N(\mathbf{Z}_2)/\mathbf{Z}_2 = \mathbf{Z}_2$$

and $\text{rank}(\mathbf{Z}_2) = 0$. Therefore, generically, relative equilibria are groups orbits of equilibria, as in Example 6.1. Moreover, if bifurcation from an equilibrium with $\Sigma_0 = \mathbf{Z}_2$ breaks the $\mathbf{Z}_2$ symmetry, then we can expect the bifurcating solution to be a relative equilibrium. These kinds of bifurcation are discussed in more detail in the next two subsections.   ◇

Tubular Neighborhoods of Relative Equilibria. We begin the study of bifurcation from relative equilibria, by describing the flow of $\dot{x} = f(x)$ in a neighborhood of a relative equilibrium. Bifurcations to relative equilibria pose no serious problems, because they can be reduced to the behavior of a suitably defined normal vector field. Hopf bifurcations to relative periodic states are considerably more difficult to understand, and are treated later.

The first step is to introduce a convenient coordinate system near the initial relative equilibrium. To achieve this we construct a tubular neighborhood of this group orbit in $\mathbf{R}^n$ using the normal bundle. Define

$$Y_x = [T_x\Gamma x]^{\perp},$$

that is, Y_x is the orthogonal complement to the tangent space of the group orbit through x at x. The tangent spaces T_x and the normal spaces Y_x are pictured in Figure 6.12.

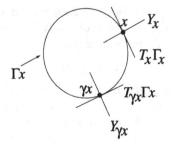

Figure 6.12: Tangent space and normal space data on a group orbit.

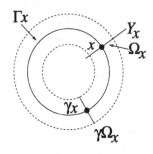

Figure 6.13: A tubular neighborhood.

Lemma 6.7 *Suppose $\gamma \in \Gamma$. Then*

$$T_{\gamma x}\Gamma = \gamma \, T_x\Gamma x \quad , \quad \gamma Y_x = Y_{\gamma x}$$

Proof. Let v be a vector in $T_x\Gamma x$. Then there is a curve γ_t in Γ such that $\gamma_0 = 1$ so that

$$v = \frac{d}{dt}\gamma_t x\Big|_{t=0} = \dot\gamma_0 x$$

Set $\hat\gamma_t = \gamma\gamma_t\gamma^{-1} \in \Gamma$ and note that $\hat\gamma_0 = 1$. Then

$$\frac{d}{dt}\hat\gamma_t(\gamma x)\Big|_{t=0} = \gamma\dot\gamma_0 x = \gamma v$$

so $\gamma v \in T_{\gamma x}\Gamma x$. □

Corollary 6.8 *For all $\sigma \in \Sigma_x$, $\sigma(Y_x) = Y_x$. Thus there exists a Γ-invariant normal bundle $\mathcal{U}$ to Γx. In particular, if $y \in \Omega_x \equiv \mathcal{U} \cap Y_x$ and $\gamma y = y$, then $\gamma \in \Sigma_x$.*

Proof. Either $\gamma\Omega_x = \Omega_x$ or $\gamma\Omega_x \cap \Omega_x = \emptyset$. If $\gamma y = y$, then $\gamma\Omega_x = \Omega_x$, so $\gamma x = x$. Thus $\gamma \in \Sigma_x$. □

The geometry underlying this proof is illustrated in Figure 6.13.

The main result of this subsection is a canonical decomposition of f into normal and tangential components:

Theorem 6.9 *On the tubular neighborhood $\mathcal{U}$ we can write*

(a) *$f = f_T + f_N$ where f_T is tangent to group orbits and f_N lies in the normal space. More precisely, let $y \in \Omega_{\gamma x}$. Then $f_T(y) \in T_y\Gamma y$ and $f_N(y) \in Y_{\gamma x}$.*

(b) *f_T and f_N are smooth.*

(c) *f_T and f_N are Γ-equivariant.*

Proof. See Krupa [329]. □

Note that zeros of f_N correspond to flow-invariant group orbits, that is, to relative equilibria. The behavior of f_T on such orbits adds extra detail (for example, do we get bifurcating relative equilibria, or does the state drift along the group orbit?).

Lemma 6.10 *Let $g = f|_{\Omega_x}$, then g is Σ_x-equivariant. If we can write $g = g_T + g_N$ where g_T and g_N are Σ_x-equivariant and defined on the appropriate spaces, then we can construct f_T and f_N.*

Proof. Let $y \in \Omega_x$ and define

$$f(\gamma y) \equiv \gamma g(y)$$

If f is well-defined, then f will be smooth. Moreover, if $g = g_T$, then $f(x) \in T_x \Gamma x$. So $f = f_T$. Similarly, if $g = g_N$, then $f = f_N$.

We now show that f is well-defined. Suppose that $\gamma_1 y = \gamma_2 y$ where $\gamma_1, \gamma_2 \in \Gamma$, and let $\sigma = \gamma_2^{-1} \gamma_1$. Then

$$\sigma y = \gamma_2^{-1} \gamma_1 y = y$$

and $\sigma \in \Sigma$. Therefore

$$
\begin{aligned}
f(\gamma_1 y) = f(\gamma_2 y) &\Leftrightarrow \gamma_1 g(y) = \gamma_2 g(y) && \text{by the definition of } g \\
&\Leftrightarrow \sigma g(y) = g(y) \\
&\Leftrightarrow g(\sigma y) = g(y) && \text{since } g \text{ is } \Sigma_x\text{-equivariant}
\end{aligned}
$$

Since $\sigma y = y$, the last equality holds, so f is well-defined. □

Next, Theorem 6.9 and Lemma 6.10 imply that every solution $x(t)$ near a relative equilibrium can be written as

$$x(t) = \gamma_t y(t) \tag{6.1}$$

where $y(t)$ is a solution to the normal vector field equation $\dot{y} = g(y)$ and $\gamma_t \in \Gamma$ is a smooth curve.

Another useful result is:

Theorem 6.11 *A relative equilibrium (that is, a zero of g_N) is normally hyperbolic if and only if the corresponding equilibrium of g_N is normally hyperbolic in Ω_x.*

Proof. See Krupa [329]. □

Center Bundles and Bifurcation. In local bifurcation theory, the reflex line of approach is to linearize the vector field in a neighborhood of the bifurcation point. We now describe the analog of this idea for bifurcation from a relative equilibrium, which makes it possible to linearize a vector field in a tubular neighborhood of a a group orbit. As before, let $\Gamma \subseteq \mathbf{O}(n)$ be a compact Lie group and let $f : \mathbf{R}^n \to \mathbf{R}^n$ be a Γ-equivariant vector field.

Suppose that the group orbit $X = \Gamma x_0$ is a relative equilibrium, and let $\Sigma_{x_0} \subseteq \Gamma$ be the isotropy subgroup of x_0. Let $T_{x_0} X$ be the tangent space to the group orbit at x_0, and note that $T_{x_0} X$ is Σ_{x_0}-invariant. By Theorem 6.9 f may be written as

$$f = f_N + f_T$$

on a neighborhood of X. It follows that $g = f_N|Y_{x_0}$ is a Σ_{x_0}-equivariant vector field. Moreover, g is 'generic' in the sense that any Σ_{x_0}-equivariant vector field $g : Y_{x_0} \to Y_{x_0}$ extends, in a neighborhood of X, to a Γ-equivariant vector field $f : \mathbf{R}^n \to \mathbf{R}^n$. If X is a relative equilibrium, then $g(x_0) = 0$.

Definition 6.12 The relative equilibrium X is a *critical group orbit* if $(dg)_{x_0}$ has eigenvalues on the imaginary axis. Let V_{x_0} be the center subspace of $(dg)_{x_0}$. We call

$$V = \bigcup_{\gamma \in \Gamma} \{\gamma V_{x_0}\}$$

the *center bundle*.

Suppose that $W_{x_0} \subseteq \Omega_{x_0}$ is a Σ_{x_0}-invariant center manifold for g. Based on (6.1), Krupa observes that $W = \bigcup\{\gamma W_{x_0}\}$ is a Γ-invariant flow-invariant center manifold for f in a neighborhood of the critical group orbit. In particular, if the noncritical eigenvalues of $(dg)_{x_0}$ all have negative real parts, then W is attracting for the dynamics of f. As is usual with center manifolds, we can project $f|W$ onto the center bundle V. Thus, we can understand bifurcations from critical group orbits by studying bifurcations from equilibria of the normal vector field g. In effect, we reduce the problem to the standard case of steady-state bifurcation of an equivariant ODE.

Example 6.13 Consider $\Gamma = \mathbf{O}(2)$. Steady-state bifurcations lead (generically) to a branch of solutions with $\mathbf{D}_k$ symmetry for some $k \geq 1$. The group orbit through x_0 of these solutions is a circle in phase space. In this example we consider the two simplest cases: $\Sigma_{x_0} = \mathbf{D}_1$ and $\Sigma_{x_0} = \mathbf{D}_2$. In either case the irreducible representations of Σ_{x_0} are one-dimensional: there are two possibile irreducible representations when $\Sigma_{x_0} = \mathbf{D}_1$, and four when $\Sigma_{x_0} = \mathbf{D}_2$.

First, suppose that $\Sigma_{x_0} = \mathbf{D}_1$. The trivial representation does not break symmetry, so we ignore it. The only nontrivial representation is $x \mapsto -x$. Suppose that we have a group orbit of equilibria and a one-dimensional kernel of $(dg)_{x_0}$ on which Σ_{x_0} acts nontrivially. Then, generically, we can suppose that the normal vector field undergoes a pitchfork bifurcation to a new steady state y with $\Sigma_y = \mathbf{1}$. Since

$$N(1)/1 \cong \mathbf{O}(2)$$

Theorem 6.4 implies that bifurcation leads to in a rotating wave.

In the case where $\Sigma_{x_0} = \mathbf{D}_2$, write $\mathbf{D}_2 = \langle \kappa, \pi \rangle$ as usual, $= \kappa$ being reflection and π being rotation. The kernel of the irreducible representation is either $\mathbf{D}_2, \mathbf{Z}_2^\kappa, \mathbf{Z}_2^\pi,$ or $\mathbf{Z}_2^{\kappa\pi}$. The first case does not break symmetry and again we ignore it. If the kernel is $\mathbf{Z}_2^{\kappa\pi}$ we get a rotating wave since the normalizer is $\mathbf{O}(2)$. In the other two cases the normalizer is $\mathbf{D}_2$, so bifurcation leads to a group orbit of equilibria. $\diamond$

This behavior has been observed in simulations of the Kuramoto-Sivashinsky equation by Kevrekidis *et al.* [306]. Here the $\mathbf{D}_2$ case leads to a group orbit of equilibria. The bifurcation diagram is pictured in Figure 6.14: see Krupa [329] for more detail.

6.3 Relative Periodic Orbits

Having dealt with steady-state bifurcation of the normal vector field, it is natural to consider what happens when there is a Hopf bifurcation. The result is a 'relative periodic orbit', that is, a trajectory that is periodic modulo the group action. A precise definition is given below. The dynamics on a relative periodic orbit is quasiperiodic, with some number k of independent frequencies. That is, the relative periodic orbit lies on an invariant torus of dimension k on which the flow is topologically conjugate to a linear flow on a standard torus. Our main interest

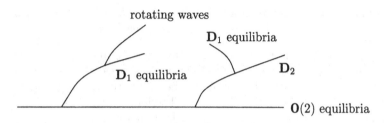

Figure 6.14: Schematic bifurcation diagram obtained by simulation of the Kuramoto-Sivashinsky equation by Kevrekidis *et al.* [306].

in this section is to bound the number k of frequencies in terms of the symmetry of the underlying periodic orbit of the normal vector field. The combination of spatio-temporal symmetry and the global nature of a relative equilibrium makes this a distinctly more subtle problem. The results described here originated with the work of Krupa [329]; they have been generalized by Field [186] and Ashwin and Melbourne [26].

We begin this section by introducing and stating a theorem of Krupa [329] which provides a sharp — indeed generically best possible — bound on the number of independent frequencies in the dynamics on a relative periodic orbit created by Hopf bifurcation in the normal vector field. We illustrate the implications of this theorem by applying it to the Couette-Taylor experiment. We then set up some basic ideas from Lie theory, and use them to prove Krupa's bound. (The proof is fairly technical and may be omitted.) Finally we indicate a variation on Krupa's approach in terms of a moving frame defined by a suitable linear flow.

We now set up the statement of Krupa's Theorem.

Using the normal/tangential decomposition of Theorem 6.9, it is not hard to show that a Hopf bifurcation in the normal vector field leads to a relative periodic orbit, in the following sense:

Definition 6.14 Let f be a Γ-equivariant vector field. Then a trajectory $x(t)$ is a *relative periodic orbit* if there exists $T > 0$ and $\gamma \in \Gamma$ such that

$$x(T) = \gamma x(0)$$

If so, it follows that $x(t + T)$ is a trajectory with initial condition $x(T)$, and so (by equivariance) is $\gamma x(t)$. By uniqueness,

$$x(t + T) = \gamma x(t)$$

for all $t \in \mathbf{R}$. $\diamond$

Another way to state this definition is that $x(t)$ becomes a periodic trajectory when projected into orbit space $\mathbf{R}^n / \Gamma$.

Suppose that Γx_0 is a relative equilibrium of the Γ-equivariant ODE

$$\frac{\mathrm{d}x}{\mathrm{d}t} = f(x) \tag{6.2}$$

and that the isotropy subgroup of x_0 is Σ_{x_0}. Let g be the normal vector field associated with f (restricted to Y_{x_0} as in the previous section), and assume that g undergoes a Hopf bifurcation at x_0 to a T-periodic solution $y(t)$ with spatial symmetry group K and spatio-temporal symmetry group H (defined initially as subgroups of Σ_{x_0}). Let $Y = \{y(t)\}$. We note in passing that H, K remain unchanged if they are defined as the corresponding subgroups of Γ:

Lemma 6.15 *The subgroup $K \subseteq \Sigma_{x_0}$ is the isotropy subgroup of $y(0)$ in Γ. The subgroup $H \subseteq \Sigma_{x_0}$ is the subgroup of Γ that leaves the set Y invariant.*

Proof. By definition K is the isotropy subgroup of $y(0)$ in Σ_{x_0}; hence K is contained in the isotropy subgroup of $y(0)$. Conversely, suppose that $\gamma y(0) = y(0)$ for some $\gamma \in \Gamma$. Since $y(0)$ is an element of the fiber of the tubular neighborhood containing x_0, γ must map that fiber into itself. Therefore, γ preserves the base point of the fiber and $\gamma x_0 = x_0$. Since $\gamma \in \Sigma_{x_0}$ it follows that $\gamma \in K$.

Similarly, if $\gamma \in \Gamma$ leaves Y invariant, then γ preserves the fiber through x_0 and $\gamma x_0 = x_0$. Therefore, $\gamma \in H$. $\qquad\square$

Krupa [329] proves that a Hopf bifurcation in the normal vector field leads to the existence of a relative periodic orbit. Specifically, if $y(t)$ is the periodic solution created by Hopf bifurcation, then he shows that there is a path γ_t in Γ, for all $t \in \mathbf{R}$, such that

$$x(t) = \gamma_t y(t) \qquad (6.3)$$

is a solution of (6.2), hence a relative periodic orbit. To prove this, substitute (6.3) into (6.2) and use the normal/tangential decomposition of Theorem 6.9 to deduce an ODE for γ_t. Moreover, he demonstrates that there are strong restrictions on the dynamics of this relative periodic orbit, related to the symmetry groups H and K.

To state these restrictions, let $P = \{x(t)\}$ be the trajectory of $x(t)$ and let $\overline{P}$ be its closure. Then it can be proved that $\overline{P}$ is a torus. Moreover, if the dimension of this torus is k, then the flow of f on $\overline{P}$ is topologically conjugate to some linear flow on a standard torus $\mathbf{R}^k/\mathbf{Z}^k$. Thus the dynamics on the relative periodic orbit P is quasiperiodic, with k independent frequencies. The main theorem of Krupa [329] places a bound on the number of these independent frequencies, in terms of the group H and its normalizer in Γ. Recall that the rank of a compact Lie group is the dimension of a maximal torus. Then the bound is given by:

Theorem 6.16 *With the above notation, $\overline{P}$ is a torus of dimension*

$$\dim \overline{P} \leq 1 + \mathrm{rank}\, \frac{N(H)}{H} \qquad (6.4)$$

Moreover, generically this bound becomes an equality.

Corollary 6.17 *If the Hopf bifurcation in the normal vector field takes place at simple eigenvalues, then $\dim \overline{P} \leq 1 + r$ where r is the dimension of the torus corresponding to the relative equilibrium x_0. In particular, if Γx_0 consists of equilibria, then P is periodic.*

Proof. Since the eigenvalues are simple, $H = \Sigma_{x_0}$. Therefore $\mathrm{rank}\, N(H)/H = \mathrm{rank}\, N(\Sigma_{x_0})/\Sigma_{x_0}$. Now appeal to Theorem 6.4. $\qquad\square$

The next example shows that Krupa's bound can be stronger than the 'elementary' bound $1 + \mathrm{rank}\, N(K)/K$ (which can be proved by an argument similar to Theorem 6.4).

Example 6.18 Let $\Gamma = \mathbf{O}(2)$, $\Sigma_{x_0} = \mathbf{Z}_2(\kappa)$, $K = 1$, and $H = \mathbf{Z}_2(\kappa)$ where κ is a reflection in $\mathbf{O}(2)$. Then Theorem 6.16 implies that $\dim \overline{P} \leq 1$, because $N(H) = \langle H, \pi \rangle$ which is finite, so $N(H)/H$ is finite. Therefore, in this case, any relative periodic orbit arising by Hopf bifurcation is actually *periodic*.

In contrast, $\dim N(K)/K = 1$, so $1 + \mathrm{rank}\, N(K)/K = 2$, so the elementary bound is too large. $\qquad\Diamond$

Hopf Bifurcation from Taylor Vortex Flow. Before sketching the proof of Krupa's Theorem, we demonstrate its utility by applying the theory of Hopf bifurcation from relative equilibria to the Couette-Taylor system. The connection is that with the modeling assumption of periodic boundary conditions (which leads to symmetry by axial translations), Taylor vortices form a group orbit of steady flows — a relative equilibrium of the velocity field of the fluid.

Based on the experimental results of Andereck *et al.* [5], Iooss [298] had a clever idea for classifying the ways in which the relative equilibrium Taylor vortices lose stability via a Hopf bifurcation to a time-periodic state. He assumed that the system spatially period doubles when it undergoes a Hopf bifurcation — an assumption that cannot be justified by reasoning directly from the Navier-Stokes equations. However, once this assumption was made, Iooss was able to recover the states that are actually observed in the experiments.

More precisely, Iooss considered two Taylor vortex pairs as the basic space unit and imposed PBC on the ends of the four-vortex unit. The symmetries of this extended system include an extra axial half-period phase shift, which we denote by ρ. Note that half-period refers to translation by one vortex pair in the two vortex pair model. The complete symmetry group is $\mathbf{O}(2) \times \mathbf{Z}_2(\rho) \times \mathbf{SO}(2)$. In this model the isotropy subgroup of Taylor vortices is the abelian subgroup $\Sigma_{x_0} = \mathbf{D}_2(\kappa, \rho) \times \mathbf{SO}(2)$.

Hopf bifurcations in the normal vector field g are determined by the irreducible representations of Σ_{x_0}, which are either one- or two-dimensional. If the representations are one-dimensional, then the azimuthal symmetry group $\mathbf{SO}(2)$ acts trivially on the bifurcating state and the new state is azimuthally symmetric at every moment of time. Such states are not observed, so we may assume (generically) that the center subspace for dg is two-dimensional and that Σ_{x_0} acts nonabsolutely irreducibly on that subspace. In particular $\mathbf{SO}(2)$ acts nontrivially, so all bifurcating solutions are rotating waves in the sense that time evolution can be identified with azimuthal rotation.

There are four irreducible representations of Σ_{x_0} determined by which of the normal subgroups $\mathbf{D}_2(\kappa, \rho)$, $\mathbf{Z}_2(\rho, \mathbf{Z}_2(\kappa)$, and $\mathbf{Z}_2(\kappa\rho)$ acts trivially on the center subspace. Each irreducible representation leads to a Hopf bifurcation where the bifurcating time-periodic solution is fixed by one of these four subgroups; that is, the isotropy subgroup Σ of a the new relative periodic orbit is one of these four subgroups. Note that κ acts with a half-period phase shift in the second and fourth cases and ρ acts with a half-period phase shift in the third and fourth cases. Therefore, in each of the four cases the group of symmetries that preserve the relative periodic orbit is $H = \mathbf{D}_2(\kappa, \rho) \times \mathbf{SO}(2)$. Moreover, $\mathrm{rank}(N(H)/H) = 0$ and Krupa's theorem implies that each of the four resulting flows in the full equations is periodic.

Iooss [298] notes that Andereck *et al.* [5] found exactly four types of periodic solution bifurcating from Taylor vortices, and the symmetries of these four types correspond exactly to the four types of $\mathbf{D}_2$ Hopf bifurcation. It is not understood why this observation works — a basic spatial period of three Taylor vortex pairs would seem an equally reasonable assumption — nevertheless, the result is striking. We end with a discussion of these four states and the realization of their symmetries in physical space.

With respect to these symmetries, wavy vortex flow is ρ-symmetric. Twisted vortices are both ρ-symmetric and κ-symmetric. Wavy outflow boundary flows have (κ, ρ) symmetry — but neither κ nor ρ symmetry by themselves. Wavy inflow

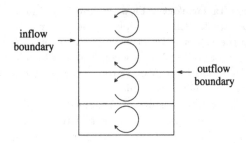

Figure 6.15: Inflow and outflow boundaries in Taylor vortex flow.

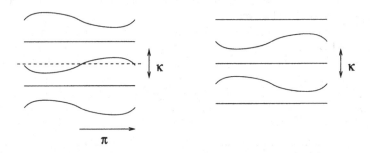

Figure 6.16: Schematic pictures of wavy outflow and wavy inflow flows.

boundary flows have κ symmetry, but not ρ symmetry. The last two flows are illustrated in Figure 6.16; the terminology *inflow* and *outflow* is set in Figure 6.15.

Lie Group Preliminaries. Having illustrated the use of Krupa's Theorem, we now move towards its proof. In this subsection we state some basic results from the theory of compact Lie groups, which will be required in the proof. For details, see Bröcker and tom Dieck [77] Sections I.2, I.3, IV.1, IV.2, IV.4. The proof itself is given in the next subsection.

Let Γ be a compact Lie group, and without loss of generality assume that $\Gamma \subseteq \mathbf{O}(n)$. Then we can identify elements of Γ with nonsingular $n \times n$ matrices. Suppose that $\{\gamma_t\}$ is a differentiable path in Γ, such that $\gamma_0 = 1$. Then we may define an element ξ of the tangent space $T_1\Gamma$ at the identity:

$$\xi = \frac{\mathrm{d}}{\mathrm{d}t}\gamma_t\Big|_{t=0}$$

The set of all such ξ is the *Lie algebra* $L(\Gamma)$ of Γ. It is a finite-dimensional vector space, of dimension equal to the dimension of Γ, and it depends only on Γ°, the connected component of the identity.

Example 6.19 Let $\Gamma = \mathbf{O}(n)$. We claim that $L(\Gamma)$ is the space of skew-symmetric $n \times n$ matrices. To see why, observe that $\gamma \in \mathbf{O}(n)$ if and only if $\gamma\gamma^T = 1$. If $\{\gamma_t\}$ is a differentiable path in $\mathbf{O}(n)$, then

$$\gamma_t\gamma_t^T = 1$$

and by differentiation,

$$\dot{\gamma}_t\gamma_t^T + \gamma_t\dot{\gamma}_t^T = 0$$

Putting $t = 0$ and letting

$$\xi = \dot{\gamma}_t|_{t=0}$$

we find that

$$\xi + \xi^T = 0$$

and ξ is skew-symmetric.

Conversely, if ξ is any skew-symmetric $n \times n$ matrix, define a path $\gamma_t = \exp(t\xi)$. Then it is easy to verify that

$$\left.\frac{\mathrm{d}}{\mathrm{d}t}\gamma_t\right|_{t=0} = \xi$$

$\diamond$

The exponential construction generalizes to any compact Lie group, and leads to the *exponential map*

$$\exp : \Gamma \to L(\Gamma)$$

Proposition 6.20 *If Γ is a compact Lie group, then the image of the exponential map is the connected component Γ° of the identity.* $\square$

Definition 6.21 Let $\xi \in L(\Gamma)$. Then the subgroup

$$\{\exp(t\xi) : t \in \mathbf{R}\}$$

is called a *one-parameter subgroup* of Γ. $\diamond$

Every one-parameter subgroup is abelian, and is isomorphic to $\mathbf{R}$ if $\exp t\xi = 1$ implies $t = 0$, or to $\mathbf{S}^1$ otherwise.

The closure of a one-parameter subgroup may have dimension larger than one: indeed, the path $\{\exp(t\xi)\}$ may wind densely through some torus of dimension 2 or more (and by compactness must do so if it does not define a circle subgroup).

Recall that a maximal torus in a compact Lie group Γ is a torus subgroup that is not contained in any larger torus subgroup, and that the rank of Γ is the dimension of any maximal torus.

Proposition 6.22 (a) *If T_1, T_2 are maximal tori of a compact Lie group Γ, then there exists $\gamma \in \Gamma^\circ$ such that*

$$\gamma^{-1}T_1\gamma = T_2$$

(b) *All maximal tori have the same dimension.*
(c) *A maximal torus is any torus of maximal dimension, and conversely.*
(d) *Every torus is contained in a maximal torus.*

Proof. See Bröcker and tom Dieck [77] Sections IV.1, IV.2. $\square$

Definition 6.23 A subgroup $M \subseteq \Gamma$ is *monogenic* (or *topologically cyclic*) if there exists an element $\gamma \in \Gamma$ such that M is the topological closure of the subgroup $\{\gamma^n : n \in \mathbf{Z}\}$ generated by γ. We call γ a *generator* of M. $\diamond$

Every torus is monogenic, but a monogenic subgroup may be disconnected. A generator of a maximal torus is called a *regular element*.

Definition 6.24 A *Cartan subgroup* of Γ is a subgroup C such that C is monogenic and $N(C)/C$ is discrete. $\diamond$

A Cartan subgroup can also be defined as a monogenic subgroup of maximal dimension. The main properties of Cartan subgroups required in the proof of Krupa's Theorem are:

Proposition 6.25 *(a) Every element $\delta \in \Gamma$ is contained in a Cartan subgroup C of Γ such that C/C° is generated by the coset δC°.*

(b) If C is a Cartan subgroup of Γ generated by γ, then any $\delta \in \Gamma^\circ \gamma$ is conjugate to an element of $C^\circ \gamma$ via an element of Γ°.

Proof. See Bröcker and tom Dieck [77] Section IV.4.   □

Proof of Krupa's Theorem. We now use the properties of Cartan subgroups to prove Theorem 6.16. Let P be the relative periodic orbit (6.3), and let $\overline{P}$ be its closure. Recall the statement of Krupa's theorem: $\overline{P}$ is a torus of dimension $\dim \overline{P} \leq 1 + \mathrm{rank}\, \frac{N(H)}{H}$. Moreover, generically this bound becomes an equality.

We prove only the inequality: for the genericity argument see Krupa [329]. (The main point in proving that generically the bound is attained is that there is sufficient freedom to define the tangential vector field to ensure maximal dimension of the invariant torus, subject to the symmetry constraints that apply.)

Let Z denote the group orbit of $y(t)$. Then $x(t) \in Z$ for all t. Indeed we can find a path $\gamma_t \subseteq \Gamma$ such that

$$x(t) = \gamma_t y(t)$$

for all $t \in \mathbf{R}$. In fact, without loss of generality we may assume that $\gamma_t \in N(K)/K$ for all $t \in \mathbf{R}$, by Lemma 6.3.

The proof now splits into two cases: $H/K \cong \mathbf{S}^1$ (the rotating wave case), and H/K finite.

Case 1: $H/K \cong \mathbf{S}^1$.

In this case, time-translation on $y(t)$ can be identified with the action of a one-parameter subgroup of Γ, so Z is a relative equilibrium. Theorem 6.4 now implies that trajectories in Z are dense in tori of dimension $\leq \mathrm{rank}\, N(K)/K$, since K is the isotropy subgroup of $y(0)$. It remains to show that

$$\mathrm{rank}\, N(K)/K = 1 + \mathrm{rank}\, N(H)/H \tag{6.5}$$

To prove this, observe that $H/K \cong \mathbf{S}^1$ is a torus of rank 1, and $H/K \subseteq N(K)/K$. Hence there exists a maximal torus $M \subseteq N(K)/K$ such that $H/K \subseteq M$. Clearly $M \subseteq N(H)/K$. Therefore M/H is a maximal torus in $N(H)/H$, and this proves the equality (6.5).

Case 2: H/K finite.

We have $H/K \cong \mathbf{Z}_m$ for some finite m. We choose a generator h for H/K such that

$$hy(t) = y(t + T/m)$$

Then

$$x(T/m) = \gamma_{T/m} y(T/m) = \gamma_{T/m} h y(0)$$

Define

$$\alpha = \gamma_{T/m} h \tag{6.6}$$

Let $A = \{\alpha^r : r \in \mathbf{Z}\}$ be the subgroup generated by α, and let $\overline{A}$ be the closure of A in Γ. Then $\overline{A}$ is compact and abelian, indeed monogenic.

Let ϕ_t be the flow of f. By definition, $\phi_{T/m} y(0) = x(T/m) = \alpha y(0)$. Therefore

$$\alpha^r y(0) = \phi_{rT/m} y(0) \tag{6.7}$$

for all $r \in \mathbf{Z}$. Let $P = \{x(t) : t \in \mathbf{R}\}$ and let $\overline{P}$ be its closure. We seek a bound on the dimension of $\overline{P}$. By (6.7)

$$Ay(0) \subseteq \overline{P}$$

so that $\phi_t Ay(0) \subseteq P$, and $\phi_t \overline{A} y(0) \subseteq \overline{P} \subseteq Z$ for all $t \in \mathbf{R}$.

Define a group action of $\Gamma \times \mathbf{R}$ on $\mathbf{R}^n$ by

$$(\gamma, s)w = \phi_t \gamma w$$

where ϕ_t is the flow of f. Then $\overline{P} = (\overline{A} \times \mathbf{R})y(0)$. Let Δ be the isotropy subgroup of $y(0)$ with respect to this action. Then $\overline{P}$ is diffeomorphic to the quotient group $(\overline{A} \times \mathbf{R})/\Delta$. This group is compact, connected, and abelian, hence it is a torus. Therefore $\overline{P}$ is diffeomorphic to a torus. The dimension of $\overline{P}$ is given by

$$\dim \overline{P} = 1 + \dim \overline{A}$$

Thus we seek a bound for $\dim \overline{A}$.

This bound is obtained using properties of Cartan subgroups. The group $\overline{A}$ is monogenic with generator α, so its dimension is bounded by the dimension of a Cartan subgroup containing α. Now α and h lie in the same connected component of the identity in $N(K)/K$, by (6.6). If C is a Cartan subgroup of $N(K)/K$, then $\dim \overline{A} \leq \dim C$ and $\dim \overline{P} \leq 1 + \dim C$.

To finish the proof, we must show that

$$\dim C \leq \operatorname{rank} N(H)/H$$

We can choose the Cartan subgroup C so that $H/K \subseteq C$. Since H/K is finite, $\operatorname{rank} N(H)/K = \operatorname{rank} N(H)/H$. Let N_H denote the normalizer of H/K in $N(K)/K$: then $N_H = N(H)/K$. Clearly $K \subseteq N_H$. We show that $\operatorname{rank} N_H = \dim C$. Suppose that $\phi \in N_H^\circ$ and let Φ be the monogenic subgroup generated by ϕ. Let $\phi_0 = \phi h$ and let Φ_0 be the monogenic subgroup generated by ϕ_0. Since $\phi \in N_H$ we have $\phi^{-1}(H/K)\phi = H/K$, which implies that

$$\phi h = h^r \phi \tag{6.8}$$

for some r. By continuity, r may be chosen to be independent of ϕ. Therefore for any positive integer j we have

$$\phi_0^j = h^s \phi_j$$

for some s that depends on j but not in ϕ. Since ϕ_0 generates Φ_0 there exists l such that $\phi_0^l \in \Phi_0^\circ$. By continuity, $\phi_0^l \in N_H^\circ$ for all $\phi \in N_H^\circ$. By (6.8) we have $h^s \phi^l \in N_H^\circ$ for some s, independent of $\phi \in N_H^\circ$. Thus for some ϕ the torus generated by $h^s \phi^l$ is a maximal torus of N_H.

Therefore

$$\dim \Phi \leq \dim \Phi_0 \leq \dim C$$

Moreover, $\dim C \geq \operatorname{rank} N_H$ because C° is a connected abelian subgroup of a compact group, hence is contained in a maximal torus. Therefore $\operatorname{rank} N_H \geq \dim K$, as required.

Moving Frames. Chossat and Lauterbach [110] approach relative periodic orbits (including those not arising from Hopf bifurcation of the normal vector field) from a different point of view, which physically amounts to reducing the relative periodic orbit to a periodic one by observing it in a moving frame defined by a *linear* flow.

Theorem 6.26 *Let $x(t)$ be any relative periodic orbit of a Γ-equivariant ODE*

$$\frac{dx}{dt} = f(x)$$

with relative period T, where Γ is compact. Then there exists an element ξ of the Lie algebra of Γ, a positive integer k, and a periodic solution $y(t)$ of the ODE

$$\frac{dy}{dt} + \xi y = f(y) \qquad such\ that \qquad x(t) = \exp(\xi t)y(t).$$

Proof. We have $x(T) = \gamma x(0)$ for some $\gamma \in \Gamma$. The order k of Γ/Γ° is finite by compactness, so that $\gamma^k \in \Gamma^\circ$. Now $x(kT) = \gamma^k x(0)$. Since $\gamma^k \in \Gamma^\circ$, there exists $\xi \in L(\Gamma)$ with

$$\gamma^k = \exp(kT\xi)$$

Define $y(t) = \exp(-\xi t)x(t)$, so that $x(t) = \exp(\xi t)y(t)$. Then

$$y(kT) = \exp(-kT\xi)x(kT) = \exp(-kT\xi)\gamma^k x(0) = y(0)$$

Finally,

$$\frac{dy}{dt} + \xi y = -\xi y(t) + \exp(-\xi t)x(t) + \xi y(t) = \exp(-\xi t)f(x(t) = f(y(t))$$

as required.   □

In fact, by Lemma 6.3, we can assume that ξ is in the Lie algebra of $N(K)/K$ where K is the isotropy subgroup of $x(0)$.

The advantage of this description over Krupa's is that it uses a one-parameter subgroup $\{\exp(\xi t)\}$ instead of a path γ_t. Its main disadvantage seems to be that we lose control over the symmetry of the vector field (which is now $f(y) - \xi y$ and is equivariant only under elements that commute with ξ modulo K). The relation between these two descriptions currently requires clarification.

A more general approach still can be found in Ashwin and Melbourne [26].

6.4 Hopf Bifurcation from Rotating Waves to Quasiperiodic Motion

Among the most important kinds of relative equilibria are *rotating waves*, which are time-periodic solutions for which time evolution is the same as spatial rotation. It follows that rotating waves occur as solutions only in systems of differential equations having at least $\mathbf{SO}(2)$ symmetry, though, in general, rotating waves occur in models containing more than $\mathbf{SO}(2)$ symmetry. Rotating waves are also the simplest form of relative equilibria that are not actually equilibria. By Case 1 of the proof of Krupa's Theorem, a relative periodic orbit bifurcating from a rotating wave via a Hopf bifurcation in the normal vector field, is actually a relative equilibrium.

For the remainder of this chapter we discuss Hopf bifurcations from rotating waves to two-frequency modulated rotating waves in Γ-equivariant systems, and following [219] make the point that the manifestation of these modulated waves in physical space depends on Γ. In our discussions we use the theory outlined in Section 6.2 and focus on three experiments where transitions from rotating to modulated rotating waves have been observed.

Rotating and Modulated Rotating Waves in Experiments. Rotating waves occur in the Couette-Taylor system as wavy vortices (see Andereck *et al.* [5] and Figure 6.2), in the Belousov-Zhabotinskii (BZ) chemical reaction as spiral waves (see Winfree [521] and Figure 6.18), and in laminar premixed flames as rotating cellular patterns (Gorman *et al.* [245] and Figure 6.17). Each of these rotating waves has also been observed numerically as a patterned solution to PDE models for the corresponding experiments: rotating waves have been observed in the Navier-Stokes equations modeling the Couette-Taylor system [373], in reaction-diffusion equations loosely modeling BZ reactions [42, 304], and in reaction-diffusion models loosely modeling combustion [54, 258].

Hopf bifurcations from rotating waves to modulated rotating waves have been observed in each experiment. In the Couette-Taylor system, wavy vortices bifurcate

  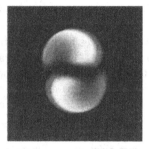

Figure 6.17: Flames on a circular burner. (Left) A circularly symmetric flame. (Center) A steady five-cell flame with $\mathbf{D}_5$ symmetry. (Right) Rotating two-cell flame. Images courtesy of M. Gorman.

to modulated wavy vortices [248, 249, 123]; in the BZ reaction, spiral waves begin to meander quasiperiodically (see Figure 6.18) and even to drift linearly [536, 537, 365, 301, 302, 477, 361]; and in laminar premixed flames, the cellular pattern appears to rotate rigidly but with an angle of rotation that depends quasiperiodically on time [245, 55, 56, 425]. In each of these experiments, the basic Hopf bifurcation from a single frequency time-periodic rotating wave to quasiperiodic motion in phase space is understood [250, 47, 56]. Even the resonant Hopf bifurcation to linear drifting spiral waves is understood [44, 528, 460, 218].

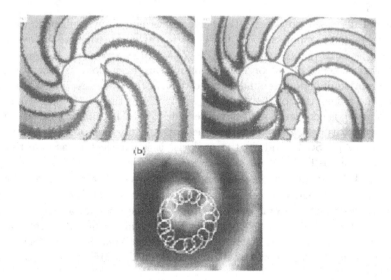

Figure 6.18: Belousov-Zhabotinskii Experiment. (Upper left) spiral waves with seven superimposed images showing the tip traversing a circle (from [477]). (Upper right) meandering spiral waves with eleven superimposed images showing the tip beginning to meander (from [477]). (Bottom) Single image of a meandering spiral wave with superimposed tip trajectory showing a *petals in* flower pattern (from [361]). Pictures courtesy of H.L. Swinney.

Hopf bifurcations from rotating waves to modulated rotating waves have been studied by a number of authors including Rand [440] and Renardy [443]. In addition, Rand classifies the various types of modulated waves that occur in systems with circular symmetry. This classification applies directly to the flame experiment (as pointed out in [56]) and less directly to the Couette-Taylor experiment (since wavy vortices have an additional glide reflection symmetry which is not taken into account in [440]). It is straightforward, as we show in Section 6.5, to complete Rand's classification of modulated wavy vortices to include the glide reflection symmetry. This extended classification accounts for additional states that have been observed in the more recent experiments of Zhang and Swinney [535], see also Coughlin and Marcus [123].

Symmetries in the Experiments. Since we view symmetry as a modeling assumption, it is important to understand how the symmetry group Γ for models of each of these experiments is determined. Some of the symmetries are clear, being based on the geometry of the apparatus and the homogeneity of the experiment; other symmetries are less transparent.

Gorman's laminar premixed flame experiment is performed on a circular burner and is modeled by a PDE with $\Gamma = \mathbf{O}(2)$ symmetry. Moreover, the transitions that are observed in this experiment are consistent with the assumption of $\mathbf{O}(2)$ symmetry, in the following sense. Steady-state bifurcation from an invariant equilibrium with $\mathbf{O}(2)$ symmetry on a circular domain leads to equilibria having a cellular structure and Hopf bifurcation leads to standing and rotating waves [227]. All of these states are observed in the flame experiment; moreover, the cellular structures and the standing waves are observed as direct transitions from a circularly symmetric flame [244].

As we discussed previously, the geometry is not always sufficient to detect all of the relevant symmetries. In the Couette-Taylor experiment, the cylindrical geometry leads to the assumption of $\Gamma = \mathbf{Z}_2 \times \mathbf{SO}(2)$ symmetry. As Taylor [501] noted, the existence of Taylor vortices bifurcating from Couette flow argues for the assumption of periodic boundary conditions in the axial direction (Figure 6.2); that is, the assumption of $\mathbf{O}(2) \times \mathbf{SO}(2)$ symmetry in the experiment. Moreover, this symmetry is consistent with the observed Hopf bifurcation from Couette flow to spiral vortices [105, 237]. However, the isotropy subgroup of wavy vortices lies inside $\Gamma = \mathbf{Z}_2 \times \mathbf{SO}(2)$ and, in the analysis, it is sufficient to assume that the symmetry group is Γ.

In the BZ reaction, the situation is even more complicated; it seems best to model these experiments by ignoring lateral boundaries. Barkley's analysis [44] of resonant linear drift of spiral waves in this experiment makes a compelling case for the assumption of full Euclidean $\Gamma = \mathbf{E}(2)$ symmetry including translations. Moreover, the more rigorous mathematical results in Wulff, Sandstede, Scheel, and Fiedler [528, 459, 460] support this assumption. See also [218] whose exposition we follow here.

The rotating waves in each of these experiments have cyclic symmetry for the instantaneous pattern. As usual, we denote the isotropy subgroup of the rotating wave at a given instant in time by Σ_{x_0}. The cellular flame pattern has k cells (Figure 6.17), and a spiral can have k arms — though one-armed spirals are what is usually observed in BZ reactions (Figure 6.18). In both cases $\Sigma_{x_0} = \mathbf{Z}_k$. The wavy vortices (Figure 6.2) have an azimuthal wave number k, and an additional symmetry — flip up and down coupled with a half wavelength azimuthal rotation (a glide reflection symmetry). The square of this symmetry is the generator of the pure azimuthal rotation symmetry — so the isotropy subgroup of wavy vortices is $\mathbf{Z}_{2k}$. Table 6.1 presents the relevant group-theoretic data for each experiment.

Experiment	Rotating Wave	Γ	Σ_{x_0}
Couette-Taylor	wavy vortices	$\mathbf{Z}_2 \times \mathbf{SO}(2)$	$\mathbf{Z}_{2k}$
flames	rotating cells	$\mathbf{O}(2)$	$\mathbf{Z}_k$
BZ reaction	1-armed spirals	$\mathbf{E}(2)$	1

Table 6.1 Symmetry data for rotating waves.

In each example $\Sigma_{x_0} = \mathbf{Z}_k$. So, generically, we expect the center subspace for Hopf bifurcation from the rotating wave in the normal direction to be two-dimensional; that is, we expect the critical eigenvalues to be simple. By Corollary 6.17, the bifurcating solution is a two-frequency motion. However, in the case of spiral waves, the symmetry group $\mathbf{E}(2)$ is not compact and Krupa's theorem does not apply. We will see in Section 6.7 that we still expect two-frequency motions to emanate from these bifurcations. So, in all cases, the transition in phase space is from a periodic rotating wave to a two-frequency modulated rotating wave. However, the spatial features of these two-frequency motions are all different — and it is these differences that we wish to discuss. The differences are summarized by three questions:

- What is the physical space interpretation of Rand's classification [440] of modulated wavy vortices?
- Why do the cellular flames appear to rotate rigidly but nonuniformly?
- Why do spirals exhibit flower-like meandering in the BZ reaction (see Figure 6.18) — and even linear drifting?

The answer to each of these questions depends on symmetry. The answer to the first question is a direct application of the theory developed in the previous section, see Section 6.5. The answer to the second question is based on a notion of inner and outer patterns in physical space, see Section 6.6. The answer to the third question requires the study of the center bundle vector fields for the Euclidean group, see Section 6.7.

6.5 Modulated Waves in Circular Domains

In this section, we consider Hopf bifurcation from rotating waves with isotropy subgroup $\Sigma_{x_0} = \mathbf{Z}_k$ ($k \geq 1$) in systems with symmetry group $\Gamma = \mathbf{SO}(2)$ or $\Gamma = \mathbf{O}(2)$. Rand [440] classified the various modulated rotating waves that occur in terms of their spatial and spatio-temporal symmetries. We rederive Rand's classification following the discussion in Golubitsky, LeBlanc, and Melbourne [219].

Hopf bifurcation from a rotating wave corresponds to Hopf bifurcation from an equilibrium for the Σ_{x_0}-equivariant normal vector field g. Since $\Sigma_{x_0} = \mathbf{Z}_k$, genericity tells us to expect a two-dimensional center subspace in the normal direction; let $z \in \mathbf{C}$ denote coordinates for this critical eigenspace. The action of the isotropy subgroup $\Sigma_{x_0} = \mathbf{Z}_k$ on $z \in \mathbf{C}$ is generated by

$$R_{\frac{2\pi}{k}} z = e^{2\pi i m/k} z \tag{6.9}$$

for some $m = 0, 1, \ldots, [k/2]$. Rand's classification of modulated rotating waves depends on the integers k and m.

One difficulty is that the integer m has no direct physical interpretation. Following Rand (though with different notation) we introduce the derived integers

$d \geq 1$ and $\alpha \in \{0, 1, \ldots, \frac{k}{d} - 1\}$ where

$$d = \gcd(k, m) \qquad \alpha m \equiv d \,(\mathrm{mod}\, k)$$

We show that the integers k, d and α are quantities that can be determined experimentally.

Hopf bifurcation in the normal vector field leads to a periodic solution $y(t)$ with spatial symmetries K and spatio-temporal symmetries H. Since the Hopf bifurcation in the normal direction occurs with simple eigenvalues, Corollary 6.17 implies that the bifurcating state is generically a two-frequency quasiperiodic state — a modulated rotating wave.

The isotropy subgroup K of the bifurcating modulated rotating wave is given by the kernel of the action (6.9) on the critical eigenspace. Hence, $K = \mathbf{Z}_d$ where $d = \gcd(k, m)$. Thus, the integers d and k correspond to the spatial symmetries ($K = \mathbf{Z}_d$) and the spatio-temporal symmetries ($H = \mathbf{Z}_k$) of the modulated rotating wave.

Remark 6.27 Note that α is the multiplicative inverse of m/d modulo k/d. Hence, given k, d and $\alpha \geq 1$ we can recover m through the equation $\alpha(m/d) \equiv 1 \,(\mathrm{mod}\, k/d)$. (When $\alpha = 0$, we have $k = d$ and $m = 0$.)

Next, we show that the integer α determines the spatio-temporal symmetry of the modulated rotating wave in a more precise way. Assume that the bifurcating periodic solution y is T-periodic. Then

$$y\left(t + \frac{m}{k}T\right) = R_{\frac{2\pi}{k}} y(t) \tag{6.10}$$

We now compute the minimal spatio-temporal symmetry of y: that is, the minimal $L > 0$ for which there exists $\gamma \in \Sigma_{x_0}$ such that $y(t + L) = \gamma y(t)$.

Proposition 6.28 *The normal vector field solution y has minimal spatio-temporal symmetry $\frac{d}{k}T$ where*

$$y\left(t + \frac{d}{k}T\right) = \left(R_{\frac{2\pi}{k}}\right)^{\alpha} y(t) \tag{6.11}$$

Proof. Since $d = \gcd(k, m)$, we have $\alpha m + \beta k = d$ for some $\beta \in \mathbf{Z}$. Hence

$$\frac{d}{k}T = \alpha\frac{m}{k}T + \beta T$$

It follows from T-periodicity and (6.10) that

$$y\left(t + \frac{d}{k}T\right) = y\left(t + \alpha\frac{m}{k}T\right) = (R_{2\pi/k})^{\alpha} y(t)$$

verifying that (6.11) is a spatio-temporal symmetry of y.

Next we show that (6.11) is the minimal spatio-temporal symmetry. Recall that in Hopf bifurcation, T is the minimal period of $y(t)$. Suppose that $y(t + S) = \gamma y(t)$ for some $S > 0$, $\gamma \in \Sigma_{x_0}$, so that

$$y(t + S) = (R_{2\pi/k})^{j} y(t)$$

where $j \geq 1$. We must show that $S \geq \frac{d}{k}T$. Now

$$y\left(t + \frac{k}{d}S\right) = (R_{2\pi/d})^{j} y(t) = y(t)$$

since $K = \mathbf{Z}_d$. Hence $\frac{k}{d}S$ is a multiple of T, and $S \geq \frac{d}{k}T$, as required. $\square$

By the results of Krupa [329], the symmetry (6.11) corresponds to an exact spatio-temporal symmetry of the full modulated rotating wave solution $x(t)$ modulo the drift along the $\mathbf{SO}(2)$-group orbit. Thus, in a suitable rotating frame, the modulated rotating wave reduces to a periodic solution, and the integer α in Proposition 6.28 determines the spatio-temporal symmetry of that periodic solution.

In numerical simulations, Bayliss, Matkowsky, and Riecke [56] obtain a number of modulated rotating waves arising through symmetry-breaking ($m > 0$ or equivalently $d < k$) bifurcations from rotating waves. They emphasize bifurcations from rotating waves with four identical cells ($k = 4$) and with seven identical cells ($k = 7$). These simulations give a number of very nice examples of the symmetries of the modulated waves described here. See also [219] for a more detailed discussion of the simulations of Bayliss *et al.* [56].

Modulated Wavy Vortices in Couette-Taylor. As discussed previously, the Couette-Taylor experiment is often modeled as having $\mathbf{O}(2) \times \mathbf{SO}(2)$ symmetry, where $\mathbf{SO}(2)$ consists of azimuthal rotations and $\mathbf{O}(2)$ consists of axial translations and an up-down flip κ. It turns out that most results are unchanged if we just assume $\mathbf{Z}_2 \times \mathbf{SO}(2)$ symmetry (the up-down flip κ and azimuthal rotations).

Recall that the isotropy subgroup of wavy vortices is given by $\Sigma_{x_0} = \mathbf{Z}_{2k}$ consisting of pure azimuthal rotations $\mathbf{Z}_k \subseteq \mathbf{SO}(2)$ generated by $\rho = (\kappa, \pi/k) \in \mathbf{Z}_2 \times \mathbf{SO}(2)$. Hence Σ_{x_0} is generated by a single glide reflection.

The representation of the isotropy subgroup $\Sigma_{x_0} = \mathbf{Z}_{2k}$ is generated by $q \mapsto e^{\pi i m / k} q$ for some $m = 0, 1, \ldots, k$. We obtain a classification of the possible types of modulated rotating waves in terms of the integers $(2k, d, \alpha)$ where $d = \gcd(2k, m)$ and $\alpha m \equiv d \pmod{2k}$.

We distinguish between modulated wavy vortices that break all the spatial glide reflection symmetry of the wavy vortices, and those that retain some of this symmetry. In the terminology of [123], *Gorman-Swinney flows* are modulated wavy vortices that break the glide reflection symmetry, while *Zhang-Swinney flows* retain some of the glide reflection symmetry. Gorman-Swinney flows occur when $2k/d$ is even and Zhang-Swinney flows occur when $2k/d$ is odd. To see this, observe that ρ^j is a glide reflection if and only if j is odd. But $K = \mathbf{Z}_d$ is generated by $\rho^{2k/d}$ and hence contains odd powers of ρ precisely when $2k/d$ is odd.

Both kinds of modulated wavy vortices have been observed in experiments. Gorman-Swinney flows are the original modulated wavy vortices of Gorman and Swinney [248, 249] and do not possess spatial glide reflection symmetry (though such symmetries must appear as spatio-temporal symmetries). Zhang-Swinney flows were obtained more recently in experiments of Zhang and Swinney [535] and possess spatial glide reflection symmetry, as noted in [123].

Rand's original classification [440] of modulated rotating waves is stated for systems with $\mathbf{SO}(2)$ symmetry. This classification is geared towards modulated wavy vortices for which the symmetry group is $\mathbf{Z}_2 \times \mathbf{SO}(2)$. In particular, the isotropy subgroup of wavy vortices is given by $\Sigma_{x_0} = \mathbf{Z}_{2k}$ and does not lie in $\mathbf{SO}(2)$, whereas Rand has $\Sigma_{x_0} = \mathbf{Z}_k \subseteq \mathbf{SO}(2)$.

6.6 Spatial Patterns

In this section we discuss why the cellular flames might appear to rotate rigidly but with a speed that varies periodically in time (so that there are two independent frequencies). Palacios *et al.* [425] call this a 'nonuniformly rotating' state. Indeed, such a state *cannot* be exactly described this way because a solution to an equivariant differential equation that lies exactly in a group orbit (in this case the group orbit given by the rotation subgroup) must produce linear flow along the group orbit, that is, the speed of rotation must be constant. In fact, careful observation of this state [425] shows that the cellular pattern itself also varies periodically in time — but by a small amount.

Let $U(X, t)$ be a solution to a PDE, or the state of an experiment where X is in some physical domain $\mathcal{D}$ and t is time. For example, in the Couette-Taylor system, U consists of the three velocity components of the fluid and the pressure variable, and $\mathcal{D} \subseteq \mathbf{R}^3$ is the region between the concentric cylinders. When we view the Couette-Taylor experiment, we look at an observable scalar quantity $u(x, t)$ which is the intensity of light reflected off aluminum platelets in the fluid; here x lies in the surface of the outer cylinder, which we denote by Ω. We call u an *observable* of the state U. Observables can be used to understand pattern when the transformation from U to u is continuous and Γ-equivariant.

In the BZ reaction the observable u is the concentration of an active chemical and $\Omega = \mathbf{R}^2$; while in the flame experiment u is the intensity of light (or heat) produced by the flame and $\Omega \subseteq \mathbf{R}^2$ is a circular disk. In PDE systems, u is usually a function of the solution vector — often one of its components.

We define a *pattern* at time t to be the region in observable space

$$\mathcal{P}(t) = \{x \in \Omega : u(x, t) \geq c\}$$

for some fixed scalar c. For example, in the BZ reaction, the pattern is the region where the observed color is red (or blue). This region consists of those points in the petri dish where a chemical concentration is larger than some critical concentration.

Note that patterns associated to rotating wave solutions have a particularly elementary structure

$$\mathcal{P}(t) = R_t \mathcal{P}(0)$$

where R_t is rotation through angle t (in appropriate units). So, for example, a spiral wave is a pattern in the concentration of a fixed chemical in physical space that rigidly rotates at constant speed as time evolves.

On bifurcation to quasiperiodic motion, the change in the pattern consists of two parts: shape change and motion change. It is our contention that in Hopf bifurcation from a rotating wave the shape change is *less important* for the observed pattern evolution than is the change in the rigid motion. To make this point precise, we introduce the notions of inner and outer patterns, which are defined using the center bundle and are valid only near bifurcation.

Inner and Outer Patterns of Modulated Rotating Waves. Let

$$v(x, t) = R_t v_0(x)$$

be an observable of a rotating wave solution $V(x, t) = R_t V_0(x)$ to the PDE

$$U_t = \mathcal{F}(U) \tag{6.12}$$

with isotropy subgroup Σ_{V_0}. The center bundle theory (Krupa [329] for compact Γ and Sandstede *et al.* [459, 460] for noncompact Γ) shows that in a neighborhood of this rotating wave the vector field $\mathcal{F}$ has the decomposition

$$\mathcal{F} = \mathcal{F}_N + \mathcal{F}_T$$

where $\mathcal{F}_T$ is tangent to group orbits, $\mathcal{F}_N$ is transverse to group orbits, and both vector fields are Γ-equivariant. Moreover, if we let $N_0 \subseteq \mathcal{H}$ be the normal section to the group orbit at V_0, then $\mathcal{F}_N$ restricts to $g : N_0 \to N_0$ where $g(V_0) = 0$. Note that g is a Σ_{V_0}-equivariant vector field on N_0.

One result of the center bundle construction is that in a neighborhood of a rotating wave, all solutions can be written as

$$U(X,t) = \gamma_t Y(X,t)$$

where $\gamma_t \in \Gamma$ is a smooth curve and $Y(x,t) \in N_0$ is a solution to the normal vector field equation

$$Y_t = g(Y,\lambda) \tag{6.13}$$

Therefore, Hopf bifurcation from a rotating wave reduces to a Hopf bifurcation from an equilibrium in (6.13) coupled with a drift along group orbits. Since the transformation from states to observables is assumed to be Γ-equivariant, we obtain the identity

$$u(x,t) = \gamma_t y(x,t) \tag{6.14}$$

where u is the observable of U and y is the observable of Y.

Suppose now that $y(x,t)$ is time-periodic with corresponding modulated rotating wave solution $u(x,t) = \gamma_t y(x,t)$, as in (6.14). We obtain a time-periodic pattern (originating from the normal equations):

$$\mathcal{P}_y(t) = \{x \in \Omega : y(x,t) \geq c\}$$

Definition 6.29 The *inner pattern* associated to the modulated rotating wave u is

$$\mathcal{Q}^{inner} = \bigcap_t \mathcal{P}_y(t) = \{x \in \Omega : \min_t y(x,t) \geq c\}$$

The *outer pattern* associated to u is

$$\mathcal{Q}^{outer} = \bigcup_t \mathcal{P}_y(t) = \{x \in \Omega : \max_t y(x,t) \geq c\}$$

Definition 6.29 implies that for every t

$$\mathcal{Q}^{inner} \subseteq \mathcal{P}_y(t) \subseteq \mathcal{Q}^{outer}$$

Since patterns for the modulated rotating wave u satisfy

$$\mathcal{P}_u(t) = \gamma_t \mathcal{P}_y(t)$$

it follows that

$$\gamma_t \mathcal{Q}^{inner} \subseteq \mathcal{P}_u(t) \subseteq \gamma_t \mathcal{Q}^{outer} \tag{6.15}$$

Thus, the pattern of the modulated rotating wave evolves in time, bounded by a time-dependent rigid motion of the region between $\mathcal{Q}^{inner}$ and $\mathcal{Q}^{outer}$.

For example, when $\Gamma = \mathbf{O}(2)$ the pattern associated with the modulated wave is bounded between two patterns that rotate rigidly with nonuniform speed. Within a small error, the modulated wave pattern itself appears to rotate rigidly with nonuniform speed, the error due to the shape change being bounded between the inner and outer patterns.

In this discussion we have suppressed the dependence of the modulated wave on the bifurcation parameter λ. We now note that when λ is near the bifurcation point, then $\mathcal{Q}^{inner}$ and $\mathcal{Q}^{outer}$ will be approximately equal to $\mathcal{P}_v(0)$ — the pattern of the rotating wave at a fixed moment in time. Indeed, these sets are all equal at the bifurcation point of λ since $y = v_0$ there.

To verify that the visible pattern associated with the modulated wave is nearly identical to the visible pattern of the rotating wave at an instant in time, we need one additional assumption. We need to assume that the level contour $\{x \in \Omega :$

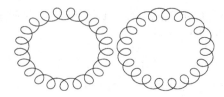

Figure 6.19: Epicycle motion of spiral tip: outward and inward petals [218].

$v_0(x) = c$} is a manifold, which is valid for almost all c. If not, the pattern can undergo a bifurcation due solely to the observation process.

6.7 Meandering of Spiral Waves

Spiral waves are a common phenomenon in PDEs with planar Euclidean symmetry, such as reaction-diffusion equations. Such waves have been observed both in experiments [519, 301, 361] and in numerical simulations [301, 47]. See Section 4.6 for a discussion of spiral waves in a disk. In suitable circumstances, the spiral can *meander* — its tip describes complicated 'epicyclic' paths in the plane, see Figures 6.18 and 6.19. (Classically, such curves are called epicycloids or hypocycloids, depending on which way the 'petals' point.) Barkley [44] showed that meandering of the spiral tip is a consequence of Euclidean symmetry.

In more detail: planar spirals rotate rigidly, so the tip of the spiral traces out a circle in the plane. Winfree [519] observed that under certain circumstances the tip can meander, creating flower-like epicyclic patterns of movement. These motions are quasiperiodic two-frequency motions, and can be thought of as an epicyclic motion superimposed on the basic spiral wave circle. When the motion on the epicycle is in the same orientation as the motion on the circle (either clockwise or counterclockwise), then the petals of the flowers point in; when the motions have the opposite orientation, the petals point out. Winfree observed both types of quasiperiodic motion and the possibility that the direction of the petals can change — we call this a change in *petality* — as a system parameter is varied.

Heuristic Description of Unbounded Tip Motion. The epicyclic motion of the spiral tip can be written phenomenologically as

$$q(t) = e^{i\omega_1 t}(z_1 + e^{-i\omega_2 t} z_2) \tag{6.16}$$

where $z_1 \in \mathbf{R}$ and $z_2 \in \mathbf{C}$. In these coordinates the change in petality occurs when $\omega_1 = \omega_2$.

In the epicyclic motion (6.16), Hopf bifurcation corresponds to the secondary amplitude $z_2 = 0$. From the standard bifurcation-theoretic viewpoint there is nothing significant about Hopf bifurcation at this critical parameter value. However, in Barkley's numerical simulation [42] and in experiments such as those by Li *et al.* [361] another phenomenon is observed. As the change in petality is approached, the amplitude of the second frequency grows unboundedly large in t. When $\omega_1 = \omega_2$, just compute

$$\int_0^t q(s)ds = \frac{1}{i\omega_1}e^{i\omega_1 t}z_1 + tz_2$$

In particular, at the point of petality change, the spiral tip appears to drift off to infinity in a straight line, see Figure 6.20. Thus, unbounded *growth* of the second frequency amplitude is a feature that seems to be connected with change in petality.

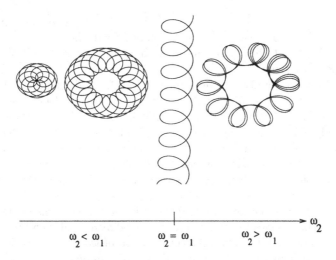

Figure 6.20: Growth of flower near change in petality [218]: path of $\int_0^t q(s)ds$, where $q(t)$ is as in (6.16) with $\omega_1 = 1$, $z_1 = 1$, $z_2 = 0.3$ and $\omega_2 = 0.61, 0.85, 1, 1.11$.

For changes in petality to be observed, the amplitude of the epicycle should be large. However, near Hopf bifurcation this amplitude must be small. This dichotomy suggests that standard Hopf bifurcation by itself cannot provide an explanation for petality change and unbounded growth. Barkley [44] made the crucial observation that Euclidean symmetry — coupled with Hopf bifurcation — is the 'real' source of the unbounded growth that accompanies changes in petality. Figure 6.21 also presents evidence that the formation of petals themselves is also based on Euclidean symmetry.

Euclidean Symmetry. Consider a reaction-diffusion system in the unbounded plane. Such equations have Euclidean symmetry. Suppose the system has a spiral wave solution and that the time-periodic spiral wave undergoes a Hopf bifurcation to a quasiperiodic motion. At the point of Hopf bifurcation, symmetry forces (at least) five eigenvalues of the linearized system to be on the imaginary axis — two generated from Hopf bifurcation and three generated by Euclidean symmetry. Assuming that some kind of 'center manifold' exists — see Section 6.2 — then the time evolution of the meandering spiral (and hence the time evolution of the meandering spiral tip) can be described by a five-dimensional Euclidean-equivariant system of ODEs. Three variables of this system represent the Euclidean group — the translation variable $p \in \mathbf{R}^2 \cong \mathbf{C}$ and the rotation variable $\varphi \in \mathbf{S}^1$ — and the variable $q \in \mathbf{C}$ represents the amplitude of the eigenfunction of Hopf bifurcation. In these variables Barkley [44] notes that translational symmetry acts by

$$T_x(p, \varphi, q) = (p + x, \varphi, q)$$

Therefore the ODE vector field is independent of p, and the (φ, q) equations decouple: see Lemmas 6.33 and 6.34.

Barkley [43] performed a numerical linear stability analysis for the basic time-periodic spiral wave solution and showed that there is a Hopf bifurcation. In particular, a simple pair of eigenvalues crosses the imaginary axis while three neutral eigenvalues lie on the imaginary axis and the remainder of the spectrum is bounded into the left-half plane. Starting from Barkley's numerical calculation, Wulff [528] proved, using Liapunov-Schmidt reduction, that resonant unbounded growth occurs in Hopf bifurcation near the codimension two point where $\omega_1 = \omega_2$. Her proof is nontrivial, because technical difficulties, such as the nonsmoothness of the group action, must be overcome.

Here we describe an alternative to the methods in [528]: an approach to bifurcations from rotating waves in Euclidean-equivariant differential equations based on center bundles. Our exposition follows Golubitsky, LeBlanc, and Melbourne [218], who also study bifurcation from many armed spirals. Biktashev $et\ al.$ [63] obtain a reduction for one-armed spirals by orbit space reduction methods. For many-armed spirals, the center bundle reduction has the advantage of not introducing singularities. Our exposition here focuses exclusively on Hopf bifurcation from one-armed spirals.

The group orbit of the rotating spiral wave is three-dimensional and the center subspace corresponding to Hopf bifurcation is two-dimensional, so we obtain a five-dimensional center bundle. With the construction of this bundle, we recover the action of the Euclidean group in Barkley's five-dimensional model. The general equivariant vector field on the center bundle can then be analyzed.

The center bundle reduction of Krupa [329] is formulated under the assumption that the symmetry group Γ is $compact$ and hence is not directly applicable to the problem of meandering spirals. However, it turns out that the main requirement is compactness of the isotropy subgroups of points on the critical group orbit. Sandstede $et\ al.$ [459, 460] and Golubitsky $et\ al.$ [218] have proved, under certain hypotheses, that Krupa's theorems are valid even when the group of symmetries is not compact and acts on an infinite-dimensional function space. In this theorem, it suffices that the isotropy subgroups are either discrete or compact, which is the case for spiral solutions since their isotropy subgroups are finite.

Group Action on the Center Bundle. We continue to use the notation from Section 6.2. In particular, x_0 is a point with isotropy Σ_{x_0} and $X = \Gamma x_0$ is a critical relative equilibrium for the Γ-equivariant vector field f on $N(X)$. Recall that the action of Γ on points $(x, v) \in N(X)$ is given by $\gamma(x, v) = (\gamma x, \gamma v)$.

Define the Σ_{x_0}-equivariant vector field $g : N_{x_0} \to N_{x_0}$ as in Section 6.2 and let $V_0 = V_{x_0}$ denote the Σ_{x_0}-invariant center subspace for g with corresponding Γ-invariant center bundle $V = \cup\{\gamma V_0\}$. Although $N(X)$ is a trivial bundle, the subbundle V is not necessarily trivial. We now give a sufficient condition for V to be a trivial bundle — and hence to have globally defined coordinates.

Lemma 6.30 $Suppose\ that\ the\ action\ \rho\ of\ \Sigma\ on\ V_0\ extends\ to\ an\ action\ \rho\ of\ \Gamma\ on$ $V_0.\ Then\ there\ is\ a\ trivialization$

$$V \cong X \times V_0$$

$The\ action\ of\ \Gamma\ on\ V\ is\ given\ by$

$$\gamma(x, v) = (\gamma x, \rho_\gamma v)$$

$where\ \gamma \in \Gamma,\ x \in X,\ v \in V_0.$

$Proof.$ Let $(x, w) \in V$, so $x \in X$ and $w \in V_x$ where V_x is the fiber over x. Write $x = \delta x_0$ where $\delta \in \Gamma$ and observe that $V_x = \delta V_0$. Hence $\delta^{-1} w \in V_0$. Define the trivialization $h : V \to X \times V_0$ by

$$h(x, w) = (x, \rho_\delta(\delta^{-1} w))$$

To show that h is well-defined, suppose that $x = \delta_1 x_0 = \delta_2 x_0$ where $\delta_1, \delta_2 \in \Gamma$. Then $\delta_2^{-1}\delta_1 = \sigma \in \Sigma$. The assumption on the action ρ ensures that $\rho_\sigma v = \sigma v$ for all $v \in V_0$. We compute that

$$\rho_{\delta_2}(\delta_2^{-1}w) = \rho_{\delta_2}\rho_\sigma(\sigma^{-1}\delta_2^{-1}w) = \rho_{\delta_2}\rho_{\delta_2^{-1}\delta_1}(\delta_1^{-1}\delta_2\delta_2^{-1}w) = \rho_{\delta_1}(\delta_1^{-1}w)$$

hence h is well-defined.

Next, we verify the action of Γ on $X \times V_0$. In other words, we show that h is Γ-equivariant with respect to the actions on V and $X \times V_0$. Let $(x, w) \in V$ as at the beginning of the proof. Then

$$
\begin{aligned}
h(\gamma(x, w)) &= h(\gamma x, \gamma w) = (\gamma x, \rho_{\gamma\delta}((\gamma\delta)^{-1}\gamma w)) \\
&= (\gamma x, \rho_\gamma\rho_\delta(\delta^{-1}w)) = \gamma(x, \rho_\delta(\delta^{-1}w)) = \gamma h(x, w)
\end{aligned}
$$

as required. $\square$

The next corollary includes the case of one-armed spirals where $\Sigma_{x_0} = 1$.

Corollary 6.31 *Suppose that Σ_{x_0} acts trivially on V_0. Then $V \cong X \times V_0$ is a trivial bundle and Γ acts as $(x, v) \to (\gamma x, v)$.*

The Center Bundle for Spirals. Now suppose that $\Gamma = \mathbf{SE}(2)$, the *special Euclidean group* consisting of rotations and translations. We suppose that $X = \mathbf{SE}(2)x_0$ is a relative equilibrium where x_0 is an 1-armed spiral. As a manifold $\mathbf{SE}(2)$ is diffeomorphic to $\mathbf{R}^2 \times \mathbf{S}^1$. The assumptions on the symmetry of x_0 imply that

$$X = \mathbf{SE}(2)x_0 \cong \mathbf{C} \times \mathbf{S}^1$$

That is, X is a cylinder with coordinates (p, φ).

Lemma 6.32 *The action of $(x, \theta) \in \mathbf{SE}(2)$ on $(p, \varphi) \in X$ is*

$$(x, \theta)(p, \varphi) = (e^{i\theta}p + x, \varphi + \theta) \tag{6.17}$$

Proof. To verify (6.17), note that the action of $\mathbf{SE}(2)$ on X is induced by the action of group multiplication in $\mathbf{SE}(2)$. Group multiplication in $\mathbf{SE}(2)$ is most easily understood through the action of $\mathbf{SE}(2)$ on $\mathbf{R}^2 = \mathbf{C}$. Let $w \in \mathbf{C}$; then

$$(x, \theta)w = e^{i\theta}w + x$$

It follows that

$$(x, \theta)(y, \psi)w = (x, \theta)(e^{i\psi}w + y) = e^{i\theta}(e^{i\psi}w + y) + x = e^{i(\theta+\psi)}w + (e^{i\theta}y + x)$$

Hence, the group multiplication on $\mathbf{SE}(2)$ induced by its action on $\mathbf{C}$ is

$$(x, \theta)(y, \psi) = (e^{i\theta}y + x, \psi + \theta)$$

Substituting (p, φ) for (y, ψ) gives the action of $\mathbf{SE}(2)$ on X. $\square$

Lemma 6.33 *Hopf bifurcation from an one-armed spiral reduces generically to Hopf bifurcation of a five-dimensional vector field on a trivial center bundle*

$$V = X \times V_0$$

that is equivariant under the action

$$
\begin{aligned}
T_x(p, \varphi, q) &= (p + x, \varphi, q) \\
R_\theta(p, \varphi, q) &= (e^{i\theta}p, \varphi + \theta, q)
\end{aligned}
\tag{6.18}
$$

Proof. Whether V is a trivial bundle is independent of the $\mathbf{C}$ factor in X. Corollary 6.31 implies that V is a trivial bundle. The action of $\mathbf{SE}(2)$ on the X-coordinates follows from (6.17). The action on the V-coordinates follows from Lemma 6.30. $\square$

Meandering and Resonant Growth of a One-Armed Spiral. In this subsection, we derive the general equivariant vector field on the center bundle and solve the resulting equations to obtain the conditions for resonant growth. We then interpret these results in the context of Hopf bifurcation in a PDE, recovering the results of Barkley and Wulff on the meandering and resonant growth of spirals.

Equivariant Vector Fields on the Center Bundle. We begin by finding the restrictions placed on a vector field on the center bundle by equivariance:

Lemma 6.34 *Let F be a system of differential equations on the center bundle V that is* **SE**(2)-*equivariant with respect to the (6.18) action. Then F has the form*

$$\begin{aligned}
\dot{p} &= e^{i\varphi} f(q) \\
\dot{\varphi} &= F^\varphi(q) \\
\dot{q} &= F^q(q)
\end{aligned} \tag{6.19}$$

Proof. Symmetry-invariance of a system of differential equations means that solutions are transformed to solutions by that symmetry. Thus, translation invariance implies that if

$$z(t) = (p(t), \varphi(t), q(t))$$

is a solution to (6.19), then so is

$$y(t) = (p(t) + x, \varphi(t), q(t))$$

for any $x \in \mathbf{C}$. Since $\dot{z}(t) = \dot{y}(t)$, it follows that

$$F(z(t)) = F(y(t))$$

for all solutions $z(t)$. In particular,

$$F(p + x, \varphi, q) = F(p, \varphi, q)$$

for all x. Hence F is independent of p and the differential equations have the form

$$\begin{aligned}
\dot{p} &= F^p(\varphi, q) \\
\dot{\varphi} &= F^\varphi(\varphi, q) \\
\dot{q} &= F^q(\varphi, q)
\end{aligned} \tag{6.20}$$

Similarly, the rotational invariance of (6.20) implies that

$$\begin{aligned}
F^p(\varphi + \theta, q) &= e^{i\theta} F^p(\varphi, q) \\
F^\varphi(\varphi + \theta, q) &= F^\varphi(\varphi, q) \\
F^q(\varphi + \theta, q) &= F^q(\varphi, q)
\end{aligned}$$

Thus, F^φ and F^q are independent of φ, which verifies the second and third equations in (6.19).

To complete this proof we must verify the equation for $\dot{p}$ in (6.19). Define

$$H(\varphi, q) = e^{-i\varphi} F^p(\varphi, q)$$

and note that

$$H(\varphi + \theta, q) = e^{-i(\varphi + \theta)} F^p(\varphi + \theta, q) = e^{-i\varphi} e^{-i\theta} e^{i\theta} F^p(\varphi, q) = H(\varphi, q)$$

It follows that $H(\varphi, q) = f(q)$ is independent of φ and that the $\dot{p}$ equation in (6.19) is valid.   □

Periodic Solutions and Resonant Growth. Suppose that $q(t)$ is a $2\pi/\omega_2$-periodic solution to

$$\dot{q} = F^q(q)$$

in the center bundle equations (6.19). Define $\omega_1 = F^\varphi(q(0))$. (These frequencies are related to, but not identical to, the frequencies ω_1 and ω_2 that appear in the introduction, see (6.25).)

We can solve

$$\dot{\varphi} = F^\varphi(q(t))$$

for $\varphi(t)$ and

$$\dot{p} = e^{i\varphi(t)} f(q(t))$$

for $p(t)$ to obtain a solution $(p(t), \varphi(t), q(t))$ to (6.19). In the next theorem we recover the resonance conditions of Barkley and Wulff for these solutions.

Theorem 6.35 *Let $(p(t), \varphi(t), q(t))$ be a solution to (6.19). Then*

$$\varphi(t) = \omega_1 t + \tilde{\varphi}(t) \tag{6.21}$$

where $\tilde{\varphi}(t)$ is $2\pi/\omega_2$-periodic. If

$$\omega_1 + k\omega_2 = 0$$

for some integer k, then generically $p(t)$ undergoes unbounded resonant growth.

Proof. The function $F^\varphi(q(t))$ is $2\pi/\omega_2$-periodic since $q(t)$ is. Therefore we can write $F^\varphi(q(t))$ as a Fourier series in t, obtaining

$$\dot{\varphi} = \sum_{n=-\infty}^{\infty} B_n e^{in\omega_2 t}$$

where $B_n \in \mathbf{C}$ and $B_{-n} = \overline{B_n}$. Every term except $n = 0$ in the Fourier series yields a periodic function on integration and hence $\varphi(t)$ has the form in (6.21) where $\omega_1 = B_0 = F^\varphi(q(0))$.

Next, consider the differential equation

$$\dot{p} = e^{i\varphi(t)} f(q(t)) = e^{i\omega_1 t} H(t) \tag{6.22}$$

where $H(t)$ is smooth and $2\pi/\omega_2$-periodic. We may write $H(t)$ as the uniformly convergent Fourier series

$$H(t) = \sum_{n=-\infty}^{\infty} D_n e^{in\omega_2 t}$$

where $D_n \in \mathbf{C}$.

Suppose that $\omega_1 + k\omega_2$ is close to zero for some nonzero integer k. Then integration of (6.22) yields

$$p(t) = \begin{cases} D_k t + \mathcal{P}(t) e^{i\omega_1 t} & \omega_1 + k\omega_2 = 0 \\ \dfrac{D_k}{i(\omega_1 + k\omega_2)} e^{i(\omega_1 + k\omega_2)t} + \mathcal{P}(t) e^{i\omega_1 t} & \omega_1 + k\omega_2 \neq 0 \end{cases}$$

where $\mathcal{P}(t)$ is a smooth bounded $2\pi/\omega_2$-periodic function. Generically, $D_k \neq 0$. Hence, by varying ω_2 so that $\omega_1 + k\omega_2$ goes through zero, the first summand in $p(t)$ undergoes unbounded resonant growth, while $\mathcal{P}(t)$ remains uniformly bounded for these values of ω_2. $\square$

Hopf Bifurcation from a One-Armed Spiral. Let $\mathcal{H}$ be a space of functions with domain $\mathbf{R}^2$ on which the Euclidean group $\mathbf{E}(2)$ acts as

$$\gamma u(z) = u(\gamma^{-1}z)$$

where $u \in \mathcal{H}$ and $\gamma \in \mathbf{E}(2)$. Consider a PDE

$$u_t = \mathcal{F}(u, \lambda) \tag{6.23}$$

where $\mathcal{F} : \mathcal{H} \mapsto \mathcal{H}$ is $\mathbf{E}(2)$-equivariant and λ is a real bifurcation parameter. Let R_θ denote rotation of the plane through angle θ. Suppose that

$$u(t) = R_{\omega_1 t} x_0 \tag{6.24}$$

is a rotating wave solution to (6.23) with period $2\pi/\omega_1$.

Let $X = \mathbf{SE}(2)x_0$ be the connected component of the group orbit of $u(t)$ in phase space under the action of $\mathbf{E}(2)$. As noted by Rand [440], Renardy [443] and others, it is possible to study bifurcation from rotating waves by transferring the problem to the rotating frame. Substituting (6.24) into (6.23) yields that x_0 is an equilibrium for the equation

$$u_t = \widetilde{\mathcal{F}}(u, \lambda) = \mathcal{F}(u, \lambda) - \omega_1 \xi u$$

where

$$\xi u = \left. \frac{d}{dt} R_t u \right|_{t=0}$$

The operator $(d\widetilde{\mathcal{F}})_{x_0, \lambda}$ has three eigenvalues on the imaginary axis corresponding to the continuous group orbit $\mathbf{SE}(2)$. Barkley [43] showed numerically that the rotating wave $u(t)$ can undergo a Hopf bifurcation as an additional simple pair of eigenvalues crosses the imaginary axis. We suppose that this bifurcation occurs at $\lambda = 0$. Let $V_0 \cong \mathbf{C}$ be the corresponding center subspace.

Theorem 6.36 *There exists a reduction of (6.23) to the center bundle $V = X \times V_0$. The reduced equations have the form*

$$\dot{y} = F(y, \lambda),$$

where $y = (p, \varphi, q) \in V$ and F has the form (6.19).

Proof. See Sandstede, Scheel and Wulff [459, 460]. $\square$

The reduction procedure implies that

$$F^\varphi(0, 0) = \omega_1 \quad F^q(0, 0) = 0 \quad f(0, 0) = 0$$

In equation (6.19), the original rotating wave solution corresponds to the equilibrium $q = 0$. Also, generically the critical eigenvalues cross the imaginary axis transversely as λ varies. Consequently, the vector field $F^q(q, \lambda)$ on V_0 satisfies

$$d_q F^q(0, 0) = i\omega_2 \quad \operatorname{Re} d_q F_\lambda^q(0, 0) \neq 0.$$

Thus, there is a Hopf bifurcation in the $\dot{q}$ equation of (6.19) to an approximately $2\pi/\omega_2$-periodic solution $q(t)$. We suppose that the Hopf bifurcation is supercritical.

The amplitude and frequency of the periodic solution $q(t, \lambda)$ vary as functions of λ. To leading order, the amplitude varies as $a\sqrt{\lambda}$ and the frequency varies as $\omega_2 + b\lambda$ where a and b are real coefficients. We set $\omega_1(\lambda) = F^\varphi(q(0, \lambda))$ and define $\omega_2(\lambda)$ to be the frequency of $q(t, \lambda)$. Thus

$$\omega_1(\lambda) = \omega_1 + O(\sqrt{\lambda}) \qquad \omega_2(\lambda) = \omega_2 + O(\lambda) \tag{6.25}$$

Note that $\omega_j(0)$ coincides with ω_j as defined in this subsection and in the introduction for each $j = 1, 2$. On the other hand, the ω_j in Subsection 6.7 correspond to $\omega_j(\lambda)$ evaluated at a specific value of λ.

Theorem 6.35 implies that linear meandering occurs at $\lambda = \lambda_0$ if

$$\omega_1(\lambda_0) + k\omega_2(\lambda_0) = 0$$

for some integer k. We call this resonance a k-resonance. In particular, resonant growth occurs when $\omega_1(0) + k\omega_2(0)$ is close to zero for some integer k. We can expect unbounded growth in $p(t)$ as λ approaches the resonance point, and linear drifting in $p(t)$ at the resonance point. However, by inspection of pictures, only when $k = \pm 1$ or $k = \pm 2$ do the concepts of petality and changes in petality appear to be relevant.

Visualization of Hopf Bifurcation from a One-Armed Spiral. To illustrate resonance and petality issues, Figure 6.21 shows results of numerical integration of the equations

$$\begin{aligned} \dot{p} &= e^{i\varphi}(0.2 - 0.6i)q \\ \dot{\varphi} &= 1 \\ \dot{q} &= (\lambda - 0.95i)q - (1 - 0.1i)q|q|^2 \end{aligned} \quad (6.26)$$

for six values of λ. Each figure is a plot of (f_1, f_2) where

$$f_1 = \cos\varphi(t) + \mathrm{Re}\ p(t) \qquad f_2 = \sin\varphi(t) + \mathrm{Im}\ p(t)$$

These coordinates approximate the movements of the spiral tip in the laboratory frame. The $\dot{q}$ equation in (6.26) undergoes a supercritical Hopf bifurcation at $\lambda = 0$, and the frequency of the corresponding periodic solution is $\omega_2(\lambda) = -0.95 - 0.1\lambda$. Since $\omega_1(\lambda) \equiv 1$, there is a resonance at $\lambda = 0.5$.

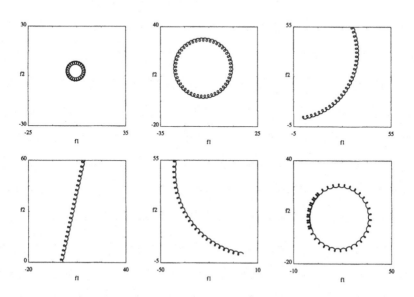

Figure 6.21: Numerical simulation of (6.26) for $\lambda = 0.1, 0.3, 0.4, 0.5, 0.6$ and 0.8 [218].

Chapter 7

Hidden Symmetry and Genericity

In Chapter 6 we discussed ways in which symmetries can be modeling assumptions in physical systems, and we framed this discussion specifically in terms of models of the Couette-Taylor experiment and the Belousov-Zhabotinskii experiment. We saw that symmetries may be exact, at least for modeling purposes, or they may be approximate. In models of the Couette-Taylor experiment, approximate symmetries appear because we assume periodic boundary conditions in the axial direction; in models of the Belousov-Zhabotinskii experiment they appear because we assume an infinitely large domain. We saw that model-independent results — those that depend only on the symmetries of the model — tell us a great deal about transitions that are actually seen in experiments.

Model-independent results are based on genericity arguments in the world of equivariant bifurcation theory. Having determined the appropriate symmetry context for analyzing a problem, we can apply mathematical techniques to work out what the generic behavior should be. A key question here is not just 'which symmetry group do we have?' but 'how does it act, and on what space?'

Our underlying philosophy is that the most common behavior should — on the whole — be 'generic'. This term has a formal technical meaning, but intuitively it means that the behavior should be 'typical' — not destroyed by a small perturbation of the equations. The reason for requiring genericity is that real systems are always subject to small perturbations, and the predictions of the model should not be seriously changed by such a perturbation. We emphasize that here we are considering perturbations of the *model*, not just of initial conditions. In other words, this issue differs from the question of the stability of a given solution, and is considerably deeper.

In this chapter we focus on a new kind of symmetry — 'hidden symmetry' — and a new twist on the issue of genericity. Hidden symmetries appear in models posed on certain domains and with certain boundary conditions. In these specific cases, the mathematical model can be embedded in a larger one with more symmetry, and it is the symmetries of the larger model that dictate the generic behavior of transitions in the smaller model.

The prototypical example is one first considered by Fujii, Mimura, and Nishiura [202] and later by Armbruster and Dangelmayr [8, 135]. The example is a reaction-diffusion equation on the interval $[0, 1]$ with Neumann boundary conditions. Such equations have only one nontrivial symmetry generated by the domain reflection $x \mapsto 1 - x$. Any solution $u(x, t)$ to this equation can be extended to the interval $[-1, 1]$ by setting $u(-x, t) = u(x, t)$. On the enlarged interval the solution satisfies the same reaction-diffusion equation — but we can now impose periodic boundary conditions because the values of u at $x = -1, 1$ are the same. As we know, periodic boundary conditions introduce $O(2)$ symmetry into the model. We discuss this example in more detail in Section 7.2. In this discussion we show that the existence of $\mathbf{O}(2)$ symmetry on the enlarged interval changes the generic behavior expected of steady-state bifurcations in equations on the small interval.

Can the effects of hidden symmetries in models be seen in experiments? The answer is 'yes'. First, in the Couette-Taylor experiment, the presence of both Taylor vortices and anomalous Taylor vortices in the experiments of Mullin [412] using a short cylinder can best be understood through the existence of a hidden symmetry, see Section 7.2. Second, the theoretical work of Crawford *et al.* [130] and the experiments of Crawford, Gollub, and Lane [129] show that a number of features of the Faraday experiment in a container with a square cross section can best be understood using hidden symmetries, see Sections 7.1 and 7.3. It is perhaps remarkable that symmetries appearing only in mathematical extensions of a model have effects that are actually seen in experiments.

The remainder of the chapter considers the effects of hidden symmetries in other bifurcations (mode interactions, Sections 7.4 and 7.5) and on different domains (hemispheres, Section 7.6).

7.1 The Faraday Experiment

Consider a layer of fluid in a square tray resting on a device that vibrates vertically with frequency ω and amplitude A, as in Figure 7.1. Faraday originally suggested this experiment as a model for studying surface waves. For small amplitude vibrations the surface is undeformed, but as A increases past some critical value A_0 the surface begins to deform.

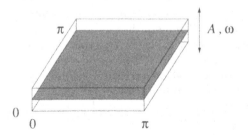

Figure 7.1: Faraday experiment: A = vibrational amplitude, ω = frequency.

What are the symmetries for this problem? Certainly the shape of the container induces $\mathbf{D}_4$ symmetry. However, there is more symmetry, and to see why, we set up a mathematical description of how the surface changes as a result of the forcing.

Definition 7.1 Let S denote a surface. Let $\mathcal{S}(S)$ denote the surface after one period of forcing. The map $\mathcal{S}$ from surfaces to surfaces is the *stroboscopic map*.

After bifurcation a deformed surface S_0 exists, as shown in Figure 7.2. Experiments show that for this surface, $S_1 = \mathcal{S}(S_0)$ is different from S_0, while $\mathcal{S}(S_1) = S_0$. That is, the period of oscillation of the surface is twice that of the forcing. It is therefore reasonable to assume that a period-doubling bifurcation (see Section 4.9) has taken place.

Experimentally these deformed surfaces may be categorized by two integers, (m, n), which represent the wave numbers of S_0 in the x and y directions, see [113],

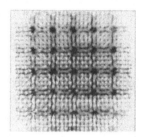

Figure 7.2: Deformed surface for Faraday experiment in a square container [211]. Picture courtesy of J.P. Gollub.

[476], and Figure 7.3. If $m \neq n$, then the kernel has dimension 2. The waveform on S_0 is of the form $\cos(mx)\cos(ny)$, so in this case

$$\mathcal{K} \supseteq \{a\cos(mx)\cos(ny) + b\cos(nx)\cos(my) : a, b \in \mathbf{R}\}$$

and generically these space are equal, which we now assume. If, for example, we choose $(m, n) = (4, 2)$, then the symmetry

$$\gamma_1 : (x, y) \mapsto (y, x)$$

acts on $\mathcal{K}$ by

$$\gamma_1(a, b) = (b, a)$$

In addition, the symmetry

$$\gamma_2 : (x, y) \mapsto (\pi - x, y)$$

also acts on $\mathcal{K}$ by

$$\gamma_2(a, b) = (a, b)$$

That is, γ_2 acts trivially. Similarly $(x, y) \mapsto (x, \pi - y)$ acts trivially on (a, b). Therefore $\mathbf{D}_4$ acts on $\mathcal{K}$ as $\mathbf{Z}_2 = \langle \gamma_1 \rangle$ with irreducible subspaces defined by $a = b$ and $a = -b$.

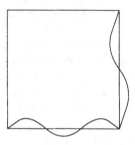

Figure 7.3: Picture of mode with $(m, n) = (3, 2)$.

Thus we have found a surprise — the symmetry group $\mathbf{D}_4$ does not act irreducibly at the point of transition, a nongeneric phenomenon. Why does this happen? The source of this apparent nongenericity is discussed in Section 7.3, but first we develop a theory of 'hidden symmetry' in certain PDEs.

7.2 Hidden Symmetry in PDEs

It is convenient to develop the ideas of hidden symmetry for a class of PDEs that is especially well-behaved. We therefore work with a reaction-diffusion equation

$$\Delta u + f(u, \lambda) = 0$$

We consider three types of boundary condition:

- Periodic Boundary Conditions (PBC) on the interval $[-\pi, \pi]$:

$$u(-\pi) = u(\pi)$$
$$u'(-\pi) = u'(\pi)$$

- Neumann Boundary Conditions (NBC) on the interval $[0, \pi]$:

$$u'(0) = u'(\pi) = 0$$

- Dirichlet Boundary Conditions (DBC) on the interval $[0, \pi]$:

$$u(0) = u(\pi) = 0$$

While analyzing an ecological model, Fujii et al. [202] realized that, for reaction-diffusion equations on an interval, NBC can give rise to additional symmetries. Armbruster and Dangelmayr [8, 135] used singularity theory to deduce the implications for one-parameter steady-state bifurcations of low codimension, and showed that the Liapunov-Schmidt reduced bifurcation equations take a particular 'prenormal form'. Later Crawford et al. [130] extended these observations to boundary value problems on multidimensional rectangular domains, with suitable mixed Neumann or Dirichlet boundary conditions. Gomes [239] and Gomes and Stewart [241] showed that the corresponding steady-state bifurcations, in the abstract, reduce to the same class of singularities that was studied by Dangelmayr and Armbruster [135]. Gomes [239] also developed algorithms for finding prenormal forms for domains with additional symmetry (boxes with square cross-section, cubes) and interactions of more than two modes. For multidimensional rectangular domains Gomes and Stewart [242] subsequently extended the ideas to Hopf bifurcation, a process begun by Castro [99], who also developed an analogous theory for DBC. Field et al. [192] applied the method to bifurcation on hemispheres (see Section 7.6), described a general class of domains to which the method applies, and proved a basic theorem on the regularity of extensions by reflection, which underpins the method. Ashwin [14] applied these ideas to the Kuramoto-Sivashinsky equation, a fourth order PDE with 'generalized Neumann' boundary conditions. Independently Healey and Kielhöfer [268, 269, 270] discovered the same basic principle in the buckling of a rectangular plate, and combined it with a positivity argument to prove a striking property of the geometry of nodal lines. Many other authors have observed various consequences of the hidden symmetries of Neumann/Dirichlet boundary value problems, especially linear degeneracies. Crawford [125, 126, 128] developed the theory for square domains, especially with regard to the Faraday experiment, and also discovered in [127] that hidden *rotational* symmetries can occur. Crawford et al. [129] obtained experimental confirmation of predictions for the Faraday experiment in a square container based on the presence or absence of hidden symmetries. The effect of hidden symmetry is thus widespread.

The main point behind all of the above work is that bifurcations with NBC have a surprising structure which changes the appropriate notion of genericity. To see why, view the PDE in operator form

$$\mathcal{P}(u, \lambda) = \Delta u + f(u, \lambda) = 0 \qquad (7.1)$$

where $\mathcal{P} : \mathcal{B} \times \mathbf{R} \to \mathcal{B}$, for some Banach space $\mathcal{B}$ of functions u on the interval $[0, \pi]$.

If we ignore the boundary of the interval and view the operator $\mathcal{P}$ as being defined on the whole of $\mathbf{R}$, then $\mathcal{P}$ has Euclidean symmetry given by translations and flips. More precisely, if we define the actions of θ, κ by

$$\begin{aligned} \theta u(x) &= u(x - \theta) \\ \kappa u(x) &= u(-x) \end{aligned}$$

then

$$\begin{aligned} \theta \mathcal{P}(u, \lambda) &= \mathcal{P}(\theta u, \lambda) \\ \kappa \mathcal{P}(u, \lambda) &= \mathcal{P}(\kappa u, \lambda) \end{aligned}$$

Periodic Boundary Conditions. The translational and flip symmetries of $\mathcal{P}$ suggest *extending* the functions $u \in \mathcal{B}$ to a space of 2π-periodic functions $\hat{u}$ on $\mathbf{R}$. To do this, define

$$\hat{u}(x) = u(-x) \quad (x \in [-\pi, 0])$$

and then extend $\hat{u}$ periodically to the whole of $\mathbf{R}$ with period 2π, obtaining a function that we continue to call $\hat{u}$. Thanks to NBC, it turns out that if u is smooth on $[0, \pi]$ and satisfies the reaction-diffusion equation, then $\hat{u}$ is smooth on $\mathbf{R}$ and also satisfies the reaction-diffusion equation (see Lemma 7.2). So instead of considering the original problem with NBC on $[0, \pi]$ we can consider the 'same' problem with PBC on $[-\pi, \pi]$.

With PBC the symmetries of $\mathcal{P}$ are all symmetries of the domain (which, by PBC, is the interval $[-\pi, \pi]$ *with ends identified*, that is, a circle). These are translations modulo the period ($\theta \in \mathbf{SO}(2)$) together with the flips $\kappa\theta$, so the overall symmetry group is $\mathbf{O}(2)$. This is considerably richer than the 'obvious' $\mathbf{Z}_2$ symmetry of the original domain given by $x \mapsto \pi - x$.

In particular, we can apply the representation theory of $\mathbf{O}(2)$. The irreducible representations of $\mathbf{O}(2)$ are either 1 or 2-dimensional. Breaking the translational symmetry implies a 2-dimensional bifurcation. The standard action of $\mathbf{O}(2)$ on $\mathbf{C} \cong \mathbf{R}^2$ is

$$\theta z = e^{i\theta} z$$

$$\kappa z = \bar{z}$$

Other actions, in which θ acts by $e^{ik\theta}$ for $k > 1$, reduce to the standard action if the kernel $\mathbf{Z}_k$ is factored out, so we concentrate on the standard action. The Equivariant Branching Lemma and $\mathrm{Fix}(\kappa) = \mathbf{R}$ imply that there exist solutions; the $\mathbf{O}(2)$-equivariance of $f : \mathbf{C} \times \mathbf{R} \to \mathbf{C}$ implies that

$$f(z, \lambda) = a(|z|^2, \lambda) z$$

Generically, this leads to a pitchfork of revolution bifurcation as in Figure 7.4, with circles of equilibria existing for $\lambda > 0$.

Neumann Boundary Conditions. In this case the only symmetry of the domain is a reflectional symmetry $\tau(x) = \pi - x$. Genericity implies a pitchfork bifurcation if $\mathbf{Z}_2$ acts nontrivially on the kernel and a limit point bifurcation if $\mathbf{Z}_2$ acts trivially. However the latter case does not happen here: see Corollary 7.3 below.

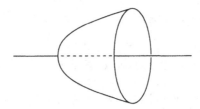

Figure 7.4: Pitchfork of revolution.

Lemma 7.2 *Bifurcations of (7.1) with NBC can be extended to bifurcations with PBC. Extensions of smooth solutions are smooth.*

Proof. Suppose $u(x)$ is a smooth solution of (7.1) on $[0, \pi]$ with NBC. As above, define $\hat{u}(\pm x) = u(x), x \in [0, \pi]$. Thus $\hat{u}$ is defined on $[-\pi, \pi]$ and $\hat{u}$ is even. We claim that the extended function $\hat{u}$ is also smooth, and satisfies (7.1) on $[-\pi, \pi]$ with PBC.

It is clear that $\hat{u}$ satisfies (7.1) on $[0, \pi]$ and that $\hat{u}$ is 2π-periodic and continuous. If $x \in [-\pi, 0]$ we compute:

$$\mathcal{P}(\hat{u}, \lambda) = \Delta u(-x) + f(u(-x), \lambda) = 0$$

since $-x \in [0, \pi]$ and Δ is invariant under reflection κ. Clearly $\hat{u}$ is smooth, except possibly when x is an integer multiple of π; by periodicity it suffices to prove $\hat{u}$ is smooth at $x = 0, \pi$. Consider x near 0: similar methods work near $x = \pi$. Denote left- and right-hand limits as $x \to 0$ by $\hat{u}(0^-), \hat{u}(0^+)$. Clearly

$$\hat{u}(0^-) = \hat{u}(0^+) = u(0)$$

and

$$\hat{u}'(0^-) = \hat{u}'(0^+) = u'(0)$$

Equation (7.1) relates $u''(x)$ to $u(x)$. Differentiating k times, we can express $u^{(k)}(x)$ in terms of $u^{(l)}(x)$ for $l < k$, whenever $k \geq 2$. Inductively, it follows that $\hat{u}^{(k)}(0^-) = \hat{u}^{(k)}(0^+)$ and smoothness is proved.

Finally, since $\hat{u}$ is 2π-periodic, it satisfies PBC on $[-\pi, \pi]$. □

Corollary 7.3 *With NBC on $[0, \pi]$, the generic steady-state bifurcation from the trivial solution is always a pitchfork.*

Proof. Consider the extended problem with PBC. This has a reflectional symmetry $x \mapsto -x$, so the generic bifurcation is a pitchfork. When we restrict back to $[0, \pi]$ the pitchfork structure remains. See Figure 7.5. □

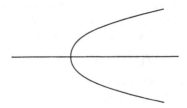

Figure 7.5: Pitchfork bifurcation with Neumann boundary conditions.

We can use a symmetry condition to determine which solutions of the extended problem with PBC are of the form $\hat{u}$ where u satisfies NBC on $[0, \pi]$.

Lemma 7.4 *Let v be a solution of the extended problem with PBC. Then $v = \hat{u}$ where u satisfies NBC if and only if v is fixed by κ.*

Proof. By construction every function $\hat{u}$ is fixed by κ. Conversely, suppose that v is fixed by κ. Then $v = \hat{v}|_{[0,\pi]}$. Moreover, the identity $v'(-x) = -v'(x)$ implies that $v'(0) = -v'(0)$, so $v'(0) = 0$, and similarly $v'(\pi) = 0$.   □

Thus any steady-state bifurcation with NBC can be extended to a bifurcation with PBC, and the NBC bifurcation can be recovered from the PBC one by using symmetry.

Remark 7.5 The reflection $\tau : x \mapsto \pi - x$ preserves the original domain $[0,\pi]$. Its action on a solution u of the NBC problem can either be trivial or nontrivial. If the action of τ is nontrivial (that is, $u(\pi - x) \neq u(x)$) then the two branches of the parabola are interchanged by τ. However, if τ acts trivially then the solutions on the upper branch of the pitchfork are not reflections by τ of the solutions on the lower branch. They are instead given by $x \mapsto x + \pi/2$, a quarter-period translation (in terms of the period of the PBC extension). See Figure 7.6. We illustrate this point in the next subsection.   ◇

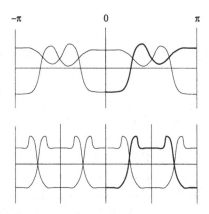

Figure 7.6: Effect of reflecting the domain $[0,\pi]$ and the two branches of the pitchfork. (Top) u not fixed by τ. (Bottom) u fixed by τ.

Taylor Vortices versus Anomalous Vortices. Consider the Couette-Taylor experiment in a short cylinder, as pictured in Figure 7.7. When Couette flow bifurcates to Taylor vortices, two flow patterns are observed in the experiments. To see this, we project the flow onto the cross-section Σ as shown in Figure 7.8. The standard Taylor vortices consist of two vortex-pairs, while the anomalous vortices have only one pair, plus two half-pairs at the ends of the cylinder. The anomalous vortices are an example of a pitchfork bifurcation where the lower branch is a quarter-period phase shift of the upper branch.

Wave Numbers. As we have seen, solutions to the NBC problem can be recovered from those to the PBC problem, since a solution $v(x)$ to the PBC problem is of the form $\hat{u}$ where u satisfies NBC on $[0,\pi]$ if and only if $v(-x) = v(x)$. This condition can be reformulated as $v \in \text{Fix}(\mathbf{Z}_2^\kappa)$ where

$$\mathbf{Z}_2^\kappa = \langle \kappa \rangle \qquad \kappa : x \mapsto -x \tag{7.2}$$

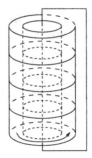

Figure 7.7: Couette-Taylor experiment in a short cylinder, and a cross-section Σ.

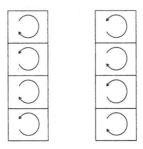

Figure 7.8: Cross-sections of flows in the Couette-Taylor experiment for a short cylinder: (left) Taylor vortices; (right) anomalous vortices.

The key point is that solutions $u(x)$ to the NBC problem have extensions $\hat{u}(x)$ to the PBC problem with nontrivial symmetries in the larger group $\mathbf{O}(2)$. The closed subgroups of $\mathbf{O}(2)$ containing $\mathbf{Z}_2^\kappa$ are the dihedral groups

$$\mathbf{D}_p = \langle \kappa, 2\pi/p \rangle \tag{7.3}$$

Because of the translational symmetries $2\pi/p$, solutions $u(x)$ to the NBC problem with symmetry group $\mathbf{D}_p$ are composed of p identical waves (that is, have *wave number* or *mode number* p). Thus the hidden symmetries of NBC are responsible, via the representation theory of $\mathbf{O}(2)$, for the occurrence of modes with well-defined wave numbers. The appropriate abstract context for the analysis of Euclidean-invariant PDEs with NBC on an interval is therefore that of the restriction of $\mathbf{O}(2)$-equivariant maps to $\mathrm{Fix}(\mathbf{Z}_2^\kappa)$. The concept of genericity (and the equivalence relation used in singularity theory) should be adapted to that context. As Dangelmayr and Armbruster [8, 135] show, the results are nontrivial and not especially intuitive. We state them in a more general context below, see Section 7.4.

Dirichlet Boundary Conditions. Castro [99] and Gomes [239] observe that a similar extension procedure applies to Dirichlet Boundary Conditions. However, we must now extend u to $\hat{u}$ by defining

$$\hat{u}(x) = \begin{cases} u(x) & \text{if } x \in [0, \pi] \\ -u(-x) & \text{if } x \in [-\pi, 0) \end{cases}$$

to preserve regularity. In order for this to work, f must satisfy an additional condition $f(-u) = -f(u)$. The appropriate symmetry group for PBC becomes a 'twisted' $\mathbf{O}(2)$ whose reflection κ' acts as

$$u(x) \mapsto -u(-x)$$

However, the $\mathbf{O}(2)$-action that results is isomorphic to the standard $\mathbf{O}(2)$ action, because minus the identity commutes with $\mathbf{O}(2)$ and has order two. Moreover, DBC solutions can be characterized group-theoretically as those that lie in $\mathrm{Fix}(\mathbf{Z}_2^{\kappa'})$. So the entire theory applies, with only minor changes in interpretation, to DBC.

7.3 The Faraday Experiment Revisited

Neumann boundary conditions are thought to be realistic for models of the Faraday experiment. In analyzing this model we can use a similar 'reflect and extend periodically' trick. If we extend the system by reflection across both horizontal coordinates we have PBC in two directions (see Figure 7.9). The symmetry group is now $\Gamma = \mathbf{D}_4 \dotplus \mathbf{T}^2$. Irreducible representations of the torus $\mathbf{T}^2$ are indexed by pairs of integers (m, n), and typically form a 4-dimensional kernel (when $m \neq n$) generated by:

$$\cos(mx)\cos(ny) \qquad \cos(mx)\sin(ny) \qquad \sin(mx)\cos(ny) \qquad \sin(mx)\sin(ny).$$

The eigenfunctions in this 4-dimensional space that satisfy NBC are invariant under $(x, y) \to (\pm x, \pm y)$. The NBC eigenfunctions form a 1-dimensional space spanned by $\cos(mx)\cos(ny)$. See [130] and [129].

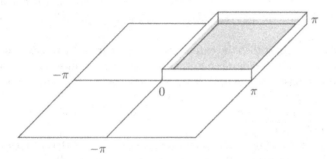

Figure 7.9: NBC extended in two directions to yield PBC in both directions.

Square $\mathbf{D}_4$ symmetry implies that when a mode (m, n) eigenfunction appears, so does mode (n, m). This is the reason for the double eigenvalue: $\mathbf{D}_4$ itself does not act irreducibly in NBC eigenspace. However, the irreducible representations of $\mathbf{D}_4 \dotplus \mathbf{T}^2$ include the (m, n) mode as well as the (n, m) mode, and they therefore bifurcate together. Indeed, the irreducible representations of $\mathbf{D}_4 \dotplus \mathbf{T}^2$ (when $m \neq n$) are 8-dimensional and the space of these eigenfunctions which satisfy NBC is 2-dimensional, generated by $\cos(mx)\cos(ny)$ and $\cos(nx)\cos(my)$.

The question of how to test this mathematical observation has been examined by Crawford *et al.* [129]. The idea is to deform the container in such a way that the $\mathbf{D}_4$ symmetry is retained but extension to PBC is not possible, see Figure 7.10. The mathematical result is that the double eigenvalues will split apart, as pictured in Figure 7.11. This result has been confirmed experimentally: see [129].

7.4 Mode Interactions and Higher-Dimensional Domains

We have seen that hidden symmetries of NBC and DBC problems can impose constraints on Liapunov-Schmidt reduced bifurcation equations, leading to apparently non-generic results. It was such a lack of genericity that showed up in Fujii *et al.* [202]. Armbruster and Dangelmayr [8, 135] pinpointed the existence of a

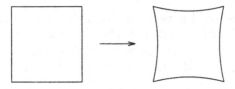

Figure 7.10: Container deformed from square while maintaining $\mathbf{D}_4$ symmetry.

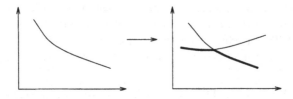

Figure 7.11: Eigenvalue splitting on breaking of torus symmetry.

PBC extension as the cause of this lack of genericity, and carried out an extensive singularity-theoretic analysis of mode interactions for NBC problems on the interval.

Their results suggest that the extension procedure from NBC/DBC to PBC can profitably be generalized to more complex domains than the interval — in particular to higher dimensions. For example NBC on a rectangle $[0, a] \times [0, b] \subseteq \mathbf{R}^2$ lead to PBC on $[-a, a] \times [-b, b] \subseteq \mathbf{R}^2$ and a symmetry group $\mathbf{O}(2) \times \mathbf{O}(2)$. It is technically harder to prove that smooth solutions extend to smooth solutions: see Field *et al.* [192] Theorem 5.18.

Again, the extended problem has continuous symmetries that give rise to mode numbers. Indeed, solutions with $\mathbf{D}_p \times \mathbf{D}_q$ symmetry satisfy NBC and are cellular patterns with p horizontal and q vertical cells — that is, nonlinear (p, q) modes with exact translational symmetries through multiples of the dimensions of a cell. The situation with DBC is similar, but again we require 'twisted' $\mathbf{O}(2)$s whose reflections change the sign of the eigenfunctions as well as reflecting their domains in the axes. For the same reason as before they are abstractly isomorphic to standard $\mathbf{O}(2)$s.

The detailed group theory and singularity theory for the r-dimensional case can be found in Gomes [239] and Gomes and Stewart [241]. In particular they derive the general form of reduced bifurcation equations for the interaction of two stationary modes of a PDE posed on an r-dimensional (generalized) rectangle (that is, a rectangular parallelepiped). Let x, y be the amplitudes of the modes, and let $m = (m_1, \ldots, m_r)$ and $n = (n_1, \ldots, n_r)$ be the vectors of mode numbers for the two modes. Define

$$m = \max_j m_j \qquad n = \max_j n_j$$

Then the reduced bifurcation equations take the form

$$\begin{aligned} a(x^2, y^2, \lambda)x &= 0 \\ b(x^2, y^2, \lambda)y &= 0 \end{aligned} \tag{7.4}$$

if not all m_j have the same parity or not all n_j have the same parity. However, if all m_j have the same parity and all n_j have the same parity then the reduced bifurcation equations take the less intuitive form

$$a(x^2, y^2, \lambda)x + c(x^2, y^2, \lambda)x^{n-1}y^m = 0$$
$$b(x^2, y^2, \lambda)y + d(x^2, y^2, \lambda)x^n y^{m-1} = 0 \tag{7.5}$$

Here a, b, c, d are smooth functions and λ is a vector of bifurcation parameters. We call these *prenormal forms* for the bifurcation equations.

In the first case we obtain the form that arises for $\mathbf{Z}_2 \times \mathbf{Z}_2$ symmetry in the plane, as studied by Schaeffer and Golubitsky [464]. In the second case, the reduced mappings effectively lead to the same prenormal forms that are studied and classified by Dangelmayr and Armbruster [135] in the one-dimensional case $r = 1$ and NBC only.

7.5 Lapwood Convection

In this section we apply the Liapunov-Schmidt procedure and the prenormalforms (7.4,7.5) to Lapwood convection (convection of a fluid in a porous medium) in a rectangular domain. Convective flows in porous media arise in many applications, including geothermal energy, oil reservoirs, waste disposal, and thermal insulation: see Bories [70] for a survey. It turns out that mode interactions occur at certain parameter values in the standard model of this problem. The modes (linearized eigenfunctions) correspond to $p \times q$ arrays of convection cells, which we call (p, q) *modes*.

The original motivation for studying this system was the numerical computation of bifurcation diagrams with unexpectedly complicated structure. Numerical calculations of the branching structure near mode interaction points lead to bifurcation diagrams for the $(3,1)/(1,1)$ and $(3,1)/(2,2)$ mode interactions which are non-generic, even when the rectangular symmetry of the domain is taken into account. When it was first noticed, this curious behavior raised questions about the accuracy of the numerical method used, a finite-element Galerkin approximation implemented using Harwell's ENTWIFE code. Impey *et al.* [297] showed that this apparent lack of genericity is partly a consequence of hidden translational symmetries. We summarize their analysis for the $(3,1)/(1,1)$ interaction, where the issues turn out to be simpler.

The main message of this subsection is not about Lapwood convection — for which see the original paper of Impey *et al.* [297]. It is that the 'reflection trick' for passing from NBC or DBC to PBC can often be used on systems of PDEs (posed on a rectangle or a higher-dimensional analogue) in which different variables satisfy different boundary conditions, and where the boundary conditions may be different on different edges of the domain.

In passing, we remark that there are cases in which the symmetry group of the partial differential operator is the Euclidean group, the domain is a rectangle, and the boundary conditions on each edge are either NBC or PBC, but nevertheless the reflection trick does not extend the problem to one with PBC. An example, concerning magnetoconvection, can be found in Nagata [414]. The linearized eigenfunctions are a useful indicator here. If the extension to PBC works, then the linearized eigenfunctions are products of trigonometric functions, such as $\sin px \cos qy$ and so on, thanks to the $\mathbf{O}(2) \times \mathbf{O}(2)$ symmetry of the extended problem. (Conversely, if the linearized eigenfunctions are trigonometric, then either there has been an amazing coincidence or the reflection trick works.) However, in Nagata's system some of the linearized eigenfunctions involve hyperbolic functions, showing that no PBC extension exists.

In the standard model of Lapwood convection, the PDE involves a mixture of both Neumann and Dirichlet boundary conditions. Specifically, on the vertical sidewalls the stream function satisfies Dirichlet boundary conditions (is zero), but the temperature satisfies Neumann (no-flux) boundary conditions. Nevertheless, it turns out that the same symmetry constraints that occur for purely Neumann boundary conditions are imposed on the Liapunov-Schmidt reduced bifurcation equations, and therefore the same list of prenormal forms (7.5) is valid. The hidden symmetries therefore force certain terms in the reduced bifurcation equations to be zero and change the generic branching geometry. With the aid of MACSYMA, Impey *et al.* [297] compute a small number of low-order coefficients of the prenormal forms, enough to characterize the qualitative form of their bifurcation diagrams. In some cases the prenormal form is more degenerate than might be anticipated, but when these degeneracies are taken into account the resulting branching geometry reproduces that found in the earlier numerical approach.

The standard model here (see Riley and Winters [448]) applies to Lapwood convection in a two-dimensional porous rectangular cavity heated from below. Sutton [498] showed that the primary bifurcations are to cellular patterns with p horizontal and q vertical cells — (p, q) modes. Their streamlines and temperature contours also form cellular arrays: see Figure 7.12. In the linearization the flow and temperature distribution in each cell is identical, except for reflections across separating boundaries.

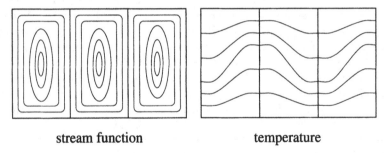

stream function   temperature

Figure 7.12: Cellular structure of primary modes of fluid stream function and temperature in Lapwood convection.

The linear analysis demonstrates the presence of numerous mode interaction points (Figure 7.15) for particular values of the Rayleigh number of the fluid and the aspect ratio of the cavity. A central objective of [448] is to study the bifurcations that arise through competition between primary modes near such parameter values. The resulting bifurcation diagrams for the $(2,1)/(1,1)$ and $(3,1)/(1,1)$ mode interactions are shown in Figures 7.13.

The geometric structure of Figure 7.13, the $(2,1)/(1,1)$ mode interaction, corresponds to that of a generic mode interaction in a system with $\mathbf{Z}_2 \times \mathbf{Z}_2$ symmetry, Schaeffer and Golubitsky [464]. However, the computed geometry of the $(3,1)/(1,1)$ mode interaction in Figure 7.13(right) is quite different from a generic $\mathbf{Z}_2 \times \mathbf{Z}_2$ mode interaction. The computed $(3,1)/(2,2)$ interaction (not shown) is different from both of these.

We show below that the Lapwood convection problem has hidden symmetry, and that the reduced bifurcation equations for the interaction of two modes must correspond to one of the prenormal forms (7.4, 7.5). For Lapwood convection in a

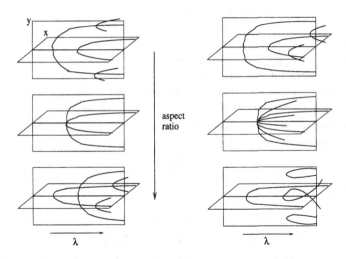

Figure 7.13: Bifurcation diagrams computed using ENTWIFE. (Left) $(2,1)/(1,1)$ mode interaction: the result is identical to a generic $\mathbf{Z}_2 \times \mathbf{Z}_2$ mode interaction. (Right) $(3,1)/(1,1)$ mode interaction: the result differs considerably from a generic $\mathbf{Z}_2 \times \mathbf{Z}_2$ mode interaction.

rectangular cell, where $r = 2$, both prenormal forms (7.4) and (7.5) occur. Computations show that the $((2,1)/(1,1)$ mode interaction has prenormal form (7.4) and so is generic for $\mathbf{Z}_2 \times \mathbf{Z}_2$ symmetry. However, the $(3,1)/(2,2)$ and $(3,1)/(1,1)$ mode interactions give rise to the prenormal form (7.5) with $(m,n) = (3,2)$ and $(3,1)$ respectively. The analysis follows the typical pattern for bifurcations of PDEs, as set up in Chapter 5: we now summarize the relevant calculations.

Linear Stability Analysis. In this subsection we describe the model equations for Lapwood convection and recall the linear stability analysis of Sutton [498], following the approach of Riley and Winters [448].

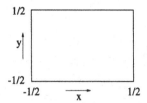

Figure 7.14: Domain and variables for Lapwood convection.

Let $\Omega \subseteq \mathbf{R}^2$ be a rectangular cavity of height H and width W, as in Figure 7.14. We call $h = W/H$ the *aspect ratio*. In the Boussinesq approximation, with large Prandtl-Darcy number, the equations for time-independent Lapwood convection in Ω may be written as

$$\Delta\psi = -\frac{\text{Ra}}{h}\frac{\partial\theta}{\partial x} \tag{7.6}$$

$$\Delta\theta = \frac{1}{h}\frac{\partial\psi}{\partial x} + \frac{1}{h}\left[\frac{\partial\psi}{\partial y}\frac{\partial\theta}{\partial x} - \frac{\partial\psi}{\partial x}\frac{\partial\theta}{\partial y}\right] \tag{7.7}$$

Here

- (x, y) are Cartesian coordinates based at the center of the cavity
- $\Delta = \frac{1}{h^2}\frac{\partial^2}{\partial x^2} + \frac{\partial^2}{\partial y^2}$ is a scaled Laplacian
- ψ is the stream function
- θ is the temperature (considered as the deviation from pure conduction)
- Ra is the Rayleigh number

and quantities have been non-dimensionalized. With the scalings used in [448], the domain Ω becomes the unit square

$$\Omega = [-\tfrac{1}{2}, \tfrac{1}{2}] \times [-\tfrac{1}{2}, \tfrac{1}{2}]$$

The boundary conditions are

$$\psi = 0, \quad \frac{\partial \theta}{\partial x} = 0 \text{ on } x = \pm\tfrac{1}{2}, |y| < \tfrac{1}{2} \tag{7.8}$$

$$\psi = 0, \theta = 0 \text{ on } y = \pm\tfrac{1}{2}, |x| < \tfrac{1}{2}. \tag{7.9}$$

That is, ψ satisfies DBC on the entire boundary, whereas θ satisfies NBC on the sidewalls but DBC at the top and bottom. Sutton [498] carried out the linear stability analysis of (7.6,7.7), showing that eigenmodes exist whenever the Rayleigh number satisfies

$$\text{Ra} = \text{Ra}_{pq} = \frac{\pi^2}{p^2 h^2}(q^2 h^2 + p^2)^2 \tag{7.10}$$

for integers $p > 0, q > 0$. The corresponding eigenmodes are

$$\psi = \Psi_{pq} \quad = \quad \sin(p\pi\bar{x})\sin(q\pi\bar{y}) \tag{7.11}$$

$$\theta = \Theta_{pq} \quad = \quad -(\text{Ra})^{-\frac{1}{2}}\cos(p\pi\bar{x})\sin(q\pi\bar{y}) \tag{7.12}$$

where

$$\bar{x} = x + \tfrac{1}{2} \qquad \bar{y} = y + \tfrac{1}{2} \tag{7.13}$$

These eigenmodes represent cellular patterns with p horizontal cells and q vertical cells, which, as previously indicated, we refer to as (p, q) modes.

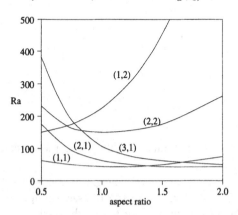

Figure 7.15: Plots of the critical Rayleigh number Ra_{pq} against aspect ratio h for low values of (p, q). Observe the mode interaction points, at which the curves cross.

Fig. 7.15 shows how the critical Rayleigh number Ra_{pq} varies with aspect ratio h for the lowest values of (p, q). A number of mode interaction points are visible where two distinct curves cross.

The system (7.6,7.7) has obvious symmetries, namely those of the domain Ω, which are generated by two reflections

$$\sigma_x : (x,y) \mapsto (-x,y)$$
$$\sigma_y : (x,y) \mapsto (x,-y)$$

These generate a group $\mathbf{Z}_2 \times \mathbf{Z}_2$ which acts as linear transformations of $\mathbf{R}^2$, and on the variables (ψ,θ) by

$$\sigma_x : \begin{bmatrix} \psi(x,y) \\ \theta(x,y) \end{bmatrix} \mapsto \begin{bmatrix} -\psi(-x,y) \\ \theta(-x,y) \end{bmatrix}$$

$$\sigma_y : \begin{bmatrix} \psi(x,y) \\ \theta(x,y) \end{bmatrix} \mapsto \begin{bmatrix} -\psi(x,-y) \\ -\theta(x,-y) \end{bmatrix}$$

The results obtained for the (2,1)/(1,1) mode interaction correspond to a generic mode interaction in a system with the above $\mathbf{Z}_2 \times \mathbf{Z}_2$ symmetry. However, as already noted, the (3,1)/(2,2) and (3,1)/(1,1) mode interactions are different.

Hidden Symmetry for Lapwood Convection. Even though the boundary conditions in (7.6, 7.7) are a mixture of DBC and NBC, the extension method described in Section 7.2 still applies to this problem. Instead of a function u defined on an interval, we have a mapping (ψ,θ) defined on the rectangle $\Omega = [-\frac{1}{2},\frac{1}{2}] \times [-\frac{1}{2},\frac{1}{2}]$. We extend a solution (ψ,θ) on Ω to $(\hat{\psi},\hat{\theta})$ on $\hat{\Omega} = [-\frac{1}{2},\frac{3}{2}] \times [-\frac{1}{2},\frac{3}{2}]$ as follows:

$$
\begin{array}{llll}
\hat{\psi}(x,y) &=& \psi(x,y) & \text{if } -\frac{1}{2} \le x \le \frac{1}{2}, \quad -\frac{1}{2} \le y \le \frac{1}{2} \\
\hat{\psi}(x,y) &=& -\psi(1-x,y) & \text{if } \frac{1}{2} \le x \le \frac{3}{2}, \quad -\frac{1}{2} \le y \le \frac{1}{2} \\
\hat{\psi}(x,y) &=& -\psi(x,1-y) & \text{if } -\frac{1}{2} \le x \le \frac{1}{2}, \quad \frac{1}{2} \le y \le \frac{3}{2} \\
\hat{\psi}(x,y) &=& \psi(1-x,1-y) & \text{if } \frac{1}{2} \le x \le \frac{3}{2}, \quad \frac{1}{2} \le y \le \frac{3}{2} \\
\hat{\theta}(x,y) &=& \theta(x,y) & \text{if } -\frac{1}{2} \le x \le \frac{1}{2}, \quad -\frac{1}{2} \le y \le \frac{1}{2} \\
\hat{\theta}(x,y) &=& \theta(1-x,y) & \text{if } \frac{1}{2} \le x \le \frac{3}{2}, \quad -\frac{1}{2} \le y \le \frac{1}{2} \\
\hat{\theta}(x,y) &=& -\theta(x,1-y) & \text{if } -\frac{1}{2} \le x \le \frac{1}{2}, \quad \frac{1}{2} \le y \le \frac{3}{2} \\
\hat{\theta}(x,y) &=& -\theta(1-x,1-y) & \text{if } \frac{1}{2} \le x \le \frac{3}{2}, \quad \frac{1}{2} \le y \le \frac{3}{2}.
\end{array}
$$

Then we extend $\hat{\theta}$ and $\hat{\psi}$ periodically to the whole plane, with period 2 in each direction (and retain the same notation). Observe that the sign of the function changes upon reflection across Dirichlet boundaries, but stays the same on reflection across Neumann boundaries: see Fig. 7.16.

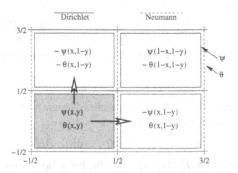

Figure 7.16: For the Lapwood convection model, reflections across boundaries of the domain, coupled with sign changes at Dirichlet boundaries, extend solutions to a domain of twice the size, on which periodic boundary conditions apply.

It is a simple but instructive exercise to check that the extended functions satisfy the same PDE as before, but now posed on the entire plane. Oddness (in the relevant variables) comes into play at Dirichlet boundaries. Regularity of the extended solution follows from Field *et. al* [192] Theorem 5.18. Its symmetry group is $\mathbf{O}(2) \times \mathbf{O}(2)$, and for a given choice of mode numbers (p, q) its action depends upon the boundary conditions. However, an abstract group-theoretic argument, carried out in Impey *et al.* [297], shows that the prenormal forms of Gomes and Stewart [241], obtained assuming NBC, also apply to the mixed boundary conditions of Lapwood convection. In fact, the group actions in the two cases turn out to be isomorphic. The low-order terms in the prenormal forms can be found by Liapunov-Schmidt reduction. For the (3,1)/(1,1) interaction we obtain

$$(\lambda_1 + x^2 + 5y^2)x + (-\lambda_1 + \tfrac{44}{35}\lambda_2)y^3 = 0$$

$$(\lambda_1 + x^2 + y^2)y + (-\tfrac{125}{63}\lambda_1 + \tfrac{710}{63}\lambda_2)xy^2 = 0$$

(7.14)

where (λ_1, λ_2) is a two-dimensional bifurcation parameter (combining a standard bifurcation parameter with a parameter that breaks up the mode interaction). Even within the world of hidden symmetries, this prenormal form turns out to be singularity-theoretically nongeneric. The additional nongenericity can be traced to certain simplifications in the modeling assumptions. Nevertheless, we can analyze (7.14), and the resulting bifurcation diagrams are in qualitative agreement with the numerically computed diagrams in all relevant details. We conclude that the numerical methods embodied in the ENTWIFE code were performing well, and that the curiosities observed in the bifurcation diagrams can be traced partly to hidden symmetries, and partly to degeneracies resulting from certain simplifications in the model.

7.6 Hemispherical Domains

The extension method, and the associated possibility of hidden symmetries, also applies to many types of non-rectangular domain. Field *et al.* [192] use it to study bifurcation of PDEs posed on a hemisphere, with NBC. Now the extended problem has $\mathbf{O}(3)$ symmetry, whereas the domain has only $\mathbf{O}(2)$ symmetry. Hemispherical domains arise in the elastic buckling of hemispherical shells (domes): see Bauer, Riess and Keller [53, 52]. For expository purposes we work with a steady-state reaction-diffusion equation, but the same results apply to a much wider class of Euclidean-invariant PDEs.

Denote coordinates on $\mathbf{R}^3$ by (x_1, x_2, x_3) and let S be the unit sphere in $\mathbf{R}^3$. Let $H = \{x \in S : x_3 \geq 0\}$ be the upper hemisphere of S, and write Δ for the Laplacian on S. Let $f : \mathbf{R}^2 \to \mathbf{R}$ be a smooth map. Consider the reaction-diffusion equation defined on H by:

$$\Delta u + f(u, \lambda) = 0 \tag{7.15}$$

where $u : H \to \mathbf{R}$. Assume that (7.15) satisfies Neumann boundary conditions on $\partial H = \{(x_1, x_2, x_3) \in H | x_3 = 0\}$, that is,

$$u_{x_3}(x_1, x_2, 0) = 0 \quad \text{on} \quad \partial H \tag{7.16}$$

Solutions of (7.15) on H that satisfy the boundary conditions (7.16) can be extended to solutions of (7.15) on S by defining u on the lower hemisphere by reflection. More precisely, let $\tau : S \to S$ be the reflection across ∂H defined by

$$\tau(x_1, x_2, x_3) = (x_1, x_2, -x_3) \tag{7.17}$$

Then we define u on the lower hemisphere by:

$$u(\tau(x)) \equiv u(x)$$

for all $x \in H$.

We shall say that a function u on S is τ-*invariant* if

$$u(\tau(x)) = u(x)$$

for all $x \in S$. The extension u that we have defined on S is τ-invariant. It can be shown that the extended solution is smooth: see Theorem 5.18 of Field *et al.* [192].

Conversely, suppose that u is a τ-invariant solution to the reaction-diffusion equation (7.15) on S. Then $u|_H$ is a solution to the Neumann boundary value problem (7.16) on H. We can therefore find solutions to the Neumann problem on the hemisphere by finding τ-invariant solutions to the extended problem on S. This extended problem has symmetry group $\mathbf{O}(3)$, and this enlarged symmetry group has far-reaching consequences. In particular, it has a substantial effect on the critical eigenspace. Indeed, suppose that (7.15) has the trivial steady state $u = 0$. Let V be the kernel of the linearization of (7.15), and let V_S be the kernel of the linearization of (7.15) on the full sphere. The symmetry group of the equation leaves V invariant; moreover, generically we would expect its action on V is (absolutely) irreducible. The irreducible representations of $\mathbf{O}(2)$ are either one- or two-dimensional. A direct application of the standard theory would imply that generically we should expect $\dim(V)$ to be one or two.

In this setting, however, the space V consists of those eigenfunctions in V_S that are τ-invariant. The extension property of (7.15) implies that the action of $\mathbf{O}(3)$ on V_S should be irreducible. There are irreducible representations of $\mathbf{O}(3)$ in each odd dimension, namely, the action of $\mathbf{O}(3)$ on the spherical harmonics of order ℓ. It can be checked that the vectors in V_S that are τ-invariant form a subspace of dimension approximately $\frac{1}{2}\dim(V_S)$. Thus, the kernel of the linearized equations may be of *much* higher dimension than would have been expected if one considered the bifurcation problem only on the hemisphere.

This change of expected dimension of the kernel V is dramatic, but does it lead to solutions to the hemisphere problem that would not have been expected on the basis of $\mathbf{O}(2)$ symmetry? The answer is 'yes', as we now indicate: full details are given in Field *et al.* [192]. One interesting feature is that some group orbits of equilibria contain *multiple* equilibria with symmetry τ. Each of these equilibria then restricts to a solution of the NBC problem on the hemisphere, and the restrictions lie lie on *different* $\mathbf{O}(2)$-orbits. (Other group orbits of equilibria contain no equilibria with symmetry τ.)

General $\mathbf{O}(3)$-equivariant bifurcation theory (see Ihrig and Golubitsky [293] or Golubitsky *et al.* [237]) shows that generically, axisymmetric solutions of order ℓ are to be expected for all ℓ. Figure 7.17(left) shows the sphere deformed by an axisymmetric spherical harmonic of order $\ell = 6$. This solution can be sliced in two different ways to obtain solutions to the original equation posed on the hemisphere. Observe that the picture in Figure 7.17(middle) has circular symmetry while the picture in Figure 7.17(right) has only a reflectional symmetry.

Similarly, when $\ell = 5$, general theory predicts the existence of a solution having fivefold symmetry, Figure 7.18(a). This solution may also be sliced in two distinct ways to obtain solutions to (7.15) on H. The first slice also has fivefold symmetry and is visualized by slicing off the obscured half of the deformed sphere in Figure 7.18(a). The other slice, which has only a reflectional symmetry, is shown in Figure 7.18(b).

Figure 7.17: Axisymmetric solution when $\ell = 6$ (left) and its restrictions to the hemisphere (middle and right).

Bifurcation from O(3)-invariant Solutions. In order to state a general classification of the symmetry types of bifurcating branches in the hemisphere problem, we first review **O**(3)-equivariant bifurcation. As stated, typically we expect the kernel of the linearization about an **O**(3)-invariant equilibrium to be an (absolutely) irreducible representation of **O**(3). Because the natural domains of partial differential equations are function spaces, we expect these irreducible representations to be isomorphic to V_ℓ, the spherical harmonics of order ℓ. There are two irreducible representations of **O**(3) in each odd dimension, depending on whether $-I \in \mathbf{O}(3)$ acts trivially or as minus the identity on V_ℓ. There is, however, a natural action on V_ℓ, since V_ℓ consists of polynomial mappings $p : \mathbf{R}^3 \to \mathbf{R}$ that are homogeneous of degree ℓ. For such polynomials, the action of $-I$ is:

$$p(-x) = (-1)^\ell p(x)$$

Thus we assume that $-I$ acts trivially when ℓ is even and as minus the identity when ℓ is odd.

Figure 7.18: Solution with five-fold symmetry when $\ell = 5$ (left) and its restrictions to the hemisphere (middle and right).

It is well known (for a proof see Golubitsky *et al.* [237]) that the closed subgroups of $\mathbf{SO}(3)$ are $\mathbf{Z}_m$, $\mathbf{D}_m$, $\mathbf{T}$ (tetrahedral group), $\mathbf{O}$ (octahedral group), $\mathbf{I}$ (the icosahedral group), $\mathbf{SO}(2)$ and $\mathbf{O}(2)$. Subgroups of $\mathbf{O}(3) = \mathbf{SO}(3) \oplus \mathbf{Z}_2^c$ are of three types. Type I and type II subgroups are respectively of the form Σ and $\Sigma \oplus \mathbf{Z}_2^c$, where Σ is a subgroup of $\mathbf{SO}(3)$. Type III subgroups Σ of $\mathbf{O}(3)$ neither contain nor are contained in $\mathbf{SO}(3)$. They may be characterized by a pair $H \supseteq K$ of subgroups of $\mathbf{SO}(3)$ with K having index two in H. Indeed, we may take $K = \Sigma \cap \mathbf{SO}(3)$ and let H be the projection of Σ into $\mathbf{SO}(3)$.

Here we shall be concerned with four type III subgroups, $\mathbf{O}(3)^-$, $\mathbf{O}^-$, $\mathbf{D}_{2m}^d$ and $\mathbf{D}_m^z$, which are defined by the pairs $\mathbf{O}(3) \supseteq \mathbf{SO}(3)$, $\mathbf{O} \supseteq \mathbf{T}$, $\mathbf{D}_{2m} \supseteq \mathbf{D}_m$ and $\mathbf{D}_m \subseteq \mathbf{Z}_m$. This abstract definition of type III subgroups of $\mathbf{O}(3)$ can be given a more concrete realization. In particular, the subgroup Σ corresponding to $H \supseteq K$ can be written as

$$K \times \{I\} \cup (H \setminus K) \times \{-I\}$$

inside $\mathbf{O}(3) = \mathbf{SO}(3) \oplus \mathbf{Z}_2^c$. The group of symmetries of the hemisphere is $\mathbf{O}(2)^-$.

Chossat, Lauterbach and Melbourne [111] list those conjugacy classes of isotropy subgroups that generically support solutions for $\mathbf{O}(3)$-equivariant bifurcation problems: see Table 7.6. With the exception of the $\ell = 5$ results for $\mathbf{D}_5^z$, $\mathbf{D}_4^z$ and $\mathbf{D}_3^z$, the isotropy groups in Table 7.6 are maximal and have odd-dimensional fixed point spaces. This follows from the computations of Ihrig and Golubitsky [293] and Golubitsky *et al.* [237]. Thus there always exist branches of solutions with the isotropy subgroups listed — and with conjugates of those isotropy subgroups.

We classify such solutions up to the action of $\mathbf{O}(2)^-$, the symmetry group of the hemisphere.

ℓ even:	
$\mathbf{O}(2) \oplus \mathbf{Z}_2^c$	all even ℓ
$\mathbf{O} \oplus \mathbf{Z}_v 2^c$	$\ell \equiv 0\ 4\ 6\ 8\ 10\ 14 \pmod{24}$, $\ell \geq 4$
$\mathbf{I} \oplus \mathbf{Z}_2^c$	$\ell \equiv 0\ 6\ 10\ 12\ 16\ 18\ 20\ 22\ 24$
	$26\ 28\ 32\ 34\ 38\ 44 \pmod{60}$, $\ell \geq 6$
ℓ odd:	
$\mathbf{O}(2)^-$	all odd ℓ
$\mathbf{O}^-$	$\ell \equiv 3\ 7\ 9\ 11\ 13\ 17 \pmod{24}$
$\mathbf{O}$	$\ell \equiv 9\ 13\ 15\ 17\ 19\ 23 \pmod{24}$
$\mathbf{I}$	$\ell \equiv 15\ 21\ 25\ 27\ 31\ 33\ 35\ 37$
	$39\ 41\ 43\ 47\ 49\ 53\ 59 \pmod{60}$
$\mathbf{D}_{2m}^d$	$3 \leq m \leq \ell$
$\mathbf{D}_4^d$	$\ell \equiv 1\ 5\ 15\ 19\ 21\ 23 \pmod{24}$, $\ell \geq 5$
$\mathbf{D}_5^z\ \mathbf{D}_4^z\ \mathbf{D}_2^z$	$\ell = 5$

Table 7.1 Isotropy subgroups of $\mathbf{O}(3)$ that generically support solutions

Let τ denote the reflection defining the hemisphere, equation (7.17). Suppose that u is a solution to an $\mathbf{O}(3)$- equivariant bifurcation problem with isotropy subgroup Σ. Then u restricts to a solution of the NBC problem on the hemisphere if and only if $\tau \in \Sigma$.

Since the groups $\mathbf{O}$ and $\mathbf{I}$ contain no orientation-reversing elements we have

Lemma 7.6 *The solutions to $\mathbf{O}(3)$- equivariant bifurcation problems with isotropy conjugate to $\mathbf{O}$ or $\mathbf{I}$ cannot restrict to solutions of the Neumann problem on the hemisphere.* $\qquad\square$

We may now classify the axisymmetric solutions — those that have an axis of rotation, and so have at least $\mathbf{SO}(2)$-symmetry. There are two types of isotropy subgroup that contain $\mathbf{SO}(2)$: $\mathbf{O}(2) \oplus \mathbf{Z}_2^c$ (when ℓ is even) and $\mathbf{O}(2)^-$ (when ℓ is odd). We find:

Theorem 7.7 *Each orbit of axisymmetric solutions to the sphere problem restricts to solutions of the hemisphere problem as follows:*

(a): *When ℓ is even, an isolated axisymmetric solution with the x_3-axis as axis of symmetry.*

(b): *For all ℓ, a unique circle of solutions with isotropy $\mathbf{Z}_2^-$ (inside $\mathbf{O}(2)^-$.)*

$\square$

We complete the classification by finding the possible solutions with finite isotropy subgroups (inside $\mathbf{O}(2)^-$):

Theorem 7.8 *Suppose that we have an $\mathbf{O}(2)$-orbit of solutions to the sphere problem with finite isotropy conjugate to Σ. On restriction, we obtain the following $\mathbf{O}(3)^-$-orbits of solutions to the hemisphere problem:*

$\Sigma = \mathbf{D}_{2m}^d$ *(sector solutions): When m is odd there is a unique circle of solutions with isotropy $\mathbf{D}_m^z$. For all m there is a unique circle of solutions with isotropy $\mathbf{Z}_2^-$.*

$\Sigma = \mathbf{D}_m^z$ *(sector solutions): There is a unique circle of solutions with isotropy $\mathbf{Z}_2^-$.*

$\Sigma = \mathbf{O}^-$ *(octahedral solutions): There is a unique circle of solutions with isotropy $\mathbf{Z}_2^-$.*

$\Sigma = \mathbf{O} \oplus \mathbf{Z}_2^c$ *(octahedral solutions): There is a unique circle of solutions with isotropy $\mathbf{D}_4^-$, and a unique circle of solutions with isotropy $\mathbf{D}_2^-$.*

$\Sigma = \mathbf{I} \oplus \mathbf{Z}_2^c$ *(icosahedral solutions): There is a unique circle of solutions with isotropy $\mathbf{Z}_2^-$.* $\square$

Some of the above results concerning reaction-diffusion equations on a hemisphere have been used by Nagata [414] to model the growth of a plant shoot, in connection with phyllotaxis.

Chapter 8

Heteroclinic Cycles

One of the classic pattern-forming systems of mathematical physics is the Rayleigh-Bénard experiment, in which a thin layer of fluid is heated from below. We discussed this experiment in Section 5.7 from the point of view of lattice-symmetric solutions to Euclidean-equivariant PDEs. In the classic experiment, the homogeneous conduction state becomes unstable and there is a bifurcation to a patterned state — typically rolls, although, as Bénard famously observed, hexagons can also occur.

The dynamics of Rayleigh-Bénard convection becomes more complex, and in several respects more interesting, if the fluid layer is rotated in the horizontal plane (see, for example, Ahlers and Bajaj [2]). In the presence of rotation, Coriolis forces act on the fluid and change the observed patterns. Time-dependent states occur arbitrarily close to the onset of convection. Küppers and Lortz [335] predicted that for an angular velocity of rotation above some critical value, the primary bifurcation will be to rolls that are *unstable*. The instability is strongest in response to perturbations that are plane waves whose direction is advanced, relative to the rolls, by a fixed angle Θ_{KL}. The resulting phenomenon is known as the Küppers-Lortz instability, and an instantaneous snapshot is shown in Figure 8.1. The angle Θ_{KL} is fairly close to $60°$, so at any given time the system appears to consist of domains of rolls in three different directions. These domains incessantly replace each other, in a chaotic manner, as the instability of the rolls in a given domain manifests itself and the direction of the rolls changes. The main mechanism for the change is irregular domain wall motion.

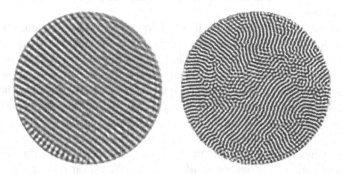

Figure 8.1: (Left) Rolls in convection from Liu and Ahlers [362]. (Right) Instantaneous snapshot of the Küppers-Lortz instability from Hu, Ecke, and Ahlers [285]. Courtesy of Günter Ahlers.

Busse and Clever [91] introduced a qualitative model for the Küppers-Lortz instability, modeling it as a competition between three modes: rolls at three angles $0°, 60°$, and $120°$. This leads to a three-dimensional system of ODEs with three fixed points, each a saddle: the stable and unstable manifolds of these fixed points create an equilateral triangle as in Figure 8.2. Such a configuration is known as a *heteroclinic cycle*. In this model, the cycle is a globally stable object, at least when approached from inside, as shown by the spiral trajectory in the figure. A state on such a trajectory will appear to cycle among the three fixed points in turn, remaining near each fixed point for an increasing length of time. The interpretation of this cycle for the Küppers-Lortz instability is that the rolls in a domain will remain at one orientation for a substantial period of time, before switching (relatively suddenly) to an orientation that is advanced by 60°. This change of direction repeats indefinitely. Busse and Heikes [92] carried out experiments and compared them to this model.

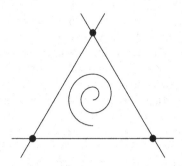

Figure 8.2: Heteroclinic cycle in the Busse-Clever model.

In generic dynamical systems, heteroclinic connections are structurally unstable: the unstable manifold of a saddle point generically does not coincide with the stable manifold of another saddle. Thus at first sight the Busse-Clever model is nongeneric. However, Field [184] pointed out that heteroclinic cycles can be structurally stable in systems with symmetry and Guckenheimer and Holmes [256] showed that this phenomenon occurs in the Busse-Clever model. The same system was also used by May and Leonard [379] to study the population dynamics of three identical species.

Heteroclinic cycles have been invoked to explain other dynamic phenomena in which a system repeatedly seems to switch between different dynamical states. Such behavior is observed, for example, in 'bursting' of neurons. A neuron may cycle between a quiescent state and one in which it fires periodically, or between two different periodic states. One popular explanation of bursting involved heteroclinic cycles, see Hoppensteadt and Izhikevich [283] and Keener and Sneyd [305].

In this chapter we discuss two different examples of heteroclinic cycles. The first example is the one introduced by Guckenheimer and Holmes, the second arises in $O(2)$ mode interactions. We also present a general method for finding heteroclinic cycles introduced by Melbourne *et al.* [392], and based on the lattice of isotropy subgroups. Finally, we show how the ideas behind heteroclinic cycles lead to one notion of bursting.

8.1 The Guckenheimer-Holmes Example

Consider the group Γ whose action on $\mathbf{R}^3$ is generated by

$$
\begin{aligned}
(x, y, z) &\mapsto (\pm x, \pm y, \pm z) \\
(x, y, z) &\mapsto (y, z, x)
\end{aligned}
\tag{8.1}
$$

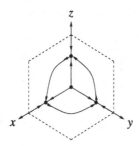

Figure 8.3: Guckenheimer-Holmes cycle in $\mathbf{R}^3$ phase space.

This group Γ has 24 elements, and is the symmetry group of the cube in $\mathbf{R}^3$. We want to study the dynamics of a system of Γ-equivariant ODEs. In particular, we want to show that the simplest bifurcations can give rise to a heteroclinic cycle. Such a system will have the following properties:

- The coordinate axes are flow-invariant. To see this, observe that

$$\mathrm{Fix}(-x, -y, z) = \mathbf{R}\{(0, 0, 1)\}$$

so the Equivariant Branching Lemma implies that generic bifurcation leads to a solution on the z-axis. We assume this bifurcation to be a supercritical pitchfork. By symmetry there also exist solutions on the x- and y-axes.

- The coordinate planes are flow-invariant since

$$\mathrm{Fix}(-x, y, z) = \mathbf{R}\{(0, 1, 0), (0, 0, 1)\}$$

By symmetry the other coordinate planes are also invariant.

Suppose we can show that

(a) There are no other equilibria in the coordinate planes.

(b) The two remaining eigenvalues of the equilibria on the axes are of opposite sign.

(c) Infinity is a source.

Then we can draw a phase portrait as in Figure 8.3.

We now consider a general Γ-equivariant system

$$F(x, y, z) = (f_1(x, y, z), f_2(x, y, z), f_3(x, y, z))$$

and show that the above assumptions are not unreasonable. First we describe the general Γ-equivariant vector field to third order. Equivariance with respect to the cyclic symmetry implies that

$$(f_2(x, y, z), f_3(x, y, z), f_1(x, y, z)) = (f_1(y, z, x), f_2(y, z, x), f_3(y, z, x))$$

In addition, the $(x, y, z) \mapsto (-x, y, z)$ symmetry implies that f_1 is odd in x and f_2 and f_3 are even in x. So to cubic order

$$
\begin{aligned}
f_1(x, y, z) &= \lambda x + (Ax^2 + By^2 + Cz^2)x \\
f_2(x, y, z) &= \lambda y + (Cx^2 + Ay^2 + Bz^2)y \\
f_3(x, y, z) &= \lambda z + (Bx^2 + Cy^2 + Az^2)z
\end{aligned}
$$

We now study the truncated vector field. If we solve for equilibria, we find that $f_1 = 0$ when $y = z = 0$ and $(\lambda + Ax^2) = 0$; thus the nontrivial equilibrium occurs when $\lambda = -Ax^2$. We require $A < 0$ in order for the equilibrium to occur supercritically (that is, for $\lambda > 0$).

We can determine the eigenvalues at this equilibrium. The linearization is

$$(df)_{x,0,0} = \begin{bmatrix} \lambda + 3Ax^2 & 0 & 0 \\ 0 & \lambda + Cx^2 & 0 \\ 0 & 0 & \lambda + Bx^2 \end{bmatrix}$$

$$= \begin{bmatrix} 2Ax^2 & 0 & 0 \\ 0 & (C-A)x^2 & 0 \\ 0 & 0 & (B-A)x^2 \end{bmatrix}$$

Since $2Ax^2 < 0$ we require that $(C-A)(B-A) < 0$ for the eigenvalues in the y- and z-directions to be of opposite sign. A simple symmetry argument provides the same result for the y and z equilibria, so the eigenvalue assumption is satisfied.

Now consider the xy-plane $\{z = 0\}$. A nontrivial equilibrium in this plane is a solution of

$$\lambda + (Ax^2 + By^2) = 0$$
$$\lambda + (Cx^2 + Ay^2) = 0$$

Subtracting, we get

$$(A - C)x^2 + (B - A)y^2 = 0$$

By the previous assumption $A - C$ and $B - A$ are of the same sign, so the only solution of the above equation is $x = y = 0$. Hence we have satisfied the condition that there be no other equilibria in the coordinate planes.

If we now assume $A \ll 0$ then the vector field will point in from infinity. To see this we need consider only the plane $z = 0$:

$$\mu = F\left(\begin{bmatrix} x \\ y \end{bmatrix}\right) = \lambda(x^2 + y^2) + A(x^4 + y^4) + (B + C)(x^2 y^2)$$

If $B+C$ is small (indeed, if $|B+C| < 2|A|$) then the fourth order terms are negative, and μ is negative when $x^2 + y^2$ is sufficiently large.

An example of this heteroclinic cycle occurs when $\lambda = 1.0$, $A = 1.0$, $B = 1.5$, and $C = 0.6$. The resulting time series is shown in Figure 8.4.

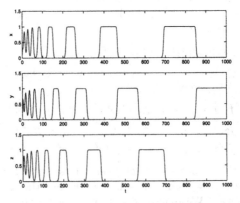

Figure 8.4: Typical time series for the Guckenheimer-Holmes system.

When symmetry is broken, the Guckenheimer-Holmes heteroclinic cycle perturbs to a periodic solution. For example, consider the perturbation

$$\dot{x} = x - (Ax^2 + By^2 + Cz^2)x + \varepsilon y$$
$$\dot{y} = y - (Cx^2 + Ay^2 + Bz^2)y + \varepsilon x$$
$$\dot{z} = z - (Bx^2 + Cy^2 + Az^2)z + \varepsilon z$$

where $A = 1.0$, $B = 1.5$, $C = 0.6$, $\varepsilon = 0.00001$. Time series are shown in Figure 8.5.

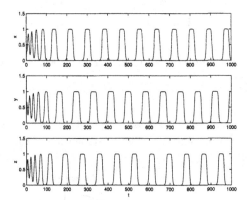

Figure 8.5: Typical time series for the Guckenheimer-Holmes system when symmetry is broken: solution becomes periodic.

8.2 Heteroclinic Cycles by Group Theory

There is a group-theoretic signature for heteroclinic cycles. As in [392], consider the isotropy lattice in Figure 8.6 where Σ_1 and Σ_2 have 1-dimensional fixed-point subspaces and T_1 and T_2 have 2-dimensional fixed-point subspaces. Suppose that the configuration of equilibria is as shown in Figure 8.7.

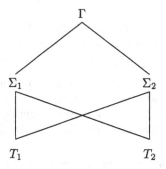

Figure 8.6: Sample lattice of isotropy subgroups leading to a possible heteroclinic cycle.

Recall that the isotropy subgroup structure in Figure 8.8 appears in the isotropy lattice for $\mathbf{O}(2)$ mode interaction in the Couette-Taylor system. This is not quite the structure we need, because the dimensions of the fixed point subspaces for ribbons, twisted vortices and wavy vortices are too large. However, recall that ribbons, twisted vortices and wavy vortices are time-periodic states. Thus by working in

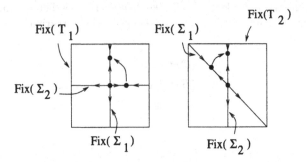

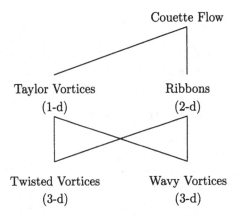

Figure 8.7: Saddle-sink connections in fixed-point subspaces leading to a hetero-clinic cycle.

phase-amplitude variables on the corresponding fixed-point subspaces, the phase equation will decouple, dropping the effective dimension of these spaces by one.

Figure 8.8: Part of the isotropy lattice in Hopf/Steady-state mode interactions, as in the Couette-Taylor experiment.

It is possible to choose coefficients in the normal form to obtain a structurally stable (and asymptotically stable) heteroclinic cycle connecting Taylor vortices and the equilibrium associated with the time-periodic solution associated with ribbons. (This choice of coefficients seems never to be realized by the actual Couette-Taylor apparatus.)

The result is rather striking numerically when the solution begins to cycle between an equilibrium state and a time-periodic state. See Figure 8.9. Melbourne *et al.* also show that heteroclinic cycles connecting different types of periodic solution exist in Hopf/Hopf mode interactions with $\mathbf{O}(2)$ symmetry.

Buono *et al.* [88, 89] show that the $\mathbf{O}(2)$ cycles also occur in systems with $\mathbf{D}_n$ symmetry — though the details are more complicated. For example, in systems with $\mathbf{D}_6$ symmetry there are two different types of equilibria and two different types of standing wave. See the lattice of isotropy subgroups pictured in Figure 8.10. A corresponding heteroclinic cycle is shown in Figure 8.11.

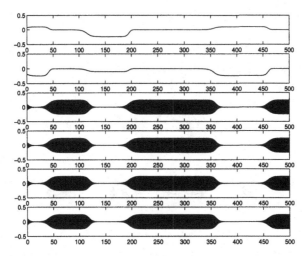

Figure 8.9: Heteroclinic cycle connecting equilibria with standing wave periodic solutions in a Hopf/Steady-state mode interaction with $\mathbf{O}(2)$ symmetry.

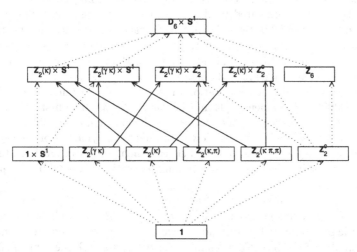

Figure 8.10: Lattice of isotropy subgroups in a Hopf/Steady-state mode interaction with $\mathbf{D}_6$ symmetry [89]. Connections involved in the cycle are shown as solid lines.

8.3 Pipe Systems and Bursting

The phenomenon of *bursting* in both biological and model neurons can be characterized, in broad terms, as a roughly periodic signal whose qualitative form 'switches' between two or more distinct types of dynamic behavior in a regular fashion (Figure 8.12). See Av-Ron *et al.* [34], Bal *et al.* [37], Bertram [61], Keener and Sneyd [305], Kopell and LeMasson [325], Maeda *et al.* [367]. For example, there may be an alternation between a series of rapid oscillations (known as *bursts*) and a (comparatively) steady signal, or the signal may switch between two distinct

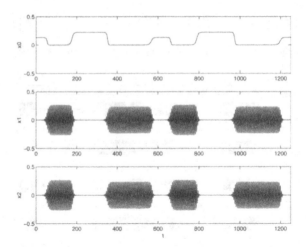

Figure 8.11: Heteroclinic cycle connecting two different types of equilibria with two different types of standing wave periodic solutions in a Hopf/Steady-state mode interaction with $\mathbf{D}_6$ symmetry [89].

kinds of oscillation. For purposes of discussion, we may describe the signal as being formed from two or more basic segments, and switching between these on a timescale that is fast compared to the durations of the segments. This bursting process is often 'periodic' in the following sense: each segment occurs repeatedly at times that are separated by approximately the same interval. However (especially in experiments) the detailed quantitative form of the signal may differ from one occurrence of a given segment to the next.

There are two standard explanation of the dynamics of bursting (see Keener and Sneyd [305] and Hoppensteadt and Izhikevich [283]). The first is in terms of singular perturbation theory, which views the phenomenon as the interaction of a fast flow and a slow flow. The slow flow determines the dynamics of the segments; the fast flow leads to rapid switching between them. The second is in terms of perturbed heteroclinic cycles connecting equilibria with other equilibria or periodic solutions.

Here we propose an alternative framework for bursting behavior: the concept of a 'pipe system', introduced by Melbourne [384, 331] to study dynamics near perturbed heteroclinic cycles. Instead of focusing on the occurrence of a small parameter, we describe a possible mechanism for dynamics with the same qualitative features as bursting, which is defined in terms of certain geometric features of the dynamics. Assume (as is standard) that the dynamics of the neuron is modeled by a system of ODEs, viewed geometrically as a flow in some phase space, which for simplicity we take to be $\mathbf{R}^n$. We shall consider the possibility of decomposing the flow, in an appropriate region of phase space, into a connected network of 'tubes' and 'joints'. Each joint is a topological ball in phase space inside which the dynamics reduces to a simple, standard form, having a unique subset that is flow-invariant for all forward and backward time — either a unique equilibrium or a unique limit cycle. A tube is a topological cylinder that connects joints to joints. The dynamics within a tube is essentially trivial: roughly speaking, each tube is arranged so that the flow is trapped inside it, and any trajectory that enters one end exits from the other end after a bounded (usually short) time. Flow through the sidewall of

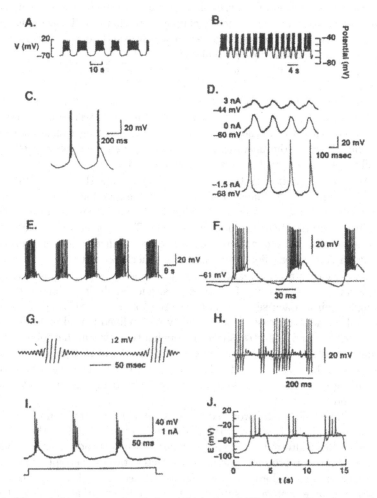

Figure 8.12: Neuronal bursting signals from Rinzel and Wang; see also Keener and Sneyd [305].

the tube is transverse to it and directed inwards. Moreover, we require that any trajectory that exits from a tube must pass into a joint, and *if* it leaves the joint, it must do so through the entrance of some tube. In other words: once a trajectory has entered the pipe system it cannot escape in forward time. Any observation of the dynamics of a pipe system will therefore decompose into a series of segments, during which it obeys the simple dynamics of a generic transient through some particular joint, followed by a rapid transition through a tube to the next joint, and so on. These segments correspond to the segments of 'regular' dynamical behavior that occur in bursting, followed by the modification of this regular behavior when exiting the joint.

For reasons that we shall explain in a moment, the dynamics within a joint will be assumed to be of saddle type, in the following sense. (For simplicity we begin with the equilibrium case.) At a joint of equilibrium type, or *E-joint*, we assume that the unstable manifold of the equilibrium is 1-dimensional and the stable manifold is $(n-1)$-dimensional, where n is the dimension of the phase space. We further assume that the dynamics is locally conjugate to that of a linear ODE determined by a matrix with distinct eigenvalues, of which exactly one (corresponding to the unstable manifold) has positive real part.

These assumptions are the simplest (but by no means the only) way to arrange for the following behavior. Suppose that a trajectory enters such a joint, passing close to its unstable manifold. It begins by converging towards the equilibrium, and for a time it remains close to that equilibrium. However, it eventually becomes captured by the outward flow near the unstable manifold, causing it to exit from the joint. Because the unstable manifold is 1-dimensional, we can easily arrange for all trajectories near it to flow into another tube.

At a joint of limit cycle type, or *C-joint*, we assume that the unstable manifold of the equilibrium is 2-dimensional and the stable manifold is $(n-2)$-dimensional. The flow transverse to the cycle is assumed to be conjugate to a linear flow of the type just described (but one dimension lower). The geometry of exiting trajectories is more complex, but again there are many circumstances in which a suitable subset of exit trajectories can flow into a suitably generalized tube.

With such an arrangement, the forward-time dynamics of any trajectory originating in a tube consists of a series of segments, during which the flow passes through a joint. Each segment is therefore close to a steady signal or a periodic one (and in the latter case there is a fairly well defined period and amplitude), but eventually makes a rapid transition (through a tube) to another joint. Suppose, for example, that a limit cycle joint links through a tube to an equilibrium joint, which in turn links by a tube back to the limit cycle joint. Then the signal will show a burst of rapid oscillations, followed by a relatively steady signal, and thereafter this alternation will repeat.

We show below that provided there is sufficient contraction of the flow along the tubes, and the flow along tubes is sufficiently rapid, then these segments of the signal must alternate in an approximately time-periodic manner. Such a pipe system therefore reproduces the main characteristics of one type of bursting.

Furthermore, it does so in a robust manner. We show that if a flow that possesses a pipe system in subjected to a small perturbation, then it continues to possess essentially the *same* pipe system. So its 'bursting' characteristics are robust under perturbations. This robustness is weaker than 'structural stability' in the sense of Smale [479], but has the same interpretation — 'reproducible in experiments to within normal experimental error'.

Pipe Systems. Melbourne [384] observed that when a heteroclinic cycle is asymptotically stable it is surrounded by a flow-invariant *pipe system* consisting of tubes and joints. The *joints* are neighborhoods of equilibria or periodic solutions in the cycle in which the system of differential equations is linearizable, and the *tubes* are neighborhoods of the connecting trajectories in the cycle. Flow-invariance (and normal hyperbolicity of the tube system) imply that the tube system persists under small perturbations of the underlying system of differential equations — even if the perturbation breaks symmetry. It follows that the intermittency associated with asymptotically stable heteroclinic cycles persists as bursting under symmetry-breaking perturbation. As we have seen in the Guckenheimer-Holmes cycle, these perturbations can lead to time-periodic states — but these states still exhibit a form of (periodic) bursting (see Figure 8.5). Melbourne used these observations to create bursting between periodic solutions in systems without symmetry by using the normal form symmetry present in a triple Hopf bifurcation to create an asymptotically stable cycle.

Example 8.1 The van der Pol oscillator (in relaxation or Liénard form) is a simple example of a fast/slow system. We show how to represent it as a pipe system. We follow the treatment of the van der Pol oscillator in Zeeman [532]. The equations we use here are:

$$\frac{dx}{dt} = y \tag{8.2}$$

$$\frac{dy}{dt} = -k(y^3 - y + x) \tag{8.3}$$

where $k > 0$ is a parameter. In the fast/slow interpretation it is assumed that $k \gg 1$ (so that $\varepsilon = 1/k$ is a small parameter suitable for singular perturbation theory). Here we take $k \geq 4$ for reasons to be explained below.

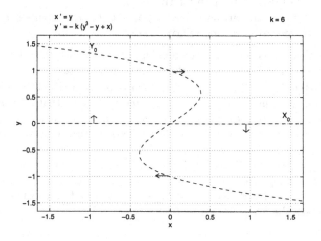

Figure 8.13: Geometry of the van der Pol oscillator.

Figure 8.13 shows the relevant geometry. There is a rotational symmetry through $180°$, $(x, y) \rightarrow (-x, -y)$, which we use to simplify the calculations. The x-nullcline (where $dx/dt = 0$) is the x-axis $y = 0$, and we call this set X_0. The y-nullcline (where $dy/dt = 0$) is the cubic curve Y_0 given by

$$y^3 - y + x = 0 \tag{8.4}$$

The 'fast/slow' intuition here is that when $k \gg 1$ there is a *fast foliation* in the y-direction, and the minus sign in (8.3) ensures that the system relaxes rapidly along the fast foliation onto the *slow manifold* Y_0. It then follows the *slow dynamic* obtained by restricting/projecting (8.2) onto Y_0. The pipe system we derive is loosely related to this picture.

For all k, the nullclines intersect only at the origin, so there is a unique equilibrium at $O = (0, 0)$. The flow across Y_0 is parallel to the x-axis ('horizontal'); flow across X_0 is parallel to the y-axis ('vertical'). Arrows show the flow directions across the nullclines ('clockwise'). The linearization of (8.2,8.3) at O is

$$\begin{bmatrix} 0 & 1 \\ -k & k \end{bmatrix}$$

with eigenvalues

$$\frac{k \pm \sqrt{k^2 - 4k}}{2}$$

So the origin is a hyperbolic source for all $k > 0$. It is of spiral type (complex eigenvalues) for $k < 4$, and of real type (real eigenvalues) for $k \geq 4$. For the sake of argument we shall assume that $k \geq 4$.

We now begin the construction of a pipe system. First, we construct two simple closed polygons A and B surrounding the origin, with A inside B, in such a manner that:

- All trajectories that start on A have a positive (or zero) component normal to A, in the outward direction relative to O.
- All trajectories that start on B have a positive (or zero) component normal to B, in the inward direction relative to O.

In other words, the flow is *trapped* in the topological annulus bounded by A and B. We shall use A, B, X_0 and Y_0 to define the required pipe system.

Since the system (8.2,8.3) has a hyperbolic source at the origin, there must exist a sufficiently small simple closed curve A' surrounding O such that all trajectories that start on A' flow *outwards*. The system also has a source at infinity, so there must also exist a sufficiently large simple closed curve B' surrounding O such that all trajectories that start on B' flow *inwards*. These curves can be used in place of A, B to define a pipe system. The explicit construction of such curves (which we will see can be defined independently of k if $k \geq 5$) is more delicate. In particular they *cannot* be taken to be circles centered at O. We shall therefore construct explicit polygons and make the necessary estimates.

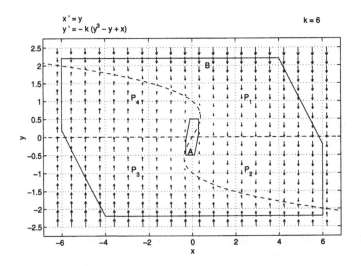

Figure 8.14: Definition of the pipe system.

The geometry we select for A and B is shown in Figure 8.14. Many arbitrary choices have gone into this figure. The main motivation is to keep the required calculations reasonably simple and to exploit useful features of the flow. The region between the two polygons could surely be reduced in area by performing a more delicate analysis.

The polygon A has vertices at

$$(0.3, 0.5) \ (0.3, 0.1) \ (0.1, -0.5) \ (-0.3, -0.5) \ (-0.3, -0.1) \ (-0.1, 0.5)$$

and B has vertices at

$$(4, 2.2) \ (6, -0.2) \ (6, -2.2) \ (-4, -2.2) \ (-6, 0.2) \ (-6, 2.2)$$

Flow on A is outwards. We claim that the normal component of the flow along each side of the polygon A is nonzero and directed along the outward normal.

Along the horizontal and vertical edges of A this is immediate by (8.2,8.3). The line segment from $(0.3, 0.1)$ to $(0.1, -0.5)$ has equation

$$x = \frac{y}{6} + \frac{1.7}{6} \quad -0.5 \leq y \leq 0.1$$

The outward normal vector is $(6, -1)$. So we need to check that the scalar product

$$S = \left(y, -k \left(y^3 - y + \frac{y + 1.7}{6} \right) \right) \cdot (6, -1) = 6y + k \left(y^3 - y + \frac{y + 1.7}{6} \right)$$

is positive for $-0.5 \leq y \leq 0.1$ and $k \geq 4$. To verify that $S > 0$ note that

$$\frac{dS}{dk} = y^3 - y + \frac{y + 1.7}{6} > 0$$

for $y \in [-0.5, 0.1]$. So we need only verify that when $k = 4$, $S > 0$ for $y \in [-0.5, 0.1]$, and this is straightforward.

By rotational symmetry, the flow is also outward along the remaining side. □

Flow on B is inwards. We claim that the normal component of the flow along each side of the polygon B is nonzero and directed along the inward normal.

Again along the horizontal and vertical edges of B this is immediate by (8.2,-8.3). The line segment from $(4, 2.2)$ to $(6, -0.2)$ has equation

$$x = \frac{-y + 7}{1.2} \quad -0.2 \leq y \leq 2.2$$

The inward normal is $(-2, -2.4)$. So we need to check that the scalar product

$$\left(y, -k \left(y^3 - y + \frac{-y + 7}{1.2} \right) \right) \cdot (-2, -2.4) = -2y + 2.4 - k \left(y^3 - y + \frac{-y + 7}{1.2} \right)$$

is positive for $-0.2 \leq y \leq 2.2$ and $k \geq 4$. This is again straightforward.

By rotational symmetry, the flow is also inward along the remaining side.

Finally we define the tubes and verify that we have a pipe system in which four tubes connect to each other in a ring: no joints.

The tubes are the four regions into which the annulus between A and B is divided by X_0 and Y_0. Call them P_1, P_2, P_3, P_4 labelled clockwise from the positive quadrant, as shown in Figure 8.14. The entrances and exits are defined by segments of nullclines (making it trivial to check the required transverse inward or outward flow). The walls are determined by the relevant segments of A and B.

We shall not define explicit axial and transverse coordinates on the pipes P_j (though this could of course be done). However, we must prove that the transit time through each pipe is bounded away from zero. We indicate the argument for P_1: that for P_2 is similar and the rest follow by rotational symmetry.

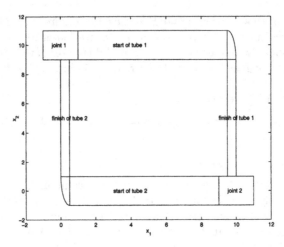

Figure 8.15: Schematic diagram for pipe system with two joints and two tubes.

Divide P_1 into the region P_{11} lying above (say) the line $y = 0.1$ and the region P_{12}, forming two 'subpipes'. In P_{11} the x-component of the flow is bounded away from zero, so after a uniformly finite time any trajectory must exit. It can do so only by crossing the line $y = 0.1$ and going into P_{12}. Inside P_{12} the y-component of the flow is bounded away from zero, so after a uniformly finite time any trajectory must exit. It can do so only by going out through the exit of pipe P_1.

Note that we cannot assert that the flow through any tube is rapid if $k \to \infty$, because all tubes include a section of slow flow. This could be adjusted by changing tube boundaries, but some tubes must have long transit times.   □

Corollary 8.2 *The annulus between A and B contains at least one periodic cycle for all $k \geq 4$.*

Proof. This follows from the Brouwer fixed-point theorem — or, more simply, from the trivial version for maps on an interval.   □

In fact, the annulus contains a *unique* cycle, which is stable, but proving this requires more delicate estimates, see Hirsch and Smale [277].

Piecewise Smooth Pipe Systems. In this section we illustrate how pipe systems can be used to create bursting solutions, and we do so in a exceptionally crude way. We experiment with a piecewise smooth two-joint two-tube systems. Indeed, there is a simple way to define a two-joint, two-tube pipe system according to the schematic diagram in Figure 8.15. Although drawn in two dimensions, this schematic easily handles extra dimensions by extending tubes and joints trivially to higher dimensional rectangles. Our initial numerical explorations were done in three dimensions.

We define the vector field on this pipe system in a piecewise smooth fashion. Each tube has three sections — a starting section, a bending section, and a finishing section. The flow in the starting and finishing sections has constant speed in the axial direction and contracts in the transverse directions. (This tube flow insures nearly periodic — in fact almost exactly periodic — signals from the pipe system.) The flow in the bend is linear elliptical flow and serves to narrow the exit of the pipe, so that it is easier to control entrance into the next tube when exiting the

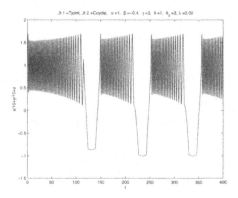

Figure 8.16: Parabolic burster from a pipe system consisting of a C-joint and a trivial E-joint and two tubes. Here $\theta_0 = 3, \alpha = 1, \beta = -0.4, \gamma = 3, \delta = 1$.

joint. No attempt is made to match the vector fields on the boundaries between segments of a tube or between tubes and joints.

An E-joint is defined by the differential equation $\dot{x} = Lx$ where

$$A = \begin{bmatrix} e & 0 \\ 0 & B \end{bmatrix}$$

where $e > 0$ is the expanding eigenvalue and B is a 2×2 matrix with eigenvalues having negative real part. We choose B to be either

$$B = \begin{bmatrix} -c_2 & 0 \\ 0 & -c_3 \end{bmatrix} \quad \text{or} \quad B = \begin{bmatrix} -\sigma & -\omega \\ \omega & -\sigma \end{bmatrix}$$

depending on whether the eigenvalues of B are real or complex conjugates. Here $c_2, c_3 > 0$ and $\sigma > 0$.

We define our C-joint by

$$
\begin{aligned}
\dot{x} &= \lambda x \\
\dot{r} &= (\alpha(r_0^2 - r^2) + \beta x)r \\
\dot{\theta} &= \theta_0 + \gamma(r_0^2 - r^2) + \delta x
\end{aligned}
\tag{8.5}
$$

When $\alpha > 0$ (8.5) has a stable limit cycle at $r = r_0$. In our integrations we have set

$$r_0 = 0.5$$

We can solve the first equation in (8.5) explicitly, obtaining

$$x(t) = x_0 e^{\lambda t}$$

Since the exit coordinate (relative to the center) is 1 in our joints, it follows that the time T spent in the joint is

$$T = \frac{1}{\lambda} \log\left(\frac{1}{x_0}\right).$$

Since we want the time in the joint to be long enough to allow for a number of circuits around the limit cycle before escape, we set the expanding eigenvalue to be

$$\lambda = 0.02$$

A sample time series from a pipe system with an E-joint and a C-joint and two tubes is given in Figure 8.16. This closely resembles the time series of a so-called *parabolic burster* in a neuron, see Figure 8.12(E). In a parabolic burster

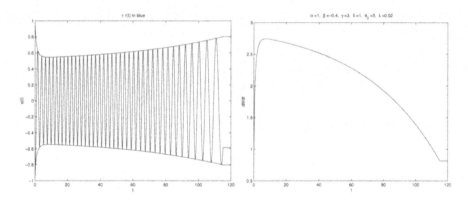

Figure 8.17: Parabolic burster from a C-joint. (Left) Graph of the time series $y(t)$ oscillating between $\pm r(t)$. (Right) Graph of $\dot{\theta}(t)$ showing parabolic shape. Here $\theta_0 = 3, \alpha = 1, \beta = -0.4, \gamma = 3, \delta = 1$.

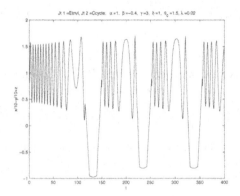

Figure 8.18: Parabolic burster from a pipe system consisting of a C-joint, a trivial E-joint and two tubes. Here $\theta_0 = 1.5, \alpha = 1, \beta = -0.4, \gamma = 3, \delta = 1$.

the periodic part of the bursting signal has a frequency that increases and then decreases before escape. In such a signal the oscillations in the time series will compress and then expand. We can measure this change by graphing $\dot{\theta}(t)$ along the solution. A parabolic burster is detected when the graph of $\dot{\theta}$ is parabolic with a unique maximum. A close-up of the transition through a C-joint yielding a parabolic burster is given in Figure 8.17.

If we slow the frequency (reduce θ_0), then it is possible to induce changes in the direction of the rotation as x and r vary, thus causing a gap in the oscillation. See Figure 8.18 and compare with Figure 8.12(F).

It is also possible to introduce rotation in the transverse directions in flow along tubes. The consequence of this rotation is that trajectories will enter the joints at different values causing variation in the length of bursts. We show an example of variable length bursting in Figure 8.19. Compare this time series with those in Figure 8.12(A-B).

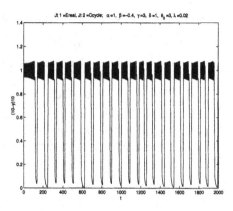

Figure 8.19: Variable length bursting in a pipe system consisting of a C-joint, a real E-joint and two tubes. Here $\theta_0 = 3, \alpha = 1, \beta = -0.4, \gamma = 3, \delta = 1$.

8.4 Cycling Chaos

As we have seen, one of the characteristic and distinguishing features of symmetric systems of differential equations is the existence of structurally stable saddle connections that would be regarded as highly degenerate in the absence of symmetry. The Guckenheimer-Holmes cycle shows that it is possible for structurally stable, asymptotically stable, cycles of saddle connections to be created by bifurcation at the loss of stability of a group invariant equilibrium in a symmetric system. The resulting heteroclinic cycle consists of a finite set of saddle points connected by trajectories. Trajectories which approach the cycle remain for long periods near each equilibrium before making a rapid transition to a neighborhood of the next equilibrium. This feature of an asymptotically stable heteroclinic cycle gives a simple mathematical model for intermittency.

In this section we follow Dellnitz et al. [144] and observe that the equilibria in the Guckenheimer-Holmes cycle may be replaced by chaotic sets. In this way, we present a phenomenological mathematical model for (spatially) cycling (temporal) chaos.

Symmetrically coupled cells. Golubitsky et al. [236] and Dionne et al. [162] observed that with an appropriate choice of coupling the Guckenheimer-Holmes example can be viewed as a system of symmetrically coupled cells. For our purposes, we regard a *cell* as a k-dimensional system of ODE

$$\dot{x} = f(x)$$

where $x \in \mathbf{R}^k$. Thus, in the Guckenheimer-Holmes system, $k = 1$ and there are three identical cells coupled in a directed ring. Generally, let x_1, x_2, x_3 be in $\mathbf{R}^k$. We consider a system of three coupled cells of the form

$$
\begin{aligned}
\dot{x}_1 &= f(x_1) + h(x_3, x_1) \\
\dot{x}_2 &= f(x_2) + h(x_1, x_2) \\
\dot{x}_3 &= f(x_3) + h(x_2, x_3)
\end{aligned}
\tag{8.6}
$$

We identify two types of symmetry in a coupled cell system of this type: *global* and *local*. The global symmetries are dictated by the pattern of coupling. In (8.6), the global symmetry group is $\mathbf{Z}_3$, generated by the cyclic permutation

$$(x_1, x_2, x_3) \mapsto (x_2, x_3, x_1)$$

Local symmetries are symmetries of f. Thus, a linear transformation σ of $\mathbf{R}^k$ is a local symmetry if

$$f(\sigma x) = \sigma f(x) \qquad (x \in \mathbf{R}^k)$$

In the Guckenheimer-Holmes system, the local symmetry group is $\mathbf{Z}_2$ and is generated by $\sigma x = -x$. Moreover, in this system, local symmetries of individual cells are symmetries of the *complete* system (8.6). That is, for all local symmetries σ we have

$$\begin{aligned} h(\sigma y, x) &= h(y, x) \\ h(y, \sigma x) &= \sigma h(y, x) \end{aligned}$$

Following Golubitsky *et al.* [1994], we call this type of coupling *wreath product coupling*. Viewed in this way, the Guckenheimer-Holmes system has coupling term given by the cubic polynomial

$$h(y, x) = \gamma |y|^2 x$$

where $\gamma \in \mathbf{R}$ represents the strength of the coupling.

The internal cell dynamics in the Guckenheimer-Holmes system are governed by the pitchfork bifurcation

$$f(x) = \lambda x - x^3$$

which is consistent with the internal symmetry. As λ varies from negative to positive through zero, a bifurcation from the trivial equilibrium ($x = 0$) to nontrivial equilibria ($x = \pm\sqrt{\lambda}$) occurs, and these bifurcating equilibria are stable in the internal cell dynamics. Guckenheimer-Holmes show that when the strength of the coupling is large and negative ($\gamma \ll 0$), an asymptotically stable heteroclinic cycle connecting these bifurcated equilibria exists. The connection between the equilibria in cell 1 to the equilibria in cell 2 occurs through a saddle-sink connection in the $x_1 x_2$-plane (which is forced by the internal symmetry to be an invariant plane for the dynamics). The global permutation symmetry guarantees connections in both the $x_2 x_3$-plane and the $x_3 x_1$-plane.

8.4.1 Cycling chaos. It turns out that the intermittent cycling of the global dynamics does not depend in an essential way on the nature of the internal dynamics, provided that f satisfies some mild restrictions (for example, the origin and infinity are repellors). Detailed mathematical analysis and generalizations will appear in Field *et al.* [1995]. In this note, we illustrate this observation by numerical simulation of the internal dynamics in a three cell system. Our first two examples have internal dynamics defined by (a modified) *Chua circuit*:

$$f(y_1, y_2, y_3) = (\alpha(y_2 - m_0 y_1 - \frac{m_1}{3} y_1^3), y_1 - y_2 + y_3, -\beta y_3)$$

where the internal variable $x = (y_1, y_2, y_3)$ and α, β, m_0, m_1 are constants. Our third example has internal dynamics governed by the *Lorenz equations*:

$$f(y_1, y_2, y_3) = (\sigma(y_2 - y_1), \rho y_1 - y_2 - y_1 y_3, -\beta y_3 + y_1 y_2)$$

where σ, ρ, β are constants. See Lorenz [364].

In Figure 1 we choose parameter values so that the internal dynamics is a *double scroll* attractor. We present a time series of a single cell in the first time history. The following three time evolutions in Figure 1 show the temporal behavior for cell 1, cell 2, and cell 3 respectively. From this figure we see that when one of the cells is active and performing the double scroll dynamics — say cell 1 — the others (cell 2 and cell 3) are quiescent (near 0). After a while cell 1 becomes quiescent while cell 2 becomes active and the transition time during which the cells interchange states is very short. The process then repeats with cell 2 and cell 3 interchanging active and quiescent states. Indeed, the process cycles forever, just as in the Guckenheimer-Holmes heteroclinic cycle, but now producing *cycling chaos*.

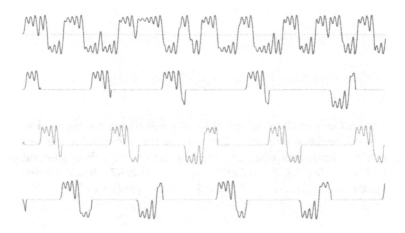

Figure 8.20: Chua circuit equations. The first row shows a single cell without coupling. The bottom three rows show the time series for each of the three cells when the coupling strength is $\gamma = -2$. Parameter values are $\alpha = 18$, $\beta = 33.136$, $m_0 = -0.230769$, $m_1 = 0.0123077$. Initial conditions are $x_1(0) = (0.01, 0.1, -0.2)$, $x_2(0) = (0.24, 0.34, -0.01)$, $x_3(0) = (0.2, -0.3, 0.1)$. [144]

In the second example, we choose parameter values in the Chua circuit that yield an asymmetric chaotic attractor. The internal symmetry forces a second conjugate attractor. Which attractor is actually observed depends on the initial conditions in the individual cell. In Figure 2, we illustrate this asymmetry. For one choice of initial conditions, the internal cell dynamics always has $y_1 > 0$; a different choice of initial conditions leads to an attractor where $y_1 < 0$ for all time. When we simulate the coupled cell system, we get the same cycling chaos but when a given cell — say cell 2 — becomes active it chooses 'randomly' which of the conjugate attractors ($y_1 < 0$ or $y_1 > 0$) it will track.

Finally, in Figure 3 we illustrate the phenomenon of cycling chaos when the internal dynamics is given by the Lorenz system. Cycling occurs even though the internal symmetry in the two examples is different. In the Chua circuit $\sigma(x) = -x$ while in the Lorenz equation $\sigma(y_1, y_2, y_3) = (-y_1, -y_2, y_3)$.

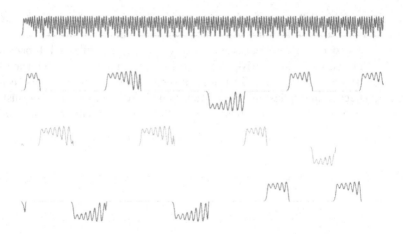

Figure 8.21: Chua circuit equations. The first row shows the temporal evolution of a single cell without coupling. The bottom three rows show the time series for each of the three cells when the coupling is $\gamma = -2$. Parameter values are $\alpha = 15$, $\beta = 33.136$, $m_0 = -0.230769$, $m_1 = 0.0123077$. Initial conditions are $x_1(0) = (0.01, 0.1, -0.2)$, $x_2(0) = (0.24, 0.34, -0.01)$, $x_3(0) = (0.2, -0.3, 0.1)$. [144]

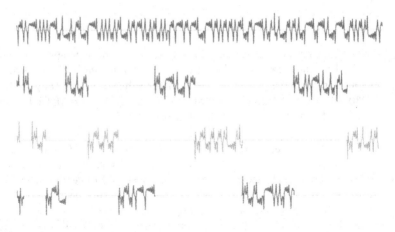

Figure 8.22: Lorenz equations. The first row shows the temporal evolution of a single cell without coupling. The bottom three rows show the time series for each of the three cells when the coupling is $\gamma = -0.025$. Parameter values are $\sigma = 15$, $\rho = 58$, $\beta = 2.4$. Initial conditions are $x_1(0) = (10, -11, 30)$, $x_2(0) = (10, -13, 20)$, $x_3(0) = (10, -12, 30)$. [144]

Chapter 9

Symmetric Chaos

In nonlinear dynamics without symmetry there is a hierarchy of increasingly complicated dynamic behavior: steady-state, equilibrium, periodic, quasiperiodic, chaotic.
See Ruelle and Takens [456] and Broer *et al.* [79]. So far we have studied analogous
behavior for steady, periodic, and quasiperiodic states of equivariant differential
equations, asking the following questions:

- What symmetry groups, in principle, might states have?
- What are the general existence and stability theorems?
- What is the interpretation in physical space of the symmetric dynamics
  found in phase space?

We now ask similar questions for chaotic solutions of equivariant dynamical systems.

We motivate our desire to answer these questions in two ways. First, on the
mathematical side, we provide evidence that symmetry and chaotic dynamics can
coexist in interesting ways. Second, we discuss an application where nontrivial symmetry is surely a modeling assumption and where chaotic dynamics is an important
issue.

Mathematical Motivation. In Section 4.9 we discussed the classical period-doubling
cascade to chaotic dynamics. We redraw the bifurcation diagram for the logistic
map

$$F(x) = \mu x(1 - x) \tag{9.1}$$

in Figure 9.1. Recall how that diagram is formed. Fix μ and iterate F with an
initial condition x_0. Iterate 100 times and then plot the next 1000 points on the $\mu =$
constant vertical line. Then increment μ and use the last computed x as the initial
condition for the next 1100 iterations, and so on. In this way we see transitions
from stable fixed points to period-two points, then to period-four points, through
to complicated chaotic dynamics.

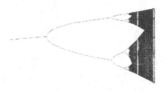

Figure 9.1: Period-doubling cascade to chaotic dynamics in logistic map (9.1) for
$2.75 \leq \mu \leq 3.75$.

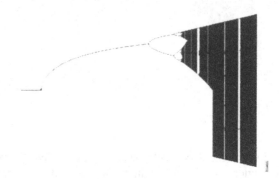

Figure 9.2: Period-doubling cascade in the odd logistic map (9.2).

Next we ask: What happens when symmetry is involved? On the line there is
one nontrivial symmetry $x \mapsto -x$, and the equivariant maps are the odd functions.
The logistic map is not odd — but we can alter it to obtain the *odd logistic map*

$$G(x) = \mu x(1 - x^2) \tag{9.2}$$

considered by Chossat and Golubitsky [104]. The bifurcation diagram for the odd
logistic map is drawn in Figure 9.2. The initial transitions are identical to those
found in the logistic map, but then there is a sudden transition from attracting
sets that are asymmetric to ones that are symmetric. We call this transition a
symmetry-increasing bifurcation.

Chossat and Golubitsky [104] showed that symmetric attractors are common
in the iterates of (noninvertible) equivariant maps of the plane. They iterated
$\mathbf{D}_m$-equivariant planar maps of the form

$$f(z) = (\lambda + \alpha|z|^2 + \beta\mathrm{Re}(z^m))z + \gamma\bar{z}^{m-1} \tag{9.3}$$

where $\lambda, \alpha, \beta, \gamma$ are real constants and $z \in \mathbf{C}$. Pictures of attractors formed by
iterating f are shown in Figure 9.3. We call $\mathbf{D}_m$-symmetric chaotic attractors
icons.

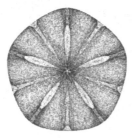

Figure 9.3: Attracting set for (9.3) where $(m, \lambda, \alpha, \beta, \gamma)$ is: (left) The Sanddollar
$(5, -2.34, 2.0, 0.2, 0.1)$; (right) Mayan Bracelet $(7, -2.08, 1.0, -0.1, 0.167)$.

Note that not only are the icons in Figure 9.3 $\mathbf{D}_m$-symmetric; they also have a rather sophisticated internal structure — the density of points obtained by iteration is far from uniform. Indeed, the probability that iterates land in a given small region of the attracting set varies from region to region. It appears that there is a naturally defined probability measure on the attracting set, and that measure can be approximated as follows. Iterate (9.3) a large number of times, count the number of times that each 'pixel' on the attracting set is visited during the iteration process, and color the pixels by number. So equal color implies equal likelihood of visitation. This process is carried out in Field and Golubitsky [189], who obtain many pictures of symmetric invariant measures in this way. Indeed, the figures shown here were produced using the program ITER developed by Mike Field.

These pictures motivate several interesting mathematical questions.

- For a given finite group, are there equivariant mappings with attractors having setwise symmetries equal to any given subgroup? For example, can a $\mathbf{D}_6$-equivariant mapping in the plane produce an attractor with exactly $\mathbf{D}_3$ symmetry?

- Attractors in symmetric systems have two kinds of symmetries — those that fix the attractor pointwise and those that leave the attracting set invariant. The visible symmetries of the icons in Figure 9.3 are setwise symmetries. Previously, in Chapter 3 this dichotomy arose in connection with periodic solutions, where the setwise symmetries corresponded to spatio-temporal symmetries of the periodic solution. What is the meaning in physical space of the setwise symmetries of chaotic attractors?

- In high dimensions (three or greater), how can we determine the setwise symmetries of an attractor? Visual inspection is generally inadequate.

- What kinds of symmetry-changing bifurcation, such as the ones shown in Figures 9.2 and 9.4, are permitted in chaotic attractors? Are there new kinds of bifurcation?

Figure 9.4: Attractors for (9.3) where $(m, \alpha, \beta, \gamma) = (3, -1.21, -0.09, -0.8)$. (Left) $\lambda = 1.50$, attractor has $\mathbf{D}_1$ symmetry. (Right) $\lambda = 1.52$, attractor has $\mathbf{D}_3$ symmetry.

Motivation from Experiments. Recall from Section 7.1 that the Faraday surface wave experiment is performed by sinusoidally vibrating a fluid layer at a fixed amplitude and frequency. When the frequency of vibration is small, the fluid layer remains flat and undeformed, but when the frequency is increased the surface deforms into waves which in small containers form patterns that seem consistent with the symmetry of the boundary. Figure 9.5 shows two images from the experiments of Gluckman *et al.* [211] — one from an experiment using a container with square cross-section and one where the container has circular cross-section.

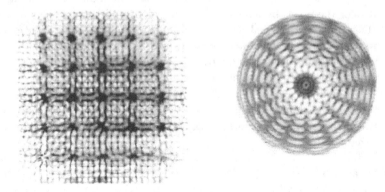

Figure 9.5: Faraday experiment performed by Gluckman *et al.* [211] near primary symmetry-breaking bifurcation in containers with square and circular cross-sections. Pictures courtesy of J.P. Gollub.

When the frequency is increased further, the surface waves bifurcate until the waves reach a state of *spatio-temporal chaos*. In such a state, both the temporal dynamics and the spatial pattern change chaotically. Pictures of spatio-temporal chaos in containers of square and circular cross-section are shown in Figure 9.6.

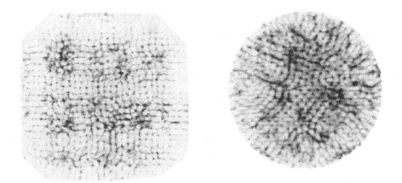

Figure 9.6: Faraday experiment performed by Gluckman *et al.* [211] in spatio-temporal chaos regime in containers with square and circular cross-sections. Pictures courtesy of J.P. Gollub.

The question that we ask concerning the spatio-temporal regime is: Where does the symmetry (pattern) that was so very visible with small frequency vibrations go when the frequency of vibration is increased?

A Brief Hint at Results. In this chapter we mix topological dynamics, ergodic theory, and symmetry to obtain our results. In Section 9.1 we prove two theorems — one concerning the topology of attractors (Theorem 9.9) and one concerning the symmetry of attractors (Theorem 9.3). A consequence of the second theorem is that planar $\mathbf{D}_6$-equivariant mappings *cannot* produce attractors with exactly $\mathbf{D}_3$ symmetry. This demonstrates that there are non-trivial restrictions on the symmetries of chaotic attractors.

In Section 9.2 we introduce ergodic theory and show that SBR measures on symmetric attractors are themselves symmetric (Theorem 9.22). This theorem helps to explain why the color pictures of symmetric attractors in Field and Golubitsky [189] are symmetric. Next, we discuss how to compute the symmetry of attractors in higher dimensional spaces. Ergodic sums and 'detectives' are used in this method (see Theorems 9.27 and 9.29). Detectives also lead to an interpretation of the setwise symmetries of attractors in phase space as symmetries of time averages in physical space. Thus, symmetries of chaotic attractors in phase space lead to patterns in the time average in physical space — we call these patterns *patterns on average*. This issue is discussed in Section 9.4 — but we introduce the topic here so that we can complete our initial discussion of the Faraday experiment.

The Brusselator and Patterns on Average. For simplicity we begin the discussion with the *Brusselator*, which is a system of reaction-diffusion equations on the interval $[0,1]$ of the form

$$
\begin{aligned}
du/dt &= \lambda^{-2}D_1 u_{xx} + u^2 v - (B+1)u + A \\
dv/dt &= \lambda^{-2}D_2 v_{xx} - u^2 v + Bu
\end{aligned}
\tag{9.4}
$$

Here u, v, A and B represent chemical concentrations and D_1, D_2 are diffusion constants. In this model A and B are constant while u and v depend on x and t. The parameter λ is the bifurcation parameter. There is a trivial solution $u = A, v = B/A$. We consider (9.4) subject to Dirichlet boundary conditions

$$
\begin{aligned}
u(0,t) &= u(1,t) = A \\
v(0,t) &= v(1,t) = B/A
\end{aligned}
$$

The Dirichlet problem possesses a reflectional symmetry given by

$$
\kappa(u(x,t), v(x,t)) = (u(1-x,t), v(1-x,t))
$$

Holodniok *et al.* [281] consider this model in detail and find multifrequency motions using numerical simulation. Presuming that chaotic dynamics will occur near multifrequency motion, Dellnitz *et al.* [146] follow [281] and set

$$
A = 2 \quad B = 5.45 \quad D_1 = 0.008 \quad D_2 = 0.004
\tag{9.5}
$$

and compute the time average

$$
F_u(x) = \lim_{T \to \infty} \left(\frac{1}{T} \int_0^T u(x,\tau) d\tau \right)
\tag{9.6}
$$

at each point in space.

Suppose that the attractor corresponding to the solution $u(x,t)$ is chaotic and symmetric. We expect u to sample the whole attractor in phase space, so the time average (9.6) should therefore be equal to a (weighted) space average over the attractor in phase space — though the rigorous proof of this point requires a theorem of ergodic type to be valid, and it is technically very difficult to prove such theorems. Presuming this, however, we expect (9.6) to be symmetric under

$x \mapsto 1 - x$ if the attractor is symmetric. Moreover, generically, we expect (9.6) to be asymmetric when the attractor is asymmetric.

The graphs of F_u both before and after a symmetry-increasing bifurcation are shown in Figure 9.7. In both cases $T = 20000$ is chosen for the numerical computations. In Figure 9.8 we indicate why we believe that this symmetry-creation is caused by collision of conjugate attractors of the type modeled in Figure 9.2. If such a collision were to occur, the difference in phase space between the union of the attractor and its conjugate attractor before collision would be approximately equal to the attractor after collision. Therefore, we expect the average

$$\tfrac{1}{2}\left(F_u(x) + F_u(1 - x)\right) \tag{9.7}$$

to vary continuously. The graph of (9.7) before symmetry-creation is given in Figure 9.8 along with the difference between (9.7) before and after symmetry-creation. Note how small this difference is.

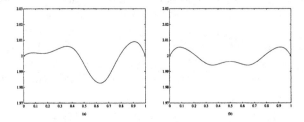

Figure 9.7: Graphs of F_u for (a) $\lambda = 1.45$, (b) $\lambda = 1.47$.

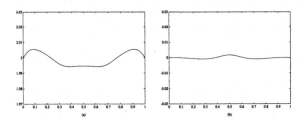

Figure 9.8: (a) The average of the graphs of F_u for the two conjugate attractors for $\lambda = 1.45$, and (b) the difference between the graph of F_u for $\lambda = 1.47$ and this graph.

An example of the phenomenon of pattern in the time average occurs in the Faraday experiment. In the experiment of [211], the time average of the intensity of transmitted light is found to be symmetric (Figure 9.9), whereas the instantaneous time picture is disordered (Figure 9.6). Thus we see how symmetry of an attractor in phase space manifests itself in physical space through time averages. Reasons why symmetry in the time average should be expected are discussed in greater detail in Section 9.4.

9.1 Admissible Subgroups

The stability issue for a chaotic solution is dealt with essentially by definition: we shall consider only solutions that lie on an attractor. There are many definitions

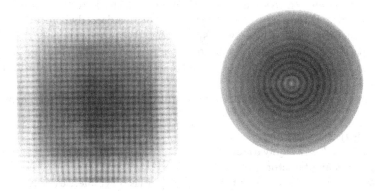

Figure 9.9: Time averages of intensity of transmitted light in the spatio-temporal chaos regime of the Faraday experiment of Gluckman *et al.* [211]. Pictures courtesy of J.P. Gollub.

of 'attractor' in the literature (see Milnor [400]): we now introduce the one we use here.

Let $f : \mathbf{R}^n \to \mathbf{R}^n$ be a continuous map and view f as a discrete dynamical system. That is, we *iterate* f, starting with an initial value x_0, by setting

$$x_{t+1} = f(x_t)$$

Then the *trajectory* of f, starting at x_0, is the sequence

$$x_0, x_1, x_2, \ldots, x_t, x_{t+1}, \ldots$$

In computer experiments, trajectories often converge to complex geometric shapes such as in Figures 9.3 and 9.4. The next definition formalizes this observation:

Definition 9.1 $A \subseteq \mathbf{R}^n$ is an *attractor* for f if

(a) A is compact.
(b) $f(A) \subseteq A$.
(c) A is (Liapunov) *stable*, that is, for every open $U \supseteq A$ there exists an open $V \supseteq A$ such that $f^m(V) \subseteq U$ for all $m \geq 0$.
(d) $A = \omega(x)$ for some $x \in \mathbf{R}^n$, where $\omega(x)$ is the limit point set of the trajectory through x. ◇

See Figure 9.10.

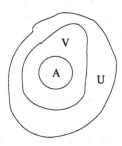

Figure 9.10: Schematic picture of an attractor.

A Symmetry Theorem. In this section we discuss the algebraic restrictions placed on a subgroup $\Sigma \subseteq \Gamma$ that insure that there is a Γ-equivariant dynamical system having an attractor with Σ as its setwise symmetries. We prove here only that the algebraic restrictions that we find on Σ are necessary, and in the proof we follow Melbourne *et al.* [393]. Sufficiency of these conditions is proved by Ashwin and Melbourne [25].

We adopt the following assumptions and notation in the statement of Theorem 9.3:

- $\Gamma \subseteq \mathbf{O}(n)$ is a finite group
- $f : \mathbf{R}^n \to \mathbf{R}^n$ is Γ-equivariant
- $A \subseteq \mathbf{R}^n$ is an attractor for f
- $\Sigma(A) = \{\gamma \in \Gamma : \gamma A = A\}$ is the group of (setwise) symmetries of A

Definition 9.2 The group element $\tau \in \Gamma$ is a *reflection* if $\text{Fix}(\tau)$ is a codimension one hyperplane. Let $R_\Gamma \subseteq \Gamma$ denote the set of reflections in Γ.   ◇

Given $\Delta \subseteq \Gamma$, let

$$L_\Delta = \bigcup_{\tau \in \Gamma \setminus \Delta, \tau \in R_\Gamma} \text{Fix}(\tau)$$

We claim that L_Δ is Δ-invariant. To see this let $\delta \in \Delta$. Observe that

$$\delta \, \text{Fix}(\tau) = \text{Fix}(\delta \tau \delta^{-1})$$

Now $\tau \notin \Delta$ implies that the reflection $\delta \tau \delta^{-1} \notin \Delta$, so $\delta \, \text{Fix}(\tau) \subseteq L_\Delta$, verifying the claim. Consequently, δ permutes the connected components of $\mathbf{R}^n \setminus L_\Delta$.

Theorem 9.3 *There exists a normal subgroup $\Delta \lhd \Sigma$ such that*

  (a) Σ/Δ *is cyclic*
  (b) Δ *fixes a connected component of* $\mathbf{R}^n \setminus L_\Delta$.

This theorem may be used to show that certain subgroups $\Sigma \subseteq \Gamma$ *cannot* be the symmetry group of an attractor for any continuous Γ-equivariant map $f : \mathbf{R}^n \to \mathbf{R}^n$. We illustrate this point with the following example:

Example 9.4 Consider $\Gamma = \mathbf{D}_6$ and $\Sigma = \mathbf{D}_3$ acting on $\mathbf{R}^2$. We ask: Does there exist an attractor A of a $\mathbf{D}_6$-equivariant map with $\mathbf{D}_3$ symmetry (that is $\Sigma(A) = \mathbf{D}_3$)? If such an A exists then there is a normal subgroup $\Delta \lhd \mathbf{D}_3$ such that $\mathbf{D}_3/\Delta$ is cyclic and Δ fixes a component of $\mathbf{R}^2 \setminus L_\Delta$. The normal subgroups of $\mathbf{D}_3$ are $\mathbf{D}_3$, $\mathbf{Z}_3$ and 1. Since $\mathbf{D}_3/1$ is not cyclic, either $\Delta = \mathbf{D}_3$ or $\Delta = \mathbf{Z}_3$. If $\Delta = \mathbf{D}_3$ then the geometry of the plane is as pictured in Figure 9.11 where C is a connected component of $\mathbf{R}^2 \setminus L_{\mathbf{D}_3}$. Note that C is not fixed under $\mathbf{D}_3$ since $\mathbf{D}_3$ contains rotation by $120°$. Similarly if $\Delta = \mathbf{Z}_3$ we have again C is not fixed under rotations in $\mathbf{Z}_3$.   ◇

The result in Example 9.4 may be abstracted as follows.

Definition 9.5 $\Sigma \subseteq \Gamma$ is Γ-*admissible* if there exists an attractor for a continuous Γ-equivariant f with $\Sigma(A) = \Sigma$.   ◇

Example 9.4 shows that $\mathbf{D}_3$ is not $\mathbf{D}_6$-admissible for the standard action of $\mathbf{D}_6$ on $\mathbf{R}^2$. Indeed more generally we have:

Corollary 9.6 $\mathbf{D}_k \subseteq \mathbf{D}_m$ *is not admissible if* $3 \le k < m$.

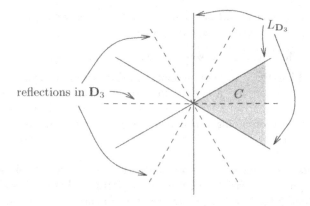

Figure 9.11: Separation of $\mathbf{R}^2$ by $\mathbf{D}_3$.

Ashwin and Melbourne [25] prove that the conditions on Γ-admissibility given in Theorem 9.3 are both necessary and sufficient. Indeed, they prove more if the group G is generated by reflections.

The subgroup $\Sigma \subseteq \Gamma$ is *strongly admissible* if there is a connected attractor A for some Γ-equivariant mapping. The following theorem of Ashwin [25] characterizes strongly admissible and admissible subgroups:

Theorem 9.7 *Let Γ be a finite reflection group.*

(a) $\Sigma \subseteq \Gamma$ *is strongly admissible if and only if Σ is an isotropy subgroup.*

(b) Σ *is admissible if and only if Σ is a cyclic extension of an isotropy subgroup.*

$\square$

In another direction, it is reasonable to ask whether subgroups are admissible for diffeomorphisms or for vector fields. It is straightforward to see that a $\mathbf{D}_3$-equivariant diffeomorphism f of the plane cannot produce a $\mathbf{D}_3$-invariant attractor. By equivariance, the diffeomorphism f leaves fixed-point subspaces invariant, so the connected components of the complements of the three one-dimensional fixed-point subspaces must be permuted by f. But absolute irreducibility of $\mathbf{D}_3$ acting on $\mathbf{R}_2$ implies that $(df)_0$ is a multiple of the identity. It follows that a given connected component C can only be taken to C or $-C$ by f. Hence f cannot start with an initial condition in C and generate an attractor with full $\mathbf{D}_3$ symmetry. Field, Melbourne, and Nicol [193] answer the question of admissibility for diffeomorphisms.

Admissibility for subgroups of non-finite compact groups is a more complicated question. We mention only one result, from Melbourne [386] (pp. 246–247 and Table 1).

Theorem 9.8 *Let A be an attractor for a system of differential equations with $\mathbf{O}(2)$ symmetry. Generically, either*

(a) *A reflectional symmetry fixes each point in A, or*

(b) $\mathbf{SO}(2)$ *leaves A invariant.*

$\square$

A Connectedness Theorem. The following theorem is proved using the same techniques that are used to prove Theorem 9.3. Note that this result is about general dynamical systems — not just symmetric ones.

Theorem 9.9 *Let A be an attractor and suppose that A contains a period-k point. Then A has at most k connected components.*

Remark 9.10 This theorem has some immediate implications.

(a) If $f : \mathbf{R}^2 \to \mathbf{R}^2$ is $\mathbf{D}_m$-equivariant, and if the origin is a fixed point of A (or indeed if A contains a fixed point), then A consists of just one connected component.

(b) Cantor sets containing periodic points cannot be attractors in the sense of Definition 9.1. $\diamond$

Preimage Sets and Connected Components. The remainder of this section is devoted to the proofs of Theorems 9.9 and 9.3. Both of these proofs use the Basic Lemma, Lemma 9.12. In order to state the Basic Lemma we require:

Definition 9.11 Let $S \subseteq \mathbf{R}^n$. Then the *preimage set* of f is

$$\mathcal{P}_s(f) = \bigcup_{m=0}^{\infty} f^{-m}(S)$$

Note that $f^{-1}(\mathcal{P}_S) \subseteq \mathcal{P}_S$, which implies $f : \mathbf{R}^n \setminus \mathcal{P}_s \to \mathbf{R}^n \setminus \mathcal{P}_S$. So f maps any connected component of $\mathbf{R}^n \setminus \mathcal{P}_S$ to a connected component. $\diamond$

Lemma 9.12 (Basic Lemma) *Suppose that A is an attractor and S is closed. If $A \cap S = \emptyset$, then there exist connected components, $C_0, \ldots, C_{r-1}$, of $\mathbf{R}^n \setminus \mathcal{P}_S$, such that*

(a) $A \subseteq C_0 \cup \cdots \cup C_{r-1}$.
(b) $f(C_1) = C_{i+1 \text{ (mod } r)}$.

Proof of Theorem 9.9. Suppose A has a period k point x_0 and

$$A = A_1 \cup \cdots \cup A_l$$

where the A_j are closed and pairwise disjoint. We prove $l \leq k$. Choose S to separate the A_j, as in Figure 9.12. Lemma 9.12 implies that there exist connected components C_j of $\mathbf{R}^n \setminus \mathcal{P}_S$ such that

$$A \subseteq C_0 \cup \cdots \cup C_{r-1}$$

Thus each A_j intersects one C_i. Moreover, since $S \subseteq \mathcal{P}_S$ separates the A_j, each A_j intersects at most one C_i. So the number of C_i is greater than or equal to the number of A_j, that is, $r \geq 1$.

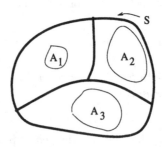

Figure 9.12: Separation of A_j in proof of Theorem 9.9.

Without loss of generality we may suppose that $x_0 \in C_0$. Then

$$f(x_0) \in C_1$$
$$f^2(x_0) \in C_2$$
$$\vdots$$
$$f^{r-1}(x_0) \in C_{r-1}$$
$$f^r(x_0) \in C_0$$

so $k \geq r$ and $r \geq l$.   □

The Basic Lemma (Lemma 9.12) is a special case of the following lemma:

Lemma 9.13 Let $x \in \mathbf{R}^n, S \subseteq \mathbf{R}^n$. Then either $\omega(x) \subseteq \overline{\mathcal{P}_S}$ or

(a) $\omega(x) \setminus \mathcal{P}_s \subseteq C_0 \cup \cdots \cup C_{r-1}$ (connected components of $\mathbf{R}^n \setminus \mathcal{P}_S$).
(b) $f(C_1) \subseteq C_{i+1 \,(\mathrm{mod}\, r)}$.
(c) $\omega(x) \subseteq \overline{C_0} \cup \cdots \cup \overline{C_{r-1}}$.

Proof of Lemma 9.12. Suppose that A is an attractor, S is closed, and $A \cap S = \emptyset$. Then we claim that $A \cap \overline{\mathcal{P}_S} = \emptyset$. To establish this claim, choose an open set $U \supseteq A$ such that $U \cap S = \emptyset$. Then there exists a set $V \supseteq A$ such that $f^m(V) \subseteq U$ for all $m \geq 0$. This implies that $V \cap \mathcal{P}_S = \emptyset$ which in turn implies that $A \cap \mathcal{P}_S = \emptyset$. Since A is an attractor, $A = \omega(x)$ for some $x \in \mathbf{R}^n$ and $\omega(x) \cap \mathcal{P}_S = \emptyset$. Thus by Lemma 9.13

$$\omega(x) \setminus \mathcal{P}_S \subseteq C_0 \cup \cdots \cup C_{r-1}$$

But $\omega(x) \cap \mathcal{P}_S = \emptyset$ which implies $\omega(x) \setminus \mathcal{P}_S = \omega(x)$. Thus $A \subseteq C_0 \cup \cdots \cup C_{r-1}$. □

Proof of Lemma 9.13. Assume $\omega(x) \not\subseteq \overline{\mathcal{P}_S}$. Choose $y \in \omega(x) \setminus \overline{\mathcal{P}_S}$ and $\varepsilon > 0$ such that

$$B_\varepsilon(y) \subseteq \mathbf{R}^n \setminus \overline{\mathcal{P}_S}$$

where $B_\varepsilon(y)$ is a ball of radius ε about y. This is possible since $x \notin \overline{\mathcal{P}_S}$. Since $y \in \omega(x)$ there exists a smallest $k \geq 0$ such that $\hat{x} = f^k(x) \in B_\varepsilon(y)$. Moreover, there exists a smallest $l > k$ such that $f^k(x) \in B_\varepsilon(y)$.

Let $r = l - k$; then $f^r(B_\varepsilon(y)) \cap B_\varepsilon(y) \neq \emptyset$. By continuity $f^r(C_0) \subseteq C_0$. The setup is pictured in Figure 9.13. Let C_i be the connected component of $\mathbf{R}^n \setminus \mathcal{P}_S$ containing $f^i(\hat{x})$. Then

$$f(C_i) \subseteq C_{i+1 \,(\mathrm{mod}\, r)}$$

This implies that

$$f^j(\hat{x}) \in C_0 \cup \cdots \cup C_{r-1}$$

for all $j \geq 0$. Thus

$$\omega(\hat{x}) = \omega(x) \subseteq \overline{C_0 \cup \cdots \cup C_{r-1}} = \overline{C_0} \cup \cdots \cup \overline{C_{r-1}}$$

This verifies (a) and (b). Finally, any limit point of $C_0 \cup \cdots \cup C_{r-1}$ is a limit point of a sequence in one of the C_j and cannot lie in another component of $\mathbf{R}^n \setminus \mathcal{P}_s$, so $\omega(x) \subseteq C_0 \cup \cdots \cup C_{r-1} \cup \mathcal{P}_s$.   □

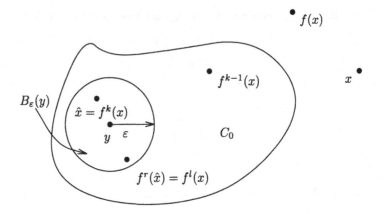

Figure 9.13: Picture of data used in proof of Lemma 9.13.

Symmetry Considerations. We now begin the proof of Theorem 9.3. This discussion starts with a basic proposition about symmetries of attractors first proved by Chossat and Golubitsky [104], which is quite useful in its own right. We follow the proof in Melbourne *et al.* [393].

Proposition 9.14 *Let* $f : \mathbf{R}^n \to \mathbf{R}^n$ *be continuous and let* $\rho \in \mathbf{O}(n)$. *Assume*

(a) $f(\rho x) = \rho f(x)$.
(b) A *is an attractor for* f.

If $A \cap \rho(A) \neq \emptyset$, *then* $\rho(A) = A$.

Proof. A is an attractor if and only if $\rho(A)$ is an attractor. We claim that if $U \supseteq A$ is open, then $U \supseteq \rho(A)$. To verify this claim, recall that if $\rho(A)$ is an attractor, then $\rho(A) = \omega(y)$. In addition, stability implies that there is an open set $V \supseteq A$ such that $f^m(V) \subseteq U$ for all $m \geq 0$. Choose

$$z \in A \cap \rho(A) \supseteq V$$

Since $\rho(A) = \omega(y)$ there exists a k such that $f^k(y) \in V$ therefore $\omega(f^k(y)) = \omega(y) = \rho(A) \subseteq U$, so $\rho(A) \subseteq U$ as claimed.

Since $U \supseteq A$ is open and arbitrary, and since $\rho(A)$ is closed, the claim implies that $\rho(A) \subseteq A$. Reversing the roles of A and $\rho(A)$ we conclude that $\rho(A) = A$. $\square$

The following lemma will be used in the proof of Theorem 9.3 to show that Σ/Δ is cyclic.

Lemma 9.15 *Let* Y *be a finite set with* $|Y| = r$. *Suppose that*

(a) $\mathbf{Z}_r$ *acts transitively on* Y.
(b) Σ *acts fixed-point freely on* Y.
(c) *The actions of* $\mathbf{Z}_r$ *and* Σ *commute.*

Then Σ *is isomorphic to a subgroup of* $\mathbf{Z}_r$. *In particular,* Σ *is cyclic.*

Proof. Suppose that $a \in \mathbf{Z}_r$ is a generator, $y \in Y$, and $\sigma \in \Sigma$. Since $\sigma y \in Y$, assumption (a) implies that there exists a unique $a^p \in \mathbf{Z}_r$ such that $\sigma y = a^p y$. Define $\chi : \Sigma \to \mathbf{Z}_r$ by

$$\chi(\sigma) = a^p$$

We show that χ is a group homomorphism. Suppose that $\chi(\sigma_1) = a^{p_1}$ and $\chi(\sigma_2) = a^{p_2}$. Then

$$
\begin{aligned}
(\sigma_1 \sigma_2) &= \sigma_1(\sigma_2 y) \\
&= \sigma_1(a^{p_2} y) \\
&= a^{p_2}(\sigma_1 y) \quad \text{by (c)} \\
&= a^{p_2} a^{p_1} y \\
&= a^{p_2 + p_1} y \quad \text{since } \mathbf{Z}_r \text{ is cyclic} \\
&= a^{p_1} a^{p_2} y
\end{aligned}
$$

Thus

$$
\chi(\sigma_1 \sigma_2) = \chi(\sigma_1) \chi(\sigma_2)
$$

Finally we show that χ is 1:1. Suppose $\chi(s) = 1$. Then $\sigma y = y$ implies $\sigma = 1$ by assumption (b). $\qquad \square$

Proof of Theorem 9.3. Let A be a Σ-symmetric attractor for f and let

$$
L = \bigcup_{\mathrm{Fix}(\tau) \cap A = \emptyset, \tau \in R_\Gamma} \mathrm{Fix}(\tau)
$$

We use L to construct the desired subgroup $\Delta \subseteq \Sigma$. First, we show that Σ acts on L. To see why, let $\sigma \in \Sigma$ and observe that

$$
\emptyset = \sigma[\mathrm{Fix}(\tau) \cap A] = \sigma[\mathrm{Fix}(\tau)] \cap \sigma(A) = \mathrm{Fix}(\sigma \tau \sigma^{-1}) \cap A
$$

Thus $\sigma[\mathrm{Fix}(\tau)] = \mathrm{Fix}(\sigma \tau \sigma^{-1}) \subseteq L$ and $\sigma(L) = L$. It follows that $\sigma(\mathcal{P}_L) = \mathcal{P}_L$, so that σ permutes the connected components of $\mathbf{R}^n \setminus \mathcal{P}_L$.

Next, by construction $L \cap A = \emptyset$, so by Lemma 9.12

$$
A \subseteq C_0 \cup \cdots \cup C_{r-1}
$$

where each C_j is a connected component of $\mathbf{R}^n \setminus \mathcal{P}_L$. Define the subgroups

$$
\Delta_j = \{\delta \in \Sigma : \delta C_j = C_j\}
$$

These subgroups are well-defined since Σ permutes the connected components of $\mathbf{R}^n \setminus \mathcal{P}_L$. We claim that for any $\delta \in \Sigma$, the equality $\delta C_i = C_j$ implies that $\delta C_{i+1} = C_{j+1}$. Since f commutes with Σ we have

$$
\delta f(C_i) = f(\delta C_1) = f(C_j) \subseteq C_{j+1}
$$

Thus $\delta C_{i+1} \cap C_{j+1} \neq \emptyset$ and connectedness implies that $\delta C_{i+1} = C_{j+1}$.
Therefore

$$
\Delta_0 \subseteq \Delta_1 \subseteq \cdots \subseteq \Delta_{r-1} \subseteq \Delta_0
$$

which implies that

$$
\Delta_0 = \cdots = \Delta_{r-1} \equiv \Delta
$$

Now Δ is the kernel of the action of Σ on $\mathcal{C} = \{C_0, \ldots, C_{r-1}\}$, so Δ is normal in Σ. It also follows that Σ/Δ acts fixed-point freely on $\mathcal{C}$.

In addition, the action of Σ/Δ commutes with the action of f on $\mathcal{C}$ (by Γ-equivariance of f), so Σ/Δ is cyclic by Lemma 9.15.

Next we show that $L_\Delta \subseteq L$. Suppose $\mathrm{Fix}(\tau) \subseteq L_\Delta$; this is true if and only if $\tau \notin \Delta$. Suppose $\mathrm{Fix}(\tau) \not\subseteq L$; this is true if and only if $\mathrm{Fix}(\tau) \cap A \neq \emptyset$. Therefore $\tau(A) \cap A \neq \emptyset$, so Proposition 9.14 implies that $\tau(A) = A$. By definition, this implies that $\tau \in \Sigma$; therefore τ permutes the C_j.

Now $\text{Fix}(\tau) \cap A \neq \emptyset$ which implies that $\text{Fix}(\tau) \cap C_j \neq \emptyset$ for some C_j. Thus $\tau(C_j) \cap C_j \neq \emptyset$ and so $\tau(C_j) = C_j$. This implies that $\tau \in \Delta_j = \Delta$ which is a contradiction. Thus $\text{Fix}(\tau) \subseteq L$.

To conclude the proof, note that by construction the subgroup $\Delta \subseteq \Sigma$ fixes some connected component of $\mathbf{R}^n \setminus \mathcal{P}_L$. Therefore, Δ fixes a connected component of $\mathbf{R}^n \setminus \mathcal{P}_L$ since $L \subseteq \mathcal{P}_L$. Similarly Δ fixes a connected component of $\mathbf{R}^n \setminus L_\Delta$ since $L_\Delta \subseteq L$.   □

9.2 Invariant Measures and Ergodic Theory

An attractor implicitly carries more structure than just its geometric form. The geometry shows which regions of phase space a recurrent trajectory visits, but it does not tell us whether those visits are common, or rare. Attractors can be equipped with 'probability measures' that answer such questions. In order to relate the measure to the dynamics, we require the measure to be invariant under the flow. The measure-theoretic aspects of attractors are complex and subtle; an entire area of mathematics, known as ergodic theory, has developed to study them. A useful general reference here is Mañé [368]. We summarize a few basic ideas and give some simple examples to help build intuition: our discussion assumes familiarity with elementary measure theory. We will make use of SBR measures and ergodic sums later in this chapter.

We abstract the important structure into a space X equipped with a measure μ. We assume furthermore that X is a *probability space*, meaning that $\mu(X) = 1$. A map $T : X \to X$ is *measure-preserving* if

$$\mu(T^{-1}(B)) = \mu(B)$$

for all measurable (or merely Borel) subsets $B \subseteq X$. We also say that μ is an *invariant measure* for T. In applications to dynamics, X is a subset of phase space and T is either a discrete dynamic or a flow.

As an example of the connection between dynamics and measure theory, we state a version of the Poincaré Recurrence Theorem. Recall that if $T : \mathbf{R}^k \to \mathbf{R}^k$ is continuous and $x \in \mathbf{R}^k$ then the *ω-limit set* $\omega(x)$ is the set of limit points of the forward trajectory $\{T^n(x) : n \geq 0\}$. We say that $x \in \mathbf{R}^k$ is *recurrent* if $x \in \omega(x)$. Intuitively this means that the trajectory starting at x returns arbitrarily close to x, infinitely often.

Theorem 9.16 (Poincaré Recurrence) *With the above notation, suppose that there exists an invariant probability measure μ on $\mathbf{R}^n$ for T. Then μ-almost all points of $\mathbf{R}^n$ are recurrent.*

Proof. See Mañé [368] Theorem I.2.3.   □

A central concept is ergodicity. This concept arose from statistical mechanics. Suppose that μ is a T-invariant probability measure on X, and let $f : X \to \mathbf{R}$ be integrable (we think of f as an *observation* of X). The time-average of f over the trajectory of a point x can be defined as the *ergodic sum*

$$\lim_{n \to \infty} \frac{f(x) + f(T(x)) + \cdots + f(T^{n-1}(x))}{n} \tag{9.8}$$

provided the limit exists and is constant almost everywhere. Birkhoff [65] proved that the limit exists almost everywhere for any T, f. Proving that the limit is constant almost everywhere required an additional condition: namely, that T is ergodic in the following sense.

Definition 9.17 The map T is *ergodic* if $T^{-1}(B) = B$ implies that $\mu(B) = 0$ or 1. That is, there are no nontrivial invariant subsets.

Theorem 9.18 (Birkhoff Ergodic Theorem) *With the above notation, assume that T is ergodic. Then*

1)
$$\lim_{n\to\infty} \frac{f(x) + f(T(x)) + \cdots + f(T^{n-1}(x))}{n}$$

exists and is constant almost everywhere.

2) *That constant is equal to*

$$\int_X f\,\mathrm{d}\mu$$

Proof. See Mañé [368] Theorem II.1.1. $\qquad\square$

We will make use of the following

Proposition 9.19 *The map T is ergodic if and only if every square-integrable $f : X \to \mathbf{R}$ that satisfies $f \circ T = f$ almost everywhere is constant almost everywhere.*

Proof. If T is not ergodic, then there exists an invariant Borel set B with $\mu(B) \neq 0, 1$. Let f be the characteristic function of B, so that $f(x) = 1$ for $x \in B$ and $f(x) = 0$ for $x \notin B$. Then f is square-integrable and $f \circ T = f$ almost everywhere. However, f is not constant almost everywhere.

The converse is similar. $\qquad\square$

We now discuss two simple examples.

Example 9.20 Let X be the circle $\mathbf{S}^1 = \mathbf{R}/\mathbf{Z}$, let $\alpha \in \mathbf{S}^1$, and define

$$T_\alpha : \; \mathbf{S}^1 \; \to \mathbf{S}^1$$
$$\theta \; \mapsto \theta + \alpha$$

so T_α is a rigid rotation through angle α. Let μ be Lebesgue measure on $\mathbf{S}^1$.

We claim that T_α is ergodic if and only if α is irrational. We prove the claim using Proposition 9.19.

It is easy to see that T_α is not ergodic if α is rational. So let $\alpha \notin \mathbf{Q}$ and consider a square-integrable $f : \mathbf{S}^1 \to \mathbf{R}$ such that $f \circ T_\alpha = f$ almost everywhere. Expand f in a Fourier series

$$f(\theta) = \sum_{n=-\infty}^{\infty} a_n e^{ni\theta}$$

Then

$$f(T_\alpha(\theta)) = \sum_{n=-\infty}^{\infty} a_n e^{ni(\theta+\alpha)}$$
$$= \sum_{n=-\infty}^{\infty} a_n e^{ni\alpha} e^{ni\theta}$$

Therefore $a_n e^{ni\alpha} = a_n$ for all n. Since α is irrational, $e^{ni\alpha} \neq 1$. Hence $a_n = 0$ for $n \neq 0$, and $f = a_0$ almost everywhere. $\qquad\diamond$

Example 9.21 In a similar vein, consider the 'Arnold's cat' map, defined as follows. Let

$$A = \begin{bmatrix} 2 & 1 \\ 1 & 1 \end{bmatrix}$$

This defines a unimodular, hence area-preserving linear transformation of $\mathbf{R}^2$, and both A and A^{-1} preserve the integer lattice $\mathbf{Z}^2$. Therefore A induces an invertible map $T : \mathbf{T}^2 \to \mathbf{T}^2$, where $T^2 = \mathbf{R}^2/\mathbf{Z}^2$ is a 2-torus. Lebesgue measure on the torus is invariant.

The discrete dynamics of T is complicated. There is a dense set of periodic points (namely those with both coefficients rational). On the other hand, there is also a dense set of homoclinic points, given by the intersections (modulo the lattice) of the stable and unstable eigenspaces of A. (The eigenvalues of A are $\frac{3-\sqrt{5}}{2} < 1$ and $\frac{3+\sqrt{5}}{2} > 1$.)

We claim that T is ergodic. Again we prove the claim using Proposition 9.19, and again we consider a square-integrable $f : \mathbf{T}^2 \to \mathbf{R}$ such that $f \circ T = f$ almost everywhere. Expand f in a Fourier series

$$f(\theta, \phi) = \sum_{m,n=-\infty}^{\infty} a_{mn} e^{i(m\theta + n\phi)}$$

By the Plancherel formula, $\sum |a_{mn}^2| < \infty$. Compute

$$\begin{aligned} f(T(\theta, \phi)) &= f(2\theta + \phi, \theta + \phi) \\ &= \sum a_{mn} e^{i((2n+m)\theta + (m+n)\phi)} \end{aligned}$$

so that

$$a_{2m+n, m+n} = a_{mn}$$

Since the eigenvalues of A are not on the unit circle, any $(m, n) \neq (0,0)$ with nonzero a_{mn} generates infinitely many (m, n) with the same value of a_{mn}. By the Plancherel formula, $a_{mn} = 0$ for all $(m, n) \neq (0,0)$. Thus f is constant almost everywhere, so T is ergodic as claimed.   ◇

In Example 9.20 the entire circle is an attractor for T_α when α is irrational. In Example 9.21 the entire torus is an attractor for T. Thus we have constructed invariant measures supported on the attractor, with the property that ergodic sums (9.8) converge to (almost everywhere) constants for *Lebesgue*-almost-all x. Such a measure is called a *Sinai-Bowen-Ruelle measure* or *SBR measure*. There is a growing fashion for the term SLYRB measure, with L = Lasota, Y = Yorke.

One of the triumphs of the Smale school of dynamical systems theory is the proof that every axiom A attractor supports an SBR measure, see Bowen [73], Ruelle [455]. However, most attractors arising in applications seem not to be axiom A. Numerical simulation strongly suggested that SBR measures are common, but only recently have proofs of this fact become available in a few cases. Benedicks and Carleson [58] proved that the Hénon map has a chaotic attractor with an SBR measure for an open set of parameter values. The work of Tucker [505] proving (with computer assistance) that the Lorenz attractor exists — that is, the Lorenz equations with the classical parameter values possess a robust chaotic attractor — seems likely to lead to an existence proof for an SBR measure on the Lorenz attractor, see Viana [512].

Symmetry in SBR Measures. Suppose that the Σ-invariant attractor A has an SBR measure ν. Then the measure ν is also Σ-invariant. Let $\sigma \in \Sigma$ and define $\nu_\sigma(B) = \nu(\sigma^{-1}(B))$.

Theorem 9.22 *If ν is an SBR measure, then $\nu_\sigma = \nu$.*

Proof. Let $\mathcal{U}$ be an attracting neighborhood of A. Then for each continuous ϕ : $\mathbf{R}^n \to \mathbf{R}$ and Lebesgue almost all $x \in \mathcal{U}$,

$$\lim_{N\to\infty} \frac{1}{N} \sum_{j=0}^{N-1} \phi(f^j(x)) = \lim_{N\to\infty} \frac{1}{N} \sum_{j=0}^{N-1} \phi(f^j(\sigma x))$$

which implies that

$$\int_A \phi \, d\nu = \int_A \phi \, d\nu_\sigma$$

Therefore $\nu = \nu_\sigma$. $\square$

Theorem 9.22 suggests why the color images of attractors pictured in Field and Golubitsky [189] are symmetric. The method used to color these attractors will approximate an SBR measure — if the attractor has an SBR measure. Since SBR measures are symmetric, so are the pictures. Although few attractors (other than Axiom A) have yet been proved to possess SBR measures, numerical simulations suggest that SBR measures — or something with similar implications — may be relatively common.

9.3 Detectives

The symmetries of chaotic attractors are discussed in Section 9.1, in terms of properties of the attractor as a subset of phase space. However, in many applications phase space is inaccessible to experiment, and what is needed is a description of the phenomena that can be observed from measurements, especially time-series of observations.

In this section we address the issues that arise when we attempt to interpret the meaning of a chaotic attractor in phase space, and its symmetry, in terms of observable quantities. In particular, can we *observe* the symmetry of a chaotic attractor by making suitable measurements? The concept of a 'detective' shows that in principle we can. We consider only finite groups, however: the continuous case poses new and currently unsolved problems.

Suppose that K is an attractor for a continuous Γ-equivariant mapping f : $\mathbf{R}^n \to \mathbf{R}^n$ where $\Gamma \subseteq \mathbf{O}(n)$ is a finite group. We address the question: what are the symmetries Σ of K? Recall that

$$\Sigma(K) = \{\gamma \in \Gamma : \gamma K = K\}$$

We want to find a method for answering this question that can be implemented on a computer. Note that when $n \geq 3$, the projection of K onto a computer screen is just a cloud of points with no indication of depth — making the determination of Σ quite difficult. In this section we follow Barany *et al.* [41] and Dellnitz *et al.* [147].

Recall Proposition 9.14: if $\gamma \in \Gamma$, then $\gamma K \cap K \neq \emptyset$ or $\gamma K = K$. We use this proposition to determine the symmetry of attractors. Since the set K can be quite complex, we instead work with the *thickened attractor*

$$A = \{x \in \mathbf{R}^n : d(x, K) < \varepsilon\}$$

which is an open set. Since Γ acts orthogonally, $\Sigma(K) \subseteq \Sigma(A)$. Proposition 9.14 implies that $\Sigma(K) = \Sigma(A)$ when $\varepsilon \ll 1$. There are two steps in determining the symmetry of the attractor K.

First, replace K with the open set A satisfying

$$\gamma A = A \quad \text{or} \quad \gamma A \cap A \neq \emptyset \tag{9.9}$$

for each $\gamma \in \Gamma$. We define $\mathcal{A}$ to be the set of A satisfying (9.9). We can replace K by A since $\Sigma(A) = \Sigma(K)$ when $\varepsilon > 0$ is sufficiently small.

Second, and this is the main idea, transfer the problem of finding the group $\Sigma(A)$ to the problem of finding the symmetries of a point in some other space W. To do this, let W be a vector space on which Γ acts and let $\phi : \mathbf{R}^n \to W$ be Γ-equivariant. Define

$$K_\phi(A) = \int_A \phi \mathrm{d}\mu$$

where μ is Lebesgue measure and $K_\phi(A)$ is a vector in W. (Note that if we were still working with K it would be possible for K to have measure zero, implying that $K_\phi(K) = 0$.) We call ϕ an *observable* and $K_\phi(A)$ an *observation*. The next lemma relates symmetries of A to symmetries of $K_\phi(A)$.

Lemma 9.23 *If* $\sigma A = A$ *then* $\sigma K_\phi(A) = K_\phi(A)$.

Proof. From the definition of $K_\phi(A)$, and using change of variables in integration,

$$
\begin{aligned}
\sigma K_\phi(A) &= \sigma \int_A \phi(x)\mathrm{d}\mu(x) &&= \int_A \phi(\sigma x)\mathrm{d}\mu(x) \\
&= \int_A \phi(\sigma x)\mathrm{d}\mu(\sigma x) &&= \int_{\sigma A} \phi(y)\mathrm{d}\mu(y) \\
&= \int_A \phi(y)\mathrm{d}\mu(y) &&= K_\phi(A)
\end{aligned}
$$

$\square$

Thus the subgroup $\Sigma(A)$ of setwise symmetries of the thickened attractor A is contained in the isotropy subgroup of the observation $K_\phi(A)$ in W. In general, the isotropy subgroup of $K_\phi(A)$ may be strictly larger than $\Sigma(A)$. We show that under reasonable assumptions on W (W needs to be large enough) and ϕ (ϕ needs to be regular enough), the two subgroups are equal — at least generically.

Statements of the Main Theorems. The notion of genericity that we use is embedded in the following definition:

Definition 9.24 The observable $\phi : \mathbf{R}^n \to W$ is a *detective* if for every $A \in \mathcal{A}$ there exists an open dense set of near-identity Γ-equivariant diffeomorphisms $\psi : \mathbf{R}^n \to \mathbf{R}^n$ such that

$$\Sigma_\phi(\psi(A)) = \Sigma(A) \tag{9.10}$$

$\diamond$

If A is a thickened attractor for f, then $\psi(A)$ is a thickened attractor for $\psi f \psi^{-1}$. Formula (9.10) states that the observed symmetries of the perturbed thickened attractor $\psi(A)$ equal the setwise symmetries of the thickened attractor A.

Next, we define what it means for W to be large enough. To do so we require the following notation:

Definition 9.25 Let $W(\Gamma) = W_1 \oplus \cdots \oplus W_s$ where the W_j for $j = 1, \ldots, s$ are all of the nontrivial irreducible representations of Γ. $\diamond$

Since Γ is a finite group, $W(\Gamma)$ is finite-dimensional. For our purposes, the space W is large enough if $W \supseteq W(\Gamma)$, in which case we can write $W = W(\Gamma) \oplus W^0$ for some Γ-invariant subspace W^0.

Definition 9.26 Assume that $W \supseteq W(\Gamma)$ and write the observable $\phi : \mathbf{R}^n \to W$ in coordinates as $\phi = (\phi_1, \ldots, \phi_s, \phi^0)$ where $\phi_j : \mathbf{R}^n \to W_j$ and $\phi^0 : \mathbf{R}^n \to W^0$ are Γ-equivariant. Then ϕ is *regular* if each ϕ_j is a nonzero polynomial for $j = 1, \ldots, s$.

The main result of Barany *et al.* [41] is:

Theorem 9.27 *Let $W \supseteq W(\Gamma)$ and let $\phi : \mathbf{R}^n \to W$ be a regular observable. Then ϕ is a detective.*

The proof of Theorem 9.27 will be given later in this section.

Example 9.28 Let $\Gamma = \mathbf{D}_3$ and let $\phi : \mathbf{C} \to \mathbf{R} \oplus \mathbf{C}$ be defined by

$$\phi(z) = (\mathrm{Im}(z^3), z)$$

The group $\mathbf{D}_3$ has two nontrivial irreducible representations; the standard two-dimensional representation and the nontrivial one-dimensional representation. It follows that $W(\Gamma) = \mathbf{R} \oplus \mathbf{C}$. The action of $\mathbf{D}_3$ on $W(\Gamma)$ is

$$\kappa(x, z) = (-x, \bar{z})$$
$$\theta(x, z) = (x, e^{i\theta} z) \qquad (\theta = 2\pi/3)$$

and ϕ is $\mathbf{D}_3$-equivariant. Therefore ϕ is a detective.   ◇

Necessity of Hypotheses Like Those in Theorem 9.27. In general, $\Sigma(A) \subsetneq \Sigma_\phi(A)$ where $\Sigma_\phi(A)$ is the isotropy subgroup of the point $K_\phi(A) \in W$. There are three reasons why this inequality might hold.

First, consider $\Gamma = \mathbf{D}_3$ acting on $\mathbf{R}^2 \equiv \mathbf{C}$ with $\phi(z) = z$ and $\Sigma(A) = \mathbf{Z}_3$, as in Figure 9.14. Geometrically, it is clear that

$$\int_A z \mathrm{d}\mu(z) = 0$$

whenever $\Sigma(A) = \mathbf{Z}_3$. However $\Sigma_\phi(0) = \mathbf{D}_3 \supsetneq \mathbf{Z}_3$. The difficulty here is that W is not large enough: some subgroups of $\mathbf{D}_3$ are not isotropy subgroups of the action on W. In particular, $\mathrm{Fix}_W(\mathbf{D}_3) = \mathrm{Fix}_W(\mathbf{Z}_3)$. So we need to assume that W is sufficiently large.

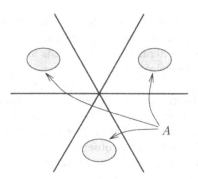

Figure 9.14: A set A with $\mathbf{Z}_3$ symmetry.

Second, suppose $\phi = 0$, then $\Sigma_\phi(A) = \Gamma$ for every A — no matter what the symmetry group of A may be. So we need to assume that the observable is sufficiently nonzero.

Figure 9.15: (Left) Set A with accidental zero integral; (right) perturbation of A.

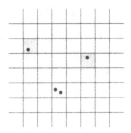

Figure 9.16: Iterates landing in grid cells.

Third, again consider $\Gamma = \mathbf{D}_3$ and let A be the set in Figure 9.15. Then the geometry of A leads to

$$\int_A z \mathrm{d}\mu(z) = 0 \tag{9.11}$$

The integral in (9.11) is accidentally zero; that is, A can easily be perturbed so that integral is no longer zero. Perturb A to $\psi(A)$ where $\psi : \mathbf{R}^2 \to \mathbf{R}^2$ is a near identity Γ-equivariant diffeomorphism and note that when ψ is a Γ-equivariant diffeomorphism, $\Sigma(\psi(A)) = \Sigma(A)$. For example, we may perturb A to the thickened attractor $\psi(A)$ pictured in Figure 9.15, and the integral over this perturbed attractor is nonzero. So we need to include some notion of genericity in the definition of a detective.

Detectives by Ergodic Sums. Let us discuss how we might calculate $K_\phi(A)$ by computer. First, form a grid, and choose an initial point x_0. Then, say that a cell in that grid is in A when an iterate of x by f lands in the cell: see Figure 9.16. Finally, integrate ϕ over A numerically.

There is one difficulty that is unavoidable with this method for systems in $\mathbf{R}^n$ with n larger than, say, 10; the number of cells in the grid becomes too large and the method cannot be used effectively because of storage limitations or insufficient computational speed. An alternative approach was proposed by Barany *et al.* [41] and proved in Dellnitz *et al.* [147]: make observations based on ergodic sums rather than the Lebesgue integral. The ergodic sum is defined as

$$K_\phi^E(A) = \lim_{N \to \infty} \frac{1}{N} \sum_{j=0}^{N-1} \phi(f^j(x_0))$$

The Birkhoff Ergodic Theorem, Theorem 9.18, implies that

$$K_\phi^E(A) = \int_A \phi \mathrm{d}\nu$$

where ν is some ergodic measure. Unfortunately, the ergodic measure ν is often very singular. One way around this difficulty is to assume the attractor has an SBR measure, which was defined in Section 9.2.

Theorem 9.29 *Let $W \supseteq W(\Gamma)$ and let $\phi : \mathbf{R}^n \to W$ be a regular observable. Then, using observations based on ergodic sums, ϕ is a detective for attractors with SBR measures.*

We refer the reader to Dellnitz *et al.* [147] for a proof of this version of the detective theorem. The storage difficulties associated with the thickened attractor version of the detective theorem disappear in the ergodic sums version.

A Detective for Coupled Cells. The results that we obtained for detectives using discrete dynamical systems apply equally to attractors in ordinary differential equations. Then the attractor is given by the limit points of the trajectory $\{x(t) : t \geq 0\}$ and the time average becomes

$$K_\phi^E(x(0)) = \lim_{T \to \infty} \frac{1}{T} \int_0^T \phi(x(t)) dt$$

As an example of detectives, we apply Theorem 9.29 to bidirectional rings of p identical coupled cells, which are $\mathbf{D}_p$ symmetric systems. We assume that each cell is a system of m ODEs, and we set $n = mp$. We present a detective for such systems and we explore numerically an example of three coupled cells ($p = 3$) consisting of two equations each ($m = 2$) so that $n = 6$.

The representation space W that we use for this detective is the space of $n \times n$ real symmetric matrices, where $\gamma \in \mathbf{D}_p$ acts by similarity transformations on W:

$$\gamma \cdot w = \gamma w \gamma^T$$

for all $w \in W$.

Proposition 9.30 *Assume that the number of cells is $p \geq 3$ and the number of equations governing each cell is $m \geq 2$. Then the mapping*

$$\phi(x) = x \cdot x^T$$

is a detective.

Proof. It is straightforward that ϕ is $\mathbf{D}_p$-equivariant, so ϕ is a polynomial observable. The proof of this theorem is given in Barany *et al.* [41]; here we prove only the case $p = 3$. We use Theorem 9.29 to prove Proposition 9.30. There are two points that must be checked. We must show that W contains every nontrivial irreducible representation of $\mathbf{D}_3$, and that this particular ϕ has a nonzero component in each of these representations. In fact, in the three-cell case, ϕ is a detective even when $m = 1$. We therefore simplify the calculations by assuming that $m = 1$. Then the matrix generators for $\mathbf{D}_3$ acting on the space of 3×3 matrices W are

$$\kappa = \begin{bmatrix} 1 & 0 & 0 \\ 0 & 0 & 1 \\ 0 & 1 & 0 \end{bmatrix} \quad \text{and} \quad \xi = \begin{bmatrix} 0 & 1 & 0 \\ 0 & 0 & 1 \\ 1 & 0 & 0 \end{bmatrix}$$

The group $\mathbf{D}_3$ has only three irreducible representations: the one-dimensional trivial representation, the one-dimensional nontrivial representation, and the two-dimensional irreducible. Since ξ acts trivially in the one-dimensional irreducible representations, the two-dimensional irreducible must be present in W if ξ acts nontrivially on W. It is a straightforward calculation to check that

$$\text{Fix}(\xi) = \begin{bmatrix} a & b & c \\ c & a & b \\ b & c & a \end{bmatrix} \quad \text{and} \quad \text{Fix}(\kappa) = \begin{bmatrix} a & b & b \\ d & e & f \\ d & f & e \end{bmatrix} \tag{9.12}$$

Thus the two-dimensional irreducible representation occurs in W. A further calculation shows that $W_1 = \mathbf{R}\{M\}$, where

$$M = \begin{bmatrix} 0 & 1 & -1 \\ -1 & 0 & 1 \\ 1 & -1 & 0 \end{bmatrix}$$

is the one-dimensional nontrivial representation of $\mathbf{D}_3$.

Finally, we verify that the coordinate functions of ϕ are nonzero. Since the image of ϕ is not contained in $\text{Fix}(\xi)$, the coordinate function of ϕ on the isotypic component of the two-dimensional irreducible representation must be nonzero. We complete the proof by showing that the coordinate function on the one-dimensional nontrivial representation is nonzero. Note that $\mathbf{D}_3$ acts orthogonally on W using the inner product

$$\langle A, B \rangle = \text{tr}(AB)$$

It follows that orthogonal projection of W onto W_1 is given by

$$w \mapsto \langle w, M \rangle M$$

A straightforward calculation shows that $\langle \phi(x), M \rangle$ is nonzero. $\square$

An Example of Three Coupled Cells. As an example of the results in the previous section we consider the following system of three coupled cells

$$\begin{aligned} \dot{x}_j &= x_{j+1} - 0.28x_j^2 x_{j+1} \\ \dot{x}_{j+1} &= -x_j - (x_j^2 - \lambda)x_{j+1} - 0.5(x_{j-1} - 2x_{j+1} + x_{j+3}) - 1.8x_j x_{j+1} \end{aligned} \tag{9.13}$$

where $j = 1, 3, 5$, $x_0 = x_6$, $x_8 = x_2$, and λ is the bifurcation parameter.

The dynamical system (9.13) possesses $\mathbf{D}_3$ symmetry and, using the notation defined previously, we have $p = 3$, $m = 2$ and $n = 6$. The equivariant polynomial observable that we use to detect symmetry is $\phi(x) = x \cdot x^*$, where to avoid confusion in the next formula we use x^* to denote the transpose instead of x^T. This choice of observable leads to the computation of the *correlation matrix*

$$K_\phi^E(x(0)) = \lim_{T \to \infty} \frac{1}{T} \int_0^T x(t) \cdot x(t)^* dt$$

Theorem 9.30 implies that this observable is a detective.

The numerical calculations include one additional stage: we must compute the isotropy subgroup Σ_K of $K = K_\phi^E(x(0))$. Since we know K only approximately, the computation is not totally straightforward. Ideally, we can test whether a symmetry $\gamma \in \Sigma_K$ by determining whether $K \in \text{Fix}(\sigma)$. Numerically, we compute the distance $d(K, \text{Fix}(\sigma))$ and ask whether this number is zero.

For the group $\Gamma = \mathbf{D}_3$, we compute the distances of K to the fixed-point subspaces of the three independent reflections κ, $\kappa\xi$, and $\kappa\xi^2$ and the distance to the fixed-point subspace of the rotation ξ, as functions of λ. It is then relative easily to see changes in these numbers corresponding to symmetry-increasing bifurcations, and hence to determine whether we have a symmetry of the attractor.

The distances between $K_\phi^E(x(0))$ and the relevant fixed-point subspaces are computed numerically in Barany et al. [41] by varying λ from -1.20 to -1.03. In Figure 9.17 we show the four distances for this range of the λ-values.

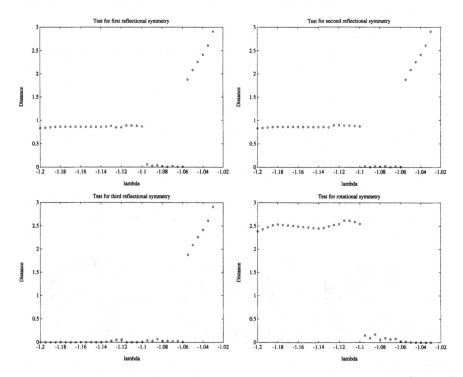

Figure 9.17: The distances to fixed-point subspaces from $K_\phi^E(x(0))$

For $-1.20 \le \lambda \le -1.10$ the attractor appears to be $\mathbf{Z}_2$-symmetric, for $-1.10 \le \lambda \le -1.06$ the attractor is $\mathbf{D}_3$-symmetric, and for $\lambda \ge -1.06$ the attractor becomes $\mathbf{Z}_3$-symmetric. For this interpretation, the distance itself is not relevant, but the change in distance that occurs at parameter values where *symmetry-creation* occurs, that is, where conjugate attractors collide and the resulting attractor has more symmetry than the single attractors had before collision.

On the other hand, at classical symmetry-breaking bifurcations the distance that we compute should vary continuously. In fact, further inspection shows that there is a period-doubling sequence occurring for $\lambda \in (-1.14, -1.11)$ in which the $\mathbf{Z}_2$ symmetry is lost. But this can hardly be seen in Figure 9.17.

We end this subsection with the numerical simulation of (9.13) for

$$\lambda = -1.18 \quad -1.13 \quad -1.08 \quad -1.03$$

The solutions are presented in Figures 9.18-9.21 in coupled cell form: cell 1 (x_1 versus x_2) on the left, cell 2 (x_3 versus x_4) in the middle, and cell 3 (x_5 versus x_6) on the right. Notice that in this presentation we cannot distinguish between attractors with $\mathbf{D}_3$ symmetry and those with $\mathbf{Z}_3$ symmetry. See Figures 9.20 and 9.21. The mathematical difficulty is the same as the one that occurred in the discussion of Figure 9.14. In order to determine exact attractor symmetries, we need to observe

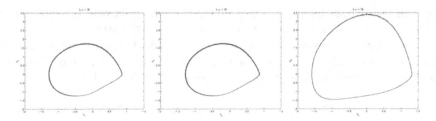

Figure 9.18: Simulation of (9.13) with $\lambda = -1.18$. Attractor has $\mathbf{Z}_2$ symmetry.

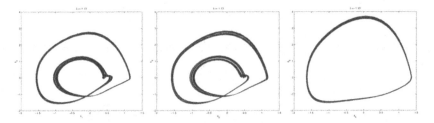

Figure 9.19: Simulation of (9.13) with $\lambda = -1.13$. Attractor has $\mathbf{Z}_2$ symmetry.

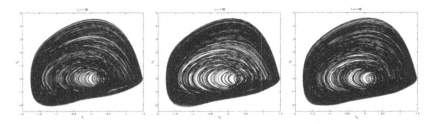

Figure 9.20: Simulation of (9.13) with $\lambda = -1.08$. Attractor has $\mathbf{D}_3$ symmetry.

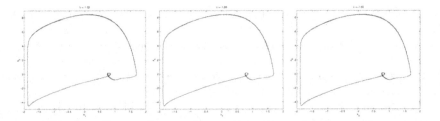

Figure 9.21: Simulation of (9.13) with $\lambda = -1.03$. Attractor is a discrete rotating wave and has $\mathbf{Z}_3$ symmetry.

the solution in a representation that distinguishes all subgroups of Γ. The most natural coupled cell form of presentation does not distinguish between $\mathbf{D}_3$ and $\mathbf{Z}_3$ symmetry.

Distinguishing Subgroups. The remainder of this section is devoted to proving the main theorem. We begin by showing that the assumption $W \supseteq W(\Gamma)$ is sufficient to prove that W is large enough to distinguish all subgroups of Γ.

Definition 9.31 The representation W *distinguishes all subgroups* of Γ if

$$\mathrm{Fix}_W(T) \subsetneq \mathrm{Fix}_W(\Delta) \text{ whenever } \Delta \subsetneq T \subseteq \Gamma$$

$\diamond$

Lemma 9.32   (a) *Suppose V distinguishes all subgroups. Then $V \oplus \hat{V}$ distinguishes all subgroups.*
    (b) *Suppose $V \oplus W$ distinguishes all subgroups and W is trivial. Then V distinguishes all subgroups.*

Proof. Observe that

$$\mathrm{Fix}_{V \oplus \hat{V}}(\Delta) = \mathrm{Fix}_V(\Delta) \oplus \mathrm{Fix}_{\hat{V}}(\Delta) \tag{9.14}$$

Therefore when $\Delta \subseteq T$,

$$\mathrm{Fix}_{V \oplus \hat{V}}(T) = \mathrm{Fix}_V(T) \oplus \mathrm{Fix}_{\hat{V}}(T)$$
$$\subseteq \mathrm{Fix}_V(\Delta) \oplus \mathrm{Fix}_{\hat{V}}(\Delta)$$

Hence, if $\mathrm{Fix}_V(T) \subsetneq \mathrm{Fix}_V(\Delta)$, then $\mathrm{Fix}_{V \oplus \hat{V}}(T) \subsetneq \mathrm{Fix}_{V \oplus \hat{V}}(\Delta)$ and (a) is valid. Identity (9.14) also implies that (b) is valid. $\qquad\square$

Proposition 9.33 *If $W \supseteq W(\Gamma)$, then W distinguishes all subgroups of Γ.*

Proof. Let V_Γ be the *left regular representation* of Γ,

$$V_\Gamma = \{h : \Gamma \to \mathbf{R}\}$$

There is a natural action of Γ on V_Γ where $\gamma \in \Gamma$ acts on $h \in V_\Gamma$ by

$$(\gamma h)(\delta) = h(\gamma^{-1}\delta)$$

It is well known that V_Γ contains every irreducible representation of Γ and hence $V_\Gamma \supseteq W(\Gamma)$. Using Lemma 9.32, it is sufficient to prove Proposition 9.33 when $W = V_\Gamma$. Consider

$$\mathrm{Fix}_{V_\Gamma} = \{h \in V_\Gamma : \delta h = h, \forall \delta \in \Delta\}$$

that is, $h(\delta^{-1}\gamma) = h(\gamma)$. This implies that h is constant on Δ. Suppose $\Delta \subsetneq T$, then $\mathrm{Fix}_{V_\Gamma}(T) \subseteq \mathrm{Fix}_{V_\Gamma}(\Delta)$. We want to construct $h \in \mathrm{Fix}_{V_\Gamma}(\Delta) \setminus \mathrm{Fix}_{V_\Gamma}(T)$. Define

$$h(\gamma) = \begin{cases} 1 & \gamma \in T \setminus \Delta \\ 0 & \text{otherwise} \end{cases}$$

Then for $\delta \in \Delta$, $h(\delta^{-1}\gamma) = h(\gamma)$ and therefore $h \in \mathrm{Fix}_{V_\Gamma}(\Delta)$. If $\tau \in T \setminus \Delta$, then $h(\tau^{-1}e) = h(\tau^{-1}) = 1$. But $h(e) = 0$, therefore $\tau h \neq h$. $\qquad\square$

Let $\mathrm{Diff}_\Gamma(\mathbf{R}^n)$ denote the group of all Γ-equivariant diffeomorphisms of $\mathbf{R}^n$. Then we may state:

Lemma 9.34 *Let $\phi : \mathbf{R}^n \to W$ be an observable. Assume that W distinguishes all subgroups. Let $\Psi_\phi^A : \mathrm{Diff}_\Gamma(\mathbf{R}^n) \to \mathrm{Fix}_W(\Sigma(A))$ be defined by*

$$\Psi_\phi^A(\psi) = K_\phi(\psi^{-1}(A))$$

and suppose for all $A \in \mathcal{A}$ there exists an open neighborhood $\mathcal{U}$ of the identity in $\mathrm{Diff}_\Gamma(\mathbf{R}^n)$ such that $\Psi_\phi^A(\mathcal{U})$ covers an open neighborhood $\mathcal{O}$ of $K_\phi(A)$ in $\mathrm{Fix}_W(\Sigma(A))$. Then ϕ is a detective.

Proof. Let $\mathcal{V}$ be the algebraic variety

$$\mathcal{V} = \bigcup_{\Delta \supseteq \Sigma(A)} \mathrm{Fix}_W(\Delta)$$

Since W distinguishes all subgroups, codim $\mathcal{V} \geq 1$ which implies that $\mathcal{O}' = \mathcal{O} \setminus \mathcal{V}$ is open and dense in $\mathcal{O}$. Therefore $(\Psi_\phi^S)^{-1}(\mathcal{O}') \cap \mathcal{U}$ is an open dense subset of $\mathcal{U}$ in $\mathrm{Diff}_\Gamma(\mathbf{R}^n)$. $\qquad\square$

Sketch of Proof of Theorem 9.27. Let X be a Γ-equivariant vector field on $\mathbf{R}^n$,

$$X = \frac{\mathrm{d}}{\mathrm{d}t}\psi_t\Big|_{t=0} \qquad \psi_t \in \mathrm{Diff}_\Gamma(\mathbf{R}^n)$$

Define $\mathcal{L}_\phi^A : C_\Gamma^\infty(\mathbf{R}^n, \mathbf{R}^n) \to \mathrm{Fix}_W(\Sigma(A))$ by

$$\mathcal{L}_\phi^A(X) = (\mathrm{d}\Psi_\phi^A)_{\psi=1}$$

If $\mathcal{L}_\phi^A$ is onto for all A, then ϕ is a detective. To compute $\mathcal{L}_\phi^A$, observe that

$$\Psi_\phi^A(\psi_t) = \int_{\psi_t(A)} \phi(x)\mathrm{d}\mu(x)$$

$$= \int_A \phi(\psi_t)(x)) \det(\mathrm{d}\psi_t)_x \mathrm{d}\mu(x).$$

Differentiating this equation with respect to t and evaluating at $t = 0$ gives

$$\mathcal{L}_\phi^A(X) = \int_A (\mathrm{d}\phi)_x(X(x)) + \phi(x)\,\frac{\mathrm{d}}{\mathrm{d}t}\det\psi_t\Big|_{t=0}\,\mathrm{d}\mu(x)$$

$$= \int_A \nabla\phi \cdot X + \phi(x)\,\mathrm{tr}(\mathrm{d}X)_x\,\mathrm{d}\mu(x)$$

$$= \int_A \nabla\phi \cdot X + \phi(x)\nabla \cdot X\,\mathrm{d}\mu(x)$$

$$= \int_{\partial A} \phi(X \cdot N)\,\mathrm{d}v$$

by Stokes' theorem, where N is the unit normal.

Now, if

$$\mathcal{F}_\phi = \{\phi(x) : x \in \partial A\} \subseteq W$$

then

$$\mathcal{L}_\phi^A(X) \subseteq \mathcal{F}_\phi$$

In fact

$$\mathrm{Im}(\mathcal{L}_\phi^A) = \mathcal{F}_\phi$$

Because ϕ is a polynomial,

$$\mathcal{F}_\phi = \{\phi(x) : x \text{ near } \partial A\} = \{\phi(x) : x \in \mathbf{R}^n\}$$

Since the ϕ_j are nonzero,

$$\mathcal{F}_\phi = \mathrm{Fix}_W(\Sigma(A))$$

$\qquad\square$

9.4 Instantaneous and Average Symmetries, and Patterns on Average

Computing detectives by use of ergodic sums provides a way to interpret the symmetries of attractors. Suppose that A is an attractor for a Γ-equivariant dynamical system and let $\Sigma(A) \subseteq \Gamma$ be its group of symmetries. Let $T(A) \subseteq \Sigma(A)$ be the subgroup that fixes every point of A. We call symmetries in $T(A)$ *instantaneous symmetries* and symmetries in $\Sigma(A)$ *average symmetries*.

To explain this terminology, suppose that the dynamical system models a physical process — such as a fluid experiment — which varies in space as well as in time. We want to understand the implications in physical space of the symmetries of A. The symmetries in $T(A)$ fix the physical state of the system at every instant in time, just as when A is an equilibrium. The effect of symmetries in $\Sigma(A)$ is more subtle.

Patterns on Average. Using Γ-equivariant observables, (9.15) shows that the time average of the these observables is a function of the space variables that is invariant under all symmetries of the attractor A in phase space. Thus, the pattern defined by this time average — its set of level contours — is invariant under all $\sigma \in \Sigma(A)$. That is, we have defined a notion of *pattern on average*.

More precisely, let A be an attractor in phase space for a PDE, and suppose that A has an SBR measure ν. Let $u(x,t)$ be a solution and suppose that $A = \omega(u(x,t))$. Let φ be any real-valued Γ-equivariant observable. Define the time average

$$U_\varphi(x) = \lim_{T \to \infty} \frac{1}{T} \int_0^T \varphi(u(x,t)) dt$$

The Birkhoff Ergodic Theorem states that

$$U_\varphi(x) = \int_A \varphi(a) d\nu(a)$$

for some ergodic measure v — which we assume is SBR. It follows that

$$U_\varphi(\sigma x) = \int_{\sigma A} \varphi(b) dv(b) = \int_{\sigma A} \varphi(\sigma a) dv(\sigma a) = \int_A \varphi(a) d\nu(a) = U_\varphi(x) \qquad (9.15)$$

Therefore, $U_\varphi(x) = U_\varphi(\sigma x)$ for all $\sigma \in \Sigma$. In particular, the time average U_φ is Σ-invariant and patterns are possible in the time average that do not exist at any instantaneous time. See Dellnitz *et al.* [146].

The Faraday Experiment Revisited. As discussed previously, Figure 9.9 shows time averages of the intensity of transmitted light at each point in space (the observable) in the Faraday experiment of Gluckman *et al.* [211]. From the numerical experiments with mappings in Chossat and Golubitsky [104] and Field and Golubitsky [189] it is reasonable to expect the symmetry group of an attractor in systems with finite symmetry to be the full symmetry group of the domain. We therefore are not surprised that the symmetry group of the time-averaged intensity function is square symmetric in experiments operated in the spatio-temporal regime in a container with square cross-section. Similarly, Theorem 9.8 of Melbourne [386] implies that experiments performed in a container with circular cross-section should produce circularly symmetric time averages, as in Figure 9.9 — provided that the time instantaneous picture has no reflection symmetries, which it does not (see Figure 9.6).

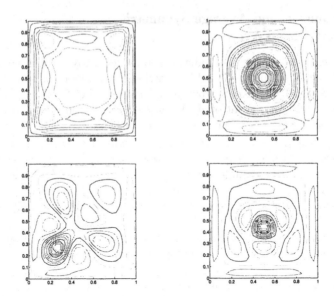

Figure 9.22: Contour plots of time average of solutions of Brusselator (9.4) with Dirichlet boundary conditions and parameters defined in (9.5): (Upper left) $\lambda = 3.75$; $\mathbf{D}_4$; (upper right) $\lambda = 2.0$; $\mathbf{Z}_4$; (lower left) $\lambda = 2.2$; $\mathbf{D}_1^p$; (lower right) $\lambda = 2.4$; $\mathbf{D}_1^s$. Courtesy of Michael Dellnitz.

There are interesting features in the Faraday experiments of [211] that cannot be explained by model-independent reasoning alone. In particular, the time-averaged pictures in the square cross-section container in Figure 9.9 have an approximate wave number with small squares appearing on a 25×25 grid. Gluckman *et al.* [211] observe that this approximate wave number in the time average changes as the frequency increases. So physically observable changes are taking place in the attractors as parameters are varied — even though there is no apparent change in the symmetry of the attractor.

The Brusselator on a Square Domain. Finally, we present numerical simulations by Michael Dellnitz of the Brusselator (9.4) that show that it is possible for average symmetries of solutions to PDEs to be any of a variety of subgroups of symmetries of the domain. These simulations were performed with Dirichlet boundary conditions on a square with $\mathbf{D}_4$ symmetry. The observable used in this numerical experiment is projection of the solution onto its first component. The simulations produced time averages with four of the eight possible subgroups of $\mathbf{D}_4$; namely, $\mathbf{D}_4$, $\mathbf{Z}_4$, $\mathbf{D}_1^s$, and $\mathbf{D}_1^p$. Recall that $\mathbf{D}_1^s$ is generated by a reflection whose line of reflection intersects midpoints of opposite sides, whereas in $\mathbf{D}_1^p$ the line of reflection contains opposite vertices. See Figure 9.22. To obtain these results, the Brusselator parameters were set at those values listed in (9.5) and the parameter λ was varied.

9.5 Synchrony of Chaotic Oscillations and Bubbling Bifurcations

The difference between instantaneous (that is, pointwise) symmetries and setwise symmetries is important in the synchronization of chaotic states of two coupled identical nonlinear cells. (A similar analysis applies to synchronization of more than two cells.) If the states of the cells are $x(t)$ and $y(t)$, we say that the cells are *synchronous* if $x(t) = y(t)$ for all t, and *asynchronous* if not. We discuss two issues. First, symmetry suggests a third kind of synchrony — which we call *synchrony on average* — and, second, the transition from synchrony to asynchrony is complex and leads to new kinds of bifurcation. We discuss the first issue only briefly.

A general two-cell system has the single transpositional symmetry $\kappa(x, y) = (y, x)$. Synchronous states are ones whose trajectories lie in $\mathrm{Fix}(\kappa)$ and that subspace is dynamically invariant by Theorem 1.17. In this context we can divide asynchronous states into two types — those whose attractors are κ-symmetric and those that are not. Suppose the attractor is κ-symmetric. It then follows directly from the ergodic theorem and the theory of detectives that any time average based on the output $x(t)$ of the first cell will be identical to that time average computed with the output from the second cell $y(t)$. So, statistically the outputs from the two cells are identical — even though the trajectories are different. Thus, we have defined *synchrony on average* between the two cells. King and Stewart [310] observe that the time series $x(t)$ and $y(t)$ that come from symmetric attractors are visually similar, whereas those that come from asymmetric attractors are visually different.

In general both synchronous and asynchronous oscillations are typical, and the choice is governed by the *stability* of synchronous states to perturbations that break synchrony. The transition between a synchronous state and an asynchronous one is surprisingly complex, and its analysis involves several different notions of stability. In more geometric language, the analysis involves the different meanings that can be given to the term 'attractor'. The full story is described in Ashwin at al. [19, 20]: here we concentrate on some of the intuition behind the formal mathematics developed in those papers. Other references include Alexander *et al.* [3], Hammer *et al.* [264], Kocarev *et al.* [319], Ott *et al.* [420, 421, 422], Pikovsky and Grassberger [433], and Platt *et al.* [434, 435].

The synchronization problem for cells is one example of a more general situation in which the same issues arise. The essential features are a dynamical system on a space M with an invariant subspace N. Suppose that the restriction of the system to N has an attractor A. The behavior of the system near A is a combination of the dynamics on N and the dynamics transverse to N. (In the coupled-cell example, N is the space of synchronized motions, and the transverse component can be interpreted as an asynchronous perturbation.) Because A is an attractor, the dynamics on A is by definition stable to perturbations in N. But when A is chaotic, having a complex spatial structure, the effect of perturbations transverse to N has a global nature that creates room for complicated dynamics. Specifically, as the point representing the state of the system moves around near A, it is subjected to *different* perturbations at different places, some of which may be repelling while others are attracting. It is the combined effect of all of these perturbations that determines the transverse stability, and this is quite a subtle matter.

The phenomena exemplified by the synchronization of cells arise more widely. For example Rand *et al.* [441] explain their relevance to ecological models of the invasion of an ecosystem by a new species. Here N is the phase space of the existing species, and M is the full phase space including new 'invading' species. These models also apply to epidemiology: here the species are species of microbes. The same phenomena are also implicated in many case of 'transition to spatio-temporal chaos'.

We begin by analyzing a simple example that includes the main features of the problem: a mapping of the plane with an invariant line upon which the dynamics is chaotic.

For simplicity we describe the ideas in the setting of discrete dynamics. A system of two identical coupled discrete-time nonlinear cells can be described by a mapping of the general form

$$x_{t+1} = f(x_t, \lambda) + K(x_t, y_t, \lambda)$$
$$y_{t+1} = f(y_t, \lambda) + K(y_t, x_t, \lambda)$$

where $x, y \in \mathbf{R}^k$ define the states of the two cells, λ is a bifurcation parameter, and for simplicity we assume that the coupling term K satisfies $K(x, x) = 0$. The state space is $M = \mathbf{R}^k \times \mathbf{R}^k$. The 'diagonal' subspace

$$N = \{(x, y) : x = y\}$$

contains all synchronous motions, and N is invariant under the dynamics.

Because K vanishes on N, the dynamics restricted to N obeys the equation

$$x_{t+1} = f(x_t, \lambda) \tag{9.16}$$

where we have identified x with (x, x). We can find all synchronous motions of the coupled system by studying this restricted equation. Attractors (in any particular sense) of (9.16) correspond to synchronous motions that are stable (in the corresponding sense) to *synchronous* perturbations. Whether such motions are fully stable depends upon their stability to asynchronous perturbations, that is, small displacements transverse to N. In particular the loss or acquisition of synchrony can be studied as a loss or gain of transverse stability.

The problem is: how do we define (and compute) a suitable quantitative measure of transverse stability? Suppose that M is a smooth manifold and $f : M \to M$ is a smooth mapping (determining a discrete dynamical system). Let N be a submanifold of M that is invariant under f, that is, $f(N) \subseteq N$. Then the restriction $g = f|_N$ defines a discrete dynamic on N. The aim is to relate the dynamics of f to that of g. In particular, suppose that A is an attractor for g on N (in some sense). When is A an attractor for f on M, and in what sense?

A naive approach to this question is the following argument, in which we suppose for simplicity of exposition that dim N = dim M - 1. Choose a global coordinate x on a neighborhood U of A in N, and a local coordinate y transverse to N on this neighborhood. Then we may write

$$f(x, y) = (f_1(x, y), f_2(x, y))$$

If y is small then we may approximate $f_1(x, y)$ by $f(x, 0)$, which is equal to $g(x)$. Since N is invariant we have

$$\left. \frac{\partial f}{\partial y} \right|_{y=0} = 0$$

so we approximate $f_2(x, y)$ by its linear part $\lambda(x)y$, where

$$\lambda : U \to \mathbf{R}.$$

Therefore we obtain an approximation to f of the form

$$\tilde{f}(x, y) = (g(x), \lambda(x)y)$$

Iterating $\tilde{f}$ we find that

$$\tilde{f}^n(x, y) = (g^n(x), \lambda(x)\lambda(g(x)) \ldots \lambda(g^{n-1}(x)))$$

This will tend towards A (implying that A is 'attracting on average') as $n \to \infty$ provided that the product

$$P_n(x) = \lambda(x)\lambda(g(x)) \ldots \lambda(g^{n-1}(x))$$

tends to 0. This will certainly be the case if the geometric mean $|P_n(x_0)|^{\frac{1}{n}} < 1$ for n large, where x_0 generates a dense orbit in A. That is, we require

$$(|\lambda(x_0)||\lambda(g(x_0))| \ldots |\lambda(g^{n-1}(x_0))|)^{\frac{1}{n}} < 1$$

for large n, or equivalently that

$$\frac{1}{n} \sum_{i=0}^{n-1} \log |\lambda(g^i(x_0))| < 0$$

for large n.

If there exists an invariant ergodic measure μ on A then this sum is equal to

$$\sigma_\mu = \int_{x \in A} \log |\lambda(x)| d\mu$$

Assuming that the approximations made are valid, the condition for transverse stability is therefore that

$$\sigma_\mu < 0$$

We interpret σ_μ as a *transverse Liapunov exponent* with respect to the invariant measure μ.

Bubbling Bifurcations. The above analysis slides over several important technical points, which turn out to have a crucial effect on the local dynamics. In particular *invariant measures on A are seldom unique*. Therefore A may be 'stable to transverse perturbations' with respect to some invariant measures but not with respect to others. For example if A contains a periodic cycle $x_0, x_1, \cdots, x_{n-1}$ then

$$\log(|\lambda(x_0)|) + \log(|\lambda(x_1)|) + \cdots + \log(|\lambda(x_{n-1})|)$$

may be positive, even though the integral of the transverse dynamics with respect to an SBR measure may be negative. If so, orbits that come close to the periodic cycle will move away from A more rapidly than orbits that do not. But since A is an attractor, most definitions imply that any orbit in A comes close to such a periodic cycle. Therefore the dynamics includes 'fluctuations' away from A and back again, as the orbit comes close to the periodic cycle and then deviates from it. We call such behavior *bubbling*. It is closely related to the phenomenon called 'riddled basins' by Alexander *et al.* [3], and also to so-called 'blowout bifurcations' and 'on-off intermittency', see Pecora and Carroll [430], Pikovsky and Grassberger [433]. perhaps the most important point to appreciate is that because of the non-uniqueness of invariant measures, the loss of transverse stability (and hence the transition from synchrony to asynchrony in a two-cell system) occurs over an *interval* of parameters, not at a single point. Indeed a typical scenario involves three values $\lambda_0 < \lambda_1 < \lambda_2$ of the bifurcation parameter λ: these are respectively where a transverse Liapunov exponent becomes positive for some invariant measure; where a transverse Liapunov exponent becomes positive for an SBR measure; and where a transverse Liapunov exponent becomes positive for all invariant measures. The transitions that occur are of the following kind:

- For $\lambda < \lambda_0$ the set A is an asymptotically stable attractor
- For $\lambda_0 < \lambda < \lambda_1$ the set A is an attractor with a 'locally riddled basin'
- For $\lambda_1 < \lambda < \lambda_2$ the set A is a chaotic saddle
- For $\lambda_2 < \lambda$ the set A is a normally repelling chaotic saddle

Bubbling is one of the phenomena that can occur in the range $\lambda_0 < \lambda < \lambda_1$. For further information see Ashwin *et al.* [19, 20].

Numerical Example. As an example, we discuss a family of planar mappings introduced in [19]. It is an extension of a cubic logistic equation $h : \mathbf{R} \to \mathbf{R}$ defined by

$$h(x) = \frac{3\sqrt{3}}{2}x(x^2 - 1)$$

to a map of the plane. Let $f_{\alpha\nu\varepsilon}$ be a 3-parameter map of $\mathbf{R}^2$ to itself that is equivariant under $\mathbf{Z}_2$ generated by $(x_1, x_2) \mapsto (-x_1, x_2)$, given by

$$f_{\alpha\nu\varepsilon}(x_1, x_2) = (\frac{3\sqrt{3}}{2}x_1(x_1^2 - 1) + \varepsilon x_1 x_2^2, \nu e^{-\alpha x_1^2}x_2 + x_2^3) \qquad (9.17)$$

The factor $\frac{3\sqrt{3}}{2}$ is such that each of the intervals $[-1, 0]$ and $[0, 1]$ is mapped onto the other in a two-to-one way (except, of course, at critical points). In fact, h has an asymptotically stable attractor $A = [-1, 1] \subseteq N = \mathbf{R} \times \{0\}$ independently of α, ν and ε. The linearization of f on N depends on α and ν but is independent of ε. The case $\varepsilon = 0$ corresponds to the existence of an invariant foliation by vertical lines.

The Jacobian at $(x_1, 0)$ is

$$d_{(x_1,0)}f = \left[\begin{array}{cc} 3\sqrt{3}(x_1^2 - \frac{1}{2}) & 0 \\ 0 & \nu e^{-\alpha x_1^2} \end{array} \right]$$

We can compute the transverse Liapunov exponent σ_μ with respect to μ, and we obtain

$$\sigma_\mu = \log|\nu| - K_\mu\alpha$$

where

$$K_\mu = \int_A x_1^2 d\mu(x_1).$$

It is possible to show that f has an absolutely continuous ergodic measure μ_{SBR} whose support is A: the coordinate change $x = \sin^3(\pi\theta/2)$ turns $f_{|A}$ into a piecewise expanding map, so by Lasota and Yorke [353] f has an ergodic absolutely continuous invariant measure. We write K_{SBR} for $K_{\mu_{SBR}}$ and K_1 for $\sup_\mu K_\mu$, where the supremum is taken over all ergodic invariant measures for f. Numerical approximation of the SBR measure from box counting $500,000$ iterates in 100 bins gives $K_{SBR} = 0.358$, and numerical evidence strongly suggests that $K_1 = 1 - \frac{2}{3\sqrt{3}} = 0.615$. Figure 9.23 shows a sequence of pictures of the basin $\mathcal{B}(A)$ in $\mathbf{R}^2$ for $\alpha = 0.7$ and increasing values of ν; Fig. 9.24 shows blow-ups of details from this figure.

Finally, we describe the nature of the invariant set A in the global phase space. For simplicity we state the results only for positive ν; note, however, that this classification is essentially independent of the sign of ν. For proofs, see Ashwin et al. [20].

Theorem 9.35 *Choose $\varepsilon \in \mathbf{R}$ and let $\alpha > 0$. Then the behavior of the map f is as follows:*

(a) *For $0 \leq \nu < 1$, A is an asymptotically stable attractor.*
(b) *For $1 < \nu < e^{K_{SBR}\alpha}$, A is a Milnor (essential) attractor whose basin is riddled with that of the attractor at infinity.*
(c) *For $e^{K_{SBR}\alpha} < \nu < e^{K_1\alpha}$, A is a (non-normally repelling) chaotic saddle.*
(d) *For $\nu > e^{K_1\alpha}$, A is a normally repelling chaotic saddle.*

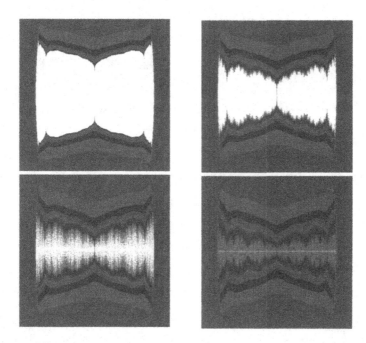

Figure 9.23: White region shows basin $\mathcal{B}(A)$ of $f_{\alpha\nu\epsilon}$ for $\alpha = 0.7$. (Top left) $\nu = 0.9$ attractor is asymptotically stable; (top right) $\nu = 1.2$ basin is riddled, but measure of holes is too small to be visible; (lower left) $\nu = 1.28$, near blow-out bifurcation, basin has visible fractal structure; (lower right) $\nu = 1.48$ basin has zero measure and A is a chaotic saddle. Figure courtesy of Peter Ashwin.

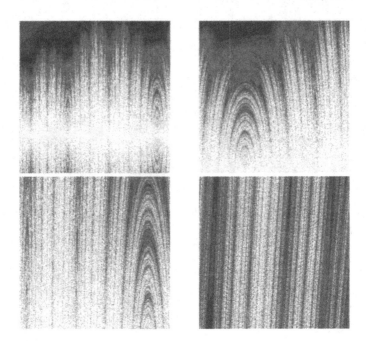

Figure 9.24: Blow-ups of basin $\mathcal{B}(A)$ for $\alpha = 0.7, \nu = 12.8$. Lower left of previous picture. Here (x_1, x_2) lies in the regions (from top left to lower right): $[0.048, 0.623] \times [-0.104, 0.317]$; $[0.204, 0.306] \times [-0.005, 0.239]$; $[0.150, 0.250] \times [0.039, 0.134]$; $[0.200, 0.210] \times [0.150, 0.160]$. Figure courtesy of Peter Ashwin.

Chapter 10

Periodic Solutions of
Symmetric Hamiltonian Systems

Until now, the dynamical systems under consideration have been dissipative ones — that is, we have not required energy to be conserved (and indeed the concept of energy has not been given any emphasis). For many models arising from mechanics, however, it is important to build in the constraint of energy-conservation. Historically this development led to a general formulation of mechanics in terms of 'Hamiltonian systems'. In this final chapter we consider effects of symmetry on Hamiltonian dynamics. This is a very broad area, and we shall discuss only certain topics within it that link to the ideas developed earlier. Our approach is closely related to the technique of 'reduction' in symmetric Hamiltonian systems — essentially the passage to the orbit space of the symmetry group — which has been extensively developed and applied: see for example Abraham and Marsden [1], Lewis [356, 357], Lewis and Ratiu [358], Lewis and Simo [359], Lewis et al. [360], Marsden [374], Marsden and Weinstein [376], Ortega and Ratiu [419], Patrick [426, 427], Roberts and Dias [451].

Many systems of physical interest are well modeled by Hamiltonian systems, and symmetries seem to be especially common in such models. In part this may be because the classically solvable Hamiltonian systems are those with extensive symmetries: the classical mathematicians were consciously or unconsciously relying on Noether's Theorem, to the effect that continuous symmetries of a Hamiltonian system generate conserved quantities. But the prevalence of symmetry in Hamiltonian systems goes deeper.

A typical example, which serves to motivate this chapter, is the problem of determining the vibrational spectrum of a molecule. Molecules can be modeled on a variety of levels — classical, semiclassical, quantum-mechanical. Many small molecules, such as water (H_2O), ammonia (NH_3), and methane (CH_4) have symmetry. So do some larger ones, notably buckminsterfullerene (C_{60}). In a classical approximation, the water molecule consists of a central oxygen atom bonded to two hydrogen atoms. The angle between the hydrogen bonds is not $180°$, so the three atoms are not collinear. The symmetry group of a molecule that is stationary in space is D_2: one generator is reflection in the common place of the three molecules, and the other is reflection in a plane at right angles to this, interchanging the two hydrogen atoms while keeping the oxygen atom fixed. Similarly, the ammonia molecule has symmetry group D_3 and the methane molecule has the symmetries of the tetrahedron. The buckminsterfullerene molecule is a cage of carbon atoms arranged in the form of a truncated icosahedron, and so has icosahedral symmetry.

We should point out immediately that molecules are more complex than the above description may suggest. For example their atoms also have their own vibrational modes. The ammonia molecule can flex between two symmetrically related states: one in which all three hydrogen bonds point downwards, another in which

Figure 10.1: Methane molecule: one carbon atom (open dot) and four hydrogen atoms (solid dots); tetrahedral symmetry.

they point upwards. The symmetry group $\mathbf{D}_3 \times \mathbf{Z}_2$ may be a more appropriate setting in such circumstances. Finally, the entire molecule is free to move in $\mathbf{R}^3$, so all states 'really' come as $\mathbf{E}(3)$ group orbits. The problem of decomposing the motion into 'spatial' and 'internal' motions is highly non-trivial: see for example Marsden [374]. For the purposes of this chapter we ignore these features, but in applications they are often important.

The dynamics of a symmetric molecule, and in particular its vibrational spectrum, are constrained by its symmetries. The symmetry also constrains the behavior of any model, be it classical or quantum-mechanical. The simplest classical models in effect treat the molecule as a collection of masses connected by springs, and factor out the spatial motions to focus on the internal ones. In its lowest energy state, such a system is at rest — an equilibrium that is invariant under the symmetry group. As its energy increases, it begins to move in some manner. Here we consider only periodic oscillations near equilibrium. (Quasiperiodic and chaotic motion are also important, but it would take us too far afield to discuss them.)

The main theoretical existence results for such oscillations are the Liapunov Center Theorem, which dates from 1895, and guarantees the existence of a family of periodic orbits near and equilibrium provided certain very stringent nondegeneracy conditions hold. Unfortunately symmetries (even finite symmetry groups) typically prevent these conditions from being valid, but a more recent result, the Moser-Weinstein Theorem, comes to the rescue. We describe work of Montaldi *et al.* [405, 406, 407, 408], deriving an equivariant analogue of the Moser-Weinstein Theorem. This theorem asserts that subject to certain technical hypotheses, certain families of symmetry-breaking periodic orbits always exist near an equilibrium of a symmetric Hamiltonian system. As in the dissipative case, the $\mathbf{S}^1$ phase shift symmetry of a periodic orbit plays a crucial role. A pleasant feature is that the relevant group theory is essentially identical to that for equivariant Hopf bifurcation. This is not entirely accidental — there are several ways to 'explain' it — but it allows us to apply existing group-theoretic calculations for Hopf bifurcation to the Hamiltonian case. These results have been significantly generalized by Roberts and Dias [451] to prove the existence of relative periodic orbits near relative equilibria. The background to their work is too technical to describe here.

We begin with mathematical generalities, motivating some of the ideas in terms of the Hénon-Heiles Hamiltonian, a system with $\mathbf{D}_3$ symmetry. As in any investigations of Hamiltonian dynamics, the symplectic structure on phase space is of central importance, and we briefly recall some pertinent concepts and discuss symplectic representations. Next we state the Equivariant Moser-Weinstein Theorem. As a non-trivial illustration we consider a slightly artificial many-body problem in which N bodies of equal mass interact via forces that depend smoothly on the square of the distance between them, making the non-physical assumption that the bodies can interpenetrate. (The methods of Roberts and Dias [451] remove the need for such an assumption.)

We then move on to the question of (linear) stability, emphasizing the new features that arise in the Hamiltonian context (because of the symplectic structure). Applications include a simplified model of the vibrations of a liquid drop, and the 1:1 resonant oscillations of a spring pendulum — which can be thought of as a simple example of the classical ball-and-spring models of molecules. We end by summarizing some results of Montaldi and Roberts [404] on the vibrations of symmetric molecules in the Born-Oppenheimer approximation, a rather more sophisticated model than the ball-and-spring one.

10.1 The Equivariant Moser-Weinstein Theorem

We recall some basic ideas concerning Hamiltonian systems, state the Liapunov Center Theorem and the Moser-Weinstein Theorem, explain why they are inadequate in the symmetric context, and state the Equivariant Moser-Weinstein Theorem.

Hamiltonian Systems. The *phase space* of a Hamiltonian system is an even-dimensional Euclidean space $P = \mathbf{R}^{2n}$ (or more generally a 'symplectic manifold', but we will not explore this generalization here). Coordinates on P are written as $(q_1 \ldots, q_n; p_1, \ldots, p_n)$: we call the q_j *position* coordinates and the p_j *momenta.* The space spanned by the q_j is *configuration space* and that spanned by the p_j is *momentum space.* The dynamics is determined by a *Hamiltonian*

$$\mathcal{H} : P \to \mathbf{R}$$

which is a generalization of energy, and we normally assume $\mathcal{H} \in C^\infty$. *Hamilton's Equations* for the dynamics are:

$$\frac{\mathrm{d}q_j}{\mathrm{d}t} = \frac{\partial \mathcal{H}}{\partial p_j} \qquad \frac{\mathrm{d}p_j}{\mathrm{d}t} = -\frac{\partial \mathcal{H}}{\partial q_j} \tag{10.1}$$

Because of the form of these equations, $\dot{\mathcal{H}} \equiv 0$, so the Hamiltonian is conserved by the flow. The level sets of $\mathcal{H}$, given by $\mathcal{H} = c$ for constant c, are called *energy levels.* See Abraham and Marsden [1] and Arrowsmith and Place [11] for further information on general Hamiltonian dynamics.

The Liapunov Center Theorem. The basic 'local bifurcation' existence theorem for periodic orbits in Hamiltonian dynamics is the Liapunov Center Theorem. The role of bifurcation parameter is played by the energy level — this is typical of Hamiltonian systems.

Suppose that $\mathcal{H}$ is a Hamiltonian on $P = \mathbf{R}^{2n}$ and let $p \in P$ be an equilibrium, so that $(\mathrm{d}\mathcal{H})_p = 0$. Assume that p is a *nondegenerate* minimum of $\mathcal{H}$, that is $(\mathrm{d}\mathcal{H})|_p = 0$ and $(\mathrm{d}^2\mathcal{H})|_p$ is positive definite. Let L be the linearization of the Hamiltonian vector field at p and let the eigenvalues of L be the purely imaginary pairs $\{\pm\lambda_1, \ldots, \pm\lambda_n\}$. In 1895 Liapunov proved:

Theorem 10.1 (Liapunov Center Theorem) *If the linearized flow at an equilibrium has a simple purely imaginary eigenvalue and some λ_i is non-resonant then there exists a smooth 2-dimensional submanifold of P, which passes through p and inter-*

sects every energy level near p in a periodic orbit, such that the period of that orbit approaches $2\pi/|\lambda_i|$ for orbits near p.

Proof. For a proof see Abraham and Marsden [1] Section 5.6.7 p. 498.   □

By 'non-resonant' we mean that λ_j is not an integer multiple of λ_i for $j \neq i$.

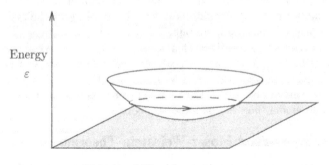

Linearized Eigenspace

Figure 10.2: Energy levels in the Liapunov Center Theorem.

Example 10.2 In the plane pendulum we have and a family of periodic orbits is visible surrounding the equilibrium at the origin (Figure 10.3).   ◇

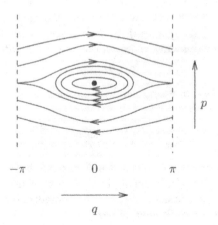

Figure 10.3: Phase portrait for the plane pendulum.

The Moser-Weinstein Theorem. Various authors have shown that the hypothesis of non-resonance can be relaxed in certain special cases. In a celebrated paper Weinstein [519] proved that even when there is resonance, there must exist at least $\frac{1}{2} \dim V_\lambda$ families of periodic solutions on each energy level near p. The proof was simplified by Moser [409], and the result has come to be known as the Moser-Weinstein Theorem:

Theorem 10.3 (Moser-Weinstein Theorem) *If λ is a resonant eigenvalue and*

  (a) $d^2\mathcal{H}_p$ *is non-degenerate*
  (b) $d^2\mathcal{H}_p|_{V_\lambda}$ *is positive definite*

then there exists at least $\frac{1}{2}\dim V_\lambda$ periodic solutions on each energy level $\varepsilon^2 + \mathcal{H}(p)$ for sufficiently small $\varepsilon \in \mathbf{R}$.

Example 10.4 Consider the spherical pendulum in Figure 10.4. There are two periodic solutions.

Standing Waves: These have $\mathbf{Z}_2$ symmetry. (Actually they have $\mathbf{Z}_2 \oplus \mathbf{Z}_2^c$ symmetry where $\mathbf{Z}_2 \oplus \mathbf{Z}_2^c = \{(0,0),(\pi,\pi)\}$.) See Figure 10.5.

Rotating Waves: These have $\mathbf{SO}(2) = \{(\theta,\theta) : \theta \in \mathbf{SO}(2)\}$ symmetry. In this case a rotation is equivalent to a phase shift. See Figure 10.5.

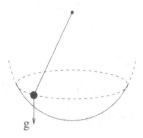

Figure 10.4: The spherical pendulum, a system with $\mathbf{O}(2)$ symmetry.

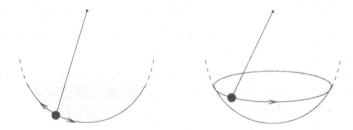

Figure 10.5: Standing wave and rotating wave in the spherical pendulum.

     Note that if $x(t)$ is a solution then so is $\gamma x(t + \theta T/2\pi)$ as solutions come in $\Gamma \times \mathbf{S}^1$ orbits. $\diamond$

The Hénon-Heiles System. One source of resonances is symmetry. This is the case in the next example, the Hénon-Heiles Hamiltonian, Hénon and Heiles [271]. We compare known results for this Hamiltonian with the generalities predicted by the Moser-Weinstein Theorem.

Example 10.5 The Hénon-Heiles Hamiltonian takes the form

$$\mathcal{H}(q_1,q_2,p_1,p_2) = \frac{1}{2}(p_1^2 + p_2^2 + q_1^2 + q_2^2) + \frac{1}{3}q_1^3 - q_1 q_2^2$$

This is the lowest order Hamiltonian that exhibits $\mathbf{D}_3$-symmetry (where the action is the standard action on the plane of q-coordinates coupled with the standard action on the plane of p-coordinates).

     A Poincaré section for this system is shown in Figure 10.6. It turns out that there are eight periodic solutions, here labeled $\Pi_1,\ldots,\Pi_8$. Three are hyperbolic (unstable) and the other five are elliptic (stable). (We discuss stability in Section 10.5.) These solutions may be grouped into three families as in Figure 10.7

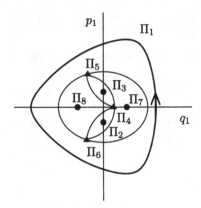

• elliptic periodic orbit

—, ▲ hyperbolic periodic orbit

Figure 10.6: Poincaré section for the Hénon-Heiles Hamiltonian showing eight periodic orbits.

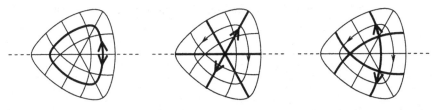

Figure 10.7: The three types of periodic solution for the Hénon-Heiles Hamiltonian. Two rotating waves Π_7, Π_8. Three type I standing waves, Π_1, Π_2, Π_3. Three type II standing waves, Π_4, Π_5, Π_6.

Rotating Waves: There are two solutions of this type: they are elliptic with symmetry $\tilde{\mathbf{Z}}_3$.

Standing Waves I: There are three solutions of this type: they are elliptic with symmetry $\mathbf{Z}_2$.

Standing Waves II: There are three solutions of this type: they are hyperbolic with symmetry $\tilde{\mathbf{Z}}_2$.

This is exactly the behavior seen in standard $\mathbf{D}_3$ Hopf bifurcation (see Example 4.11(c)), suggesting that there should be some link between Moser-Weinstein type results and Hopf bifurcation in the equivariant setting.   ◇

It is clear that the Moser-Weinstein Theorem fails to predict all periodic solutions near equilibrium in the equivariant case. In Example 10.5 the Moser-Weinstein Theorem predicts at least two periodic solutions, but we find a total of eight. Even taking the symmetry into account, we still have three group orbits of periodic solutions, not two. So it looks as if we need better machinery to obtain better estimates on the number of periodic orbits. We can develop such machinery by deriving an equivariant version of the Moser-Weinstein Theorem, as was done by Montaldi *et al.* [406] and developed in Montaldi *et al.* [407, 408]. The Equivariant Moser-Weinstein Theorem will be the main result of this chapter.

Symplectic Group Actions. We begin, as usual, by setting up the appropriate group action and investigating its implications for the linearized problem. They main new feature is the symplectic structure of Hamiltonian systems. In the equivariant setting, the Hamiltonian H is invariant under the action of a compact Lie group on phase space, but this action must satisfy a new constraint: preserving solutions of Hamilton's Equations. To formulate this requirement, we make use of a fundamental feature of Hamiltonian dynamics: the existence of a *symplectic structure*.

Definition 10.6   (a) A *symplectic vector space* over **R** is a vector space V over **R** equipped with a *symplectic form* $[,]$ — that is, a nondegenerate skew-symmetric bilinear form.

(b) A *symplectic action* of a group Γ on a symplectic vector space $V, [,]$ is an action that leaves the symplectic form invariant. That is, it must satisfy

$$[\gamma x, \gamma y] = [x, y]$$

$\diamond$

The entire theory of group representations can be extended to symplectic representations (Montaldi *et al.* [406]): in particular any symplectic representation of a compact Lie group is a direct sum of irreducible symplectic representations, and there exists a unique isotypic decomposition. Moreover, the symplectic irreducibles for compact Γ are precisely what we earlier called Γ-simple representations (a fact that is related to the complex structure on Γ-simple representations and a similar complex structure on symplectic vector spaces). In other words, symplectic irreducibles are of three types: real, complex, or quaternionic, as follows:

(a) *Real case.* Let W be absolutely irreducible for Γ, and form the space $P = W \oplus W$ with diagonal Γ-action

$$\gamma(u, v) = (\gamma u, \gamma v)$$

Here we can identify one copy of W with configuration space and the other with momentum space. Up to isomorphism, there is a unique symplectic structure compatible with this action.

(b) *Complex case.* Let W be irreducible for Γ and of complex type, so that the commuting matrices can be identified with **C**. Then there are two symplectically non-isomorphic representations of Γ on W, related to each other by complex conjugation.

(c) *Quaternionic case.* Let W be irreducible for Γ and of quaternionic type, so that the commuting matrices can be identified with the quaternions. Then up to isomorphism there is a unique symplectic structure compatible with this action.

Moser-Weinstein with Symmetry. Suppose that a compact Lie group Γ acts symplectically on P, let $p \in P$ be a fixed point for Γ, and suppose that the Hamiltonian H is Γ-invariant. This symmetry may force some of the λ_i to be equal, creating unavoidable resonances.

Let $u(t)$ be a periodic orbit of the flow of H having period T. Let $\mathbf{S}^1$ be the circle group, identified with $\mathbf{R}/2\pi\mathbf{Z}$, and consider the usual action of $\Gamma \times \mathbf{S}^1$ on the *loop space* $C^k(T)$ of k-times differentiable T-periodic functions, as in equivariant Hopf bifurcation (Section 4.2). Then $\Gamma \times \mathbf{S}^1$ acts on $u = u(t)$ by

$$(\gamma, \theta)u(t) = \gamma u(t + T\theta/2\pi)$$

Define the *symmetry group* of $u \in C^k(T)$ to be

$$\Sigma_u = \{(\gamma, \theta) \in \Gamma \times \mathbf{S}^1 | \gamma u(t + T\theta/2\pi) = u(t)\}$$

Recall that when P is a vector space over $\mathbf{R}$ and G acts linearly, $\mathrm{Fix}(\Sigma)$ is a linear subspace. Analogously, if P is a symplectic vector space over $\mathbf{R}$ and G acts linearly and symplectically, then $\mathrm{Fix}(\Sigma)$ is a symplectic linear subspace.

Let X be the vector field of H, let L be the linearization of X at p. Define the *linearized flow* to be the flow generated by the ODE

$$\dot{x} + Lx = 0$$

on the tangent space $V = T_p P$ to P at p. Let λ be a nonzero purely imaginary eigenvalue of L and define the *resonance space* $V_\lambda \subseteq V$ to be the (real part of the) sum of the generalized eigenspaces of L for eigenvalues $k\lambda$, where $k \in \mathbf{Z}$. Assume the following conditions on H:

(1) $(\mathrm{d}^2 H)_p$ is a nondegenerate quadratic form.
(2) $(\mathrm{d}^2 H)_p|_{V_\lambda}$ is positive definite.

Condition (1) is equivalent to L being nonsingular, and (2) implies that $L|_{V_\lambda}$ is semisimple (diagonalizable over $\mathbf{C}$).

Clearly L is Γ-equivariant, so V_λ is invariant under the action of Γ. It is also invariant under the linearized flow. Because $L|_{V_\lambda}$ is semisimple, the orbits of the linearized flow are all periodic with period $2\pi/|\lambda|$ and hence define an action of $\mathbf{S}^1$ on V_λ. Explicitly,

$$\theta v = \exp\left(\frac{\theta}{|\lambda|}L\right)v$$

This action commutes with the action of Γ, so together they define a $\Gamma \times \mathbf{S}^1$-action on V_λ.

We may now state the central result of this chapter:

Theorem 10.7 (Equivariant Moser-Weinstein) *Suppose that the Hamiltonian H satisfies (1) and (2). Then for every isotropy subgroup Σ of the $\Gamma \times \mathbf{S}^1$-action on V_λ, and for all sufficiently small $\varepsilon \in \mathbf{R}$, there exist at least $\frac{1}{2} \dim \mathrm{Fix}(\Sigma)$ periodic orbits of X with periods near $2\pi/|\lambda|$ and symmetry group containing Σ, on the energy surface $H(x) = H(p) + \varepsilon^2$.*

For further details see Montaldi *et al.* [406]. A rather different approach to an equivariant Liapunov Center Theorem, using the 'constrained Liapunov-Schmidt procedure', can be found in Golubitsky *et al.* [221].

Because of the symplectic structure, $\dim \mathrm{Fix}(\Sigma)$ is always even. In practice — though it is more a rule of thumb than a provable theorem — the important isotropy subgroups Σ are those for which $\dim \mathrm{Fix}(\Sigma)$ is small. The most important isotropy subgroups, and the most tractable, of all are those for which $\dim \mathrm{Fix}(\Sigma)$ attains its minimum value, namely 2. These are the **C**-axial subgroups. So the Equivariant Moser-Weinstein Theorem implies that under the usual hypotheses if Σ is **C**-axial then there exists at least one family of periodic solutions with isotropy group *equal* to Σ. This follows since $\frac{1}{2} \dim \mathrm{Fix}(\Sigma) = 1$ and Σ is a maximal isotropy subgroup, so that 'containing Σ' implies 'equal to Σ'.

The group theory involved in the $\Gamma \times \mathbf{S}^1$-action is identical to the action occurring in equivariant Hopf bifurcation. This is a consequence of the loop space technique employed in both contexts and the classification of symplectic irreducibles stated in the previous subsection. We can use this relationship to import results from equivariant Hopf bifurcation into Hamiltonian dynamics.

One interesting implication of the symplectic structure occurs when we consider the linear stability of the periodic solutions predicted by the Equivariant Moser-Weinstein Theorem. We discuss this issue further in Section 10.4 below. Stability criteria are formulated in terms of the Floquet operator, which can be defined as the linearization of a Poincaré map. The Hamiltonian case has some characteristic features, arising from the symplectic structure — for example, eigenvalues of the Floquet operator typically occur in *quartets* $(\mu^{\pm 1}, \bar{\mu}^{\pm 1})$. The solution is said to be *elliptic* (linearly stable) if all such eigenvalues lie on the unit circle, and *hyperbolic* (linearly unstable) otherwise. There exists a class of isotropy subgroups in which the combination of symmetry and symplectic structure *forces* all eigenvalues of the Floquet operator to lie on the unit circle. Such isotropy subgroups are said to be *cyclospectral*. A periodic solution with a cyclospectral isotropy subgroup is automatically elliptic.

Example 10.8   (a) For the spherical pendulum $\Gamma = \mathbf{O}(2)$ and both isotropy subgroups are cyclospectral:
    (i) $\mathbf{SO}(2)$ has two eigenvalues equal to 1 and two imaginary,
    (ii) $\mathbf{Z}_2$ has four eigenvalues equal to 1.
  (b) For the Hénon-Heiles Hamiltonian $\Gamma = \mathbf{D}_3$:
    (i) $\tilde{\mathbf{Z}}_3$ is cyclospectral with two eigenvalues equal to 1,
    (ii) $\mathbf{Z}_2$ (I) and $\mathbf{Z}_2$ (II) both have one elliptic and one hyperbolic solution with two eigenvalues equal to 1.

10.2 Many-Body Problems

The dynamical behavior of systems of point particles interacting through central forces has been of interest since the time of Newton. In the hands of Poincaré it gave rise to the qualitative theory of ODEs and modern dynamical systems theory, a recent triumph being the proof by Xia [530] that Arnold diffusion — and hence in particular chaos — occurs in the restricted 3-body problem. Despite these complexities, many results are known for special kinds of motion of an N-body system, in particular *central configurations*, see for example Meyer [398]. It is fairly common to consider the symmetric case of N equal masses, in which case many of the known solutions also possess a degree of symmetry. The aim of this section is to set up a symmetry-based context for equal-mass N-body problems, both in the plane $\mathbf{R}^2$ and in 3-space $\mathbf{R}^3$, and to illustrate how some simple general principles for equivariant dynamics might be used to organize the known results and — eventually — derive new ones. Because of technical limitations in current techniques, we will make two rather unorthodox assumptions:

Assumption 1: The potential corresponding to each body is a smooth function of (squared) distance.

Assumption 2: Bodies that collide are free to pass through each other, and several distinct bodies may occupy the same spatial position simultaneously.

Assumption 1 rules out, for example, inverse square law attraction, but it has the advantage of not creating singularities at collisions. Assumption 2 may appear unphysical, but it is central to the approach adopted here, because we will obtain 'interesting' solutions by perturbing the trivial solution in which all N bodies are coincident and stationary. Many of these perturbations do not involve collisions, and thus have physical significance. Those that do involve collisions can be further perturbed in order to remove them.

No attempt is made to 'regularize' collisions: they are simply accepted. A similar assumption is common in the dynamics of low-density distributions of celestial bodies such as globular clusters and galaxies, see Binney and Tremaine [64] p.190. Here the problem is often formulated using the collisionless Boltzmann equation; note, however, that 'collisionless' there carries the further technical implication that each body moves under the influence of the mean field potential of the others. Our model could — stretching a point — be interpreted as describing a system of N nominally identical globular clusters or galaxies, assuming (again unphysically) that these retain their general form if they pass close to or through each other. The Hamiltonian for such clusters can, in a reasonable model, be assumed to obey the smoothness condition (1), even though individual stars exert inverse-square law gravitational attraction, in the same way that Poisson's equation rather than Laplace's applies within a solid body. However, this suggestion is not intended very seriously.

We first establish the appropriate symmetry context for the N-body problem in the plane. Consider a system of N equal point particles in $\mathbf{R}^2$. Assume that each pair of particles experiences an attractive force given by a smooth function $F(d^2)$ where d is the distance between them. Let $q^1, \ldots, q^N$ be the position coordinates of the particles, and let $p^1, \ldots, p^N \in \mathbf{R}^2$ be their momentum coordinates. Then the motion is governed by a smooth Hamiltonian

$$\mathcal{H} : \mathbf{R}^{2N} \times \mathbf{R}^{2N} \to \mathbf{R}$$

with $2N$ degrees of freedom. Because the force F is attractive there is clearly an equilibrium state with all q^j and p^j zero. We will require this state to be nondegenerate, in a slightly stronger sense than Section 10.1; that is, there is a purely imaginary eigenvalue λ with $V_\lambda = V = \mathbf{R}^{2N} \times \mathbf{R}^{2N}$, and $d^2\mathcal{H}_0$ is positive definite.

An important special case arises when the group Γ acts absolutely irreducibly on configuration space and dually on momentum space. That is, the only linear maps that commute with Γ are scalar multiples of the identity. We will see below that this case applies to the N-body system. By a symplectic change of coordinates we can identify the phase space with $V \oplus V \equiv V \otimes_\mathbf{R} \mathbf{C}$ in such a way that the action of $(\gamma\theta) \in \Gamma \times \mathbf{S}^1$ is given by

$$(\gamma, \theta)(v \otimes z) = \gamma v \otimes e^{i\theta} z$$

for $v \in V, z \in \mathbf{C}$. A

Let us apply this remark to the N-body model. There are two obvious types of symmetry on configuration space:

- *Label Symmetries.* The Hamiltonian $\mathcal{H}$ is invariant under permutations of the N particles, hence under permutations of the q^j.
- *Euclidean symmetries.* The Hamiltonian $\mathcal{H}$ is invariant if the same element α of the 2-dimensional Euclidean group is applied simultaneously to all q^j.

There is a reduction of the problem that simplifies the analysis. Without loss of generality we may choose a coordinate system whose origin is at the center of mass. That is, we may restrict the dynamics to the subspace

$$q^1 + \cdots + q^N = 0$$
$$p^1 + \cdots + p^N = 0.$$

Then the Euclidean symmetries are restricted to elements of $\mathbf{O}(2)$, since the center

Let $\mathbf{S}_N$ denote the symmetric group, comprising all permutations of $\{1, \ldots, N\}$. Because label and Euclidean symmetries commute, the Hamiltonian is invariant under an action of $\Gamma = \mathbf{O}(2) \times \mathbf{S}_N$. Explicitly, this action is as follows:

$$(\gamma, \sigma)(q^1, \ldots, q^N) = (\gamma q^{\sigma^{-1}(1)}, \ldots, \gamma q^{\sigma^{-1}(N)})$$

Here we assume that $q^1 + \cdots + q^N = 0$, so there is redundancy in the coordinate system. More abstractly we can define this action as follows. Introduce a space $\mathbf{R}^N$ on which $\mathbf{S}_N$ acts by permuting a basis, and identify $\mathbf{R}^{2N} = \{(q^1, \ldots, q^N)\}$ with $\mathbf{R}^2 \otimes_{\mathbf{R}} \mathbf{R}^N$ so that $(0, \ldots, 0, q^j, 0, \ldots, 0)$ is identified with $q^j \otimes e_j$, where $e_j = (0, \ldots, 0, 1, 0, \ldots, 0)$ with the 1 in the jth position. Then $\mathbf{O}(2) \times \mathbf{S}_N$ acts on $\mathbf{R}^2 \otimes_{\mathbf{R}} \mathbf{R}^N$ by

$$(\gamma, \sigma)(v \otimes w) = \gamma v \otimes \sigma w$$

Define the subspace

$$\mathbf{R}^{N-1} = \{\sum \mu_j e_j \in \mathbf{R}^N : \sum \mu_j = 0\}$$

Then confining attention to the subspace $\{\sum q^j = 0\}$ is equivalent to restricting the action of Γ to the subspace $V = \mathbf{R}^2 \otimes_{\mathbf{R}} \mathbf{R}^{N-1}$.

It is easy to check that $\Gamma = \mathbf{O}(2) \times \mathbf{S}_N$ acts absolutely irreducibly on $V = \mathbf{R}^2 \otimes_{\mathbf{R}} \mathbf{R}^{N-1}$. This action can be identified with that on $V \oplus V \equiv V \otimes_{\mathbf{R}} \mathbf{C}$, in such a way that $\exp(\frac{1}{\omega}\theta L) \in \mathbf{S}^1$ acts as scalar multiplication by $e^{i\theta}$. But

$$V \otimes \mathbf{C} \cong (\mathbf{R}^2 \otimes_{\mathbf{R}} \mathbf{R}^{N-1}) \otimes_{\mathbf{R}} \mathbf{C} \cong \mathbf{C}^2 \otimes_{\mathbf{C}} \mathbf{C}^{N-1}$$

Thus we have:

Proposition 10.9 *The action of $\mathbf{O}(2) \times \mathbf{S}_N \times \mathbf{S}^1$ on $\mathbf{R}^{2N-2} \times \mathbf{R}^{2N-2}$ can be identified with the action on $\mathbf{C}^2 \otimes_{\mathbf{C}} \mathbf{C}^{N-1}$, where*

(a) $\gamma \in \mathbf{O}(2)$ *acts by complexifying its action on $\mathbf{R}^2$.*

(b) $\sigma \in \mathbf{S}_N$ *acts by permuting a basis of $\mathbf{C}^N \supset \mathbf{C}^{N-1}$ where*

$$\mathbf{C}^{N-1} = \{z \in \mathbf{C}^N | \sum z_j = 0\}$$

(c) $\theta \in \mathbf{S}^1$ *acts as scalar multiplication by $e^{i\theta}$.* $\qquad\square$

Dionne *et al.* [163] develop a general theory of axial and **C**-axial subgroups of a direct product $\Gamma_1 \times \Gamma_2$, reducing their existence to properties of Γ_1 and Γ_2 alone. In particular any subgroup of $\Gamma_1 \times \Gamma_2 \mathbf{S}^1$ acting on $U \otimes_{\mathbf{R}} V \otimes_{\mathbf{R}} \mathbf{C}$ generated by a **C**-axial subgroup of Γ_1 acting on $U \otimes_{\mathbf{R}} \mathbf{C}$ and a **C**-axial subgroup of Γ_2 acting on $V \otimes_{\mathbf{R}} \mathbf{C}$ is itself $\mathbf{C} - axial$. Here we can take $\Gamma_1 = \mathbf{O}(2), \Gamma_2 = \mathbf{S}_N$. The **C**-axial subgroups of $\mathbf{O}(2)$ have been discussed in Example 4.12. The **C**-axial subgroups of $\mathbf{S}_N$ are classified in Stewart [490]:

Theorem 10.10 *Suppose that $N \geq 2$. Then the axes of $\mathbf{S}_N \times \mathbf{S}^1$ acting on $\mathbf{C}^N$ have orbit representatives as follows:*

Type I:

Let $N = qk + p$ where $2 \leq k \leq N, q \geq 1, p \geq 0$. Let $\zeta = e^{2\pi i/k}$, and set

$$z = (\underbrace{1, \ldots, 1}_{q}; \underbrace{\zeta, \ldots, \zeta}_{q}; \underbrace{\zeta^2, \ldots, \zeta^2}_{q}; \ldots; \underbrace{\zeta^{k-1}, \ldots, \zeta^{k-1}}_{q}; \underbrace{0, \ldots, 0}_{p})$$

Type II:

Let $N = q + p, 1 \leq q < \frac{N}{2}$, and set

$$z = (\underbrace{1, \ldots, 1}_{q}; \underbrace{a, \ldots, a}_{p})$$

where $a = -q/p$.   □

For illustrative purposes, consider the case $N = 3$. Because $S_3 \cong D_3$, this case reduces to a system with $D_3 \times O(2)$ symmetry, which is worked out in detail as an example in Dionne *et al.* [163].

Recall that the **C**-axial subgroups of $O(2)$ are A_1, A_2 given by

$$\begin{aligned} A_1 &= \widetilde{SO(2)} = \{(\theta, \theta) \in O(2) \times S^1\} \\ A_2 &= D_1^{(\kappa,0)} \times Z_2^{(\pi,\pi)} \end{aligned}$$

By Theorem 10.10 S_3 has precisely three **C**-axial subgroups, namely

$$\Sigma_{1,0}^{I} \quad \Sigma_{1,1}^{I} \quad \Sigma_1^{II}$$

Therefore by Dionne *et al.* [163] $S_3 \times O(2)$ has at least the following six **C**-axial subgroups:

$$\begin{array}{ccc} \Sigma_{1,0}^{I} \times A_1 & \Sigma_{1,1}^{I} \times A_1 & \Sigma_1^{II} \times A_1 \\ \Sigma_{1,0}^{I} \times A_2 & \Sigma_{1,1}^{I} \times A_2 & \Sigma_1^{II} \times A_2 \end{array}$$

The complete analysis of Dionne *et al.* [163] shows that there is precisely one additional **C**-axial subgroup, namely

$$\widetilde{D}_3 = \langle (\rho, R_{2\pi/3}), (\sigma, \kappa) \rangle$$

where $\rho = (123)$ is a 3-cycle, $\sigma = (12)$ is a transposition, $R_{2\pi/3}$ is rotation through an angle $2\pi/3$, and κ is a flip.

The Equivariant Moser-Weinstein Theorem therefore implies that near the trivial (and 'unphysical') equilibrium in which all bodes coincide, there are seven classes of periodic solution:

- *Collinear, 2 coalesce.* The bodies oscillate along a line where two have coalesced.
- *Collinear, 1 fixed.* Two of the bodies oscillate along a line through the third which remains stationary.
- *Collinear. Z_3* The bodies oscillate back and forth along a line separated by phase shifts of $2\pi/3$.
- *2 Fixed, 1 orbiting.* Two bodies coalesce and all bodies orbit a common center.
- *1 Fixed, 2 orbiting.* Two bodies orbit the third at opposite ends of a diameter.
- *Rotating Triangle.* The bodies travel in circular orbits at the vertices of an equilateral triangle.
- *Colliding Triangle.* The bodies remain in a triangular arrangement, each oscillating along a line.

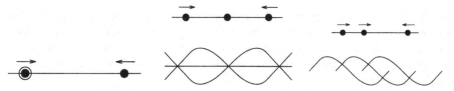

Figure 10.8: Collinear solutions.

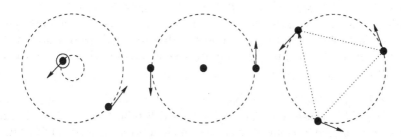

Figure 10.9: Rotating wave solutions.

Figure 10.10: Colliding triangle.

By perturbing these solutions to remove collisions or coincidences of particles, a number of physically realistic solutions may also be obtained. See Stewart [490] for details.

10.3 Spatio-Temporal Symmetries in Hamiltonian Systems

There is a Hamiltonian version of Theorem 3.4, the symmetry classification of periodic solutions of equivariant ODEs. The proof is very similar, but if anything simpler, and it leads to a slightly different result. For simplicity we restrict ourselves to phase spaces $\mathbf{R}^{2n}$ rather than general symplectic manifolds, but the generalization is immediate once the proof is described — see below.

Theorem 10.11 *Let Γ be a finite group acting symplectically on $\mathbf{R}^{2n}$. There is a periodic solution to some Γ-equivariant Hamiltonian system of ODE on $\mathbf{R}^{2n}$ with spatial symmetries K and spatio-temporal symmetries H if and only if*

    (a) *H/K is cyclic.*
    (b) *K is an isotropy subgroup.*
    (c) *$\dim \mathrm{Fix}(K) \geq 2$. If $\dim \mathrm{Fix}(K) = 2$, then either $H = K$ or $H = N(K)$.*

Moreover, when these conditions hold, there exists a smooth Γ-equivariant Hamiltonian vector field having an elliptic periodic solution with the desired symmetries.

Proof. The proof of this theorem is virtually identical to that of Theorem 3.4. As before, conditions (a)-(c) are necessary conditions. Note that condition (d) of Theorem 3.4 is superfluous in the Hamiltonian setting, since the symplectic structure implies that the codimension of $\text{Fix}(\gamma) \cap \text{Fix}(K)$ in $\text{Fix}(K)$ is at least two; hence the complement of L_K is always connected.

Conversely, choose the closed curve C with the desired symmetry properties, as in the proof of Theorem 3.4. Then choose a nonnegative Hamiltonian in a small neighborhood of C whose zero set is C. Extend the Hamiltonian to be Γ-invariant on all of $\mathbf{R}^{2n}$ in a way analogous to the construction of the vector field in the proof of Theorem 3.4. We can also assume that the Hamiltonian is chosen so that C is the trajectory of an elliptic periodic solution (see the next section for a discussion of stability).   □

The same proof works if the phase space of the system is an arbitrary symplectic manifold, because everything takes place in a tubular neighborhood of the periodic orbit; indeed the Hamiltonian can be assumed to be zero outside such a neighborhood, so the topology of the phase space is unimportant here.

10.4 Poincaré-Birkhoff Normal Form

We now consider linear stability in more detail. Given a Hamiltonian system we would like to do two things:

(1) Find all of the periodic solutions whose period is near that of the linearization.
(2) Determine the stability of these solutions.

We approach these problems by considering systems in Poincaré-Birkhoff normal form. The Hamiltonian analogue of the Normal Form Theorem says that for a Γ-invariant Hamiltonian $\mathcal{H}$, there exists a symplectic change of coordinates

$$\mathcal{H} = \underbrace{\mathcal{H}_k}_{\Gamma \times \mathbf{S}^1 - \text{invariant}} + \underbrace{O(k+1)}_{\Gamma - \text{invariant only}} .$$

If $\mathcal{H}$ is $\Gamma \times \mathbf{S}^1$-invariant (that is, $\mathcal{H}$ is in Poincaré-Birkhoff normal form to all orders) we have a simple version of the theory. Otherwise, k big enough implies that we get the same result using $\mathcal{H}_k$. (There exist singularity-theoretic tests for 'big enough'.)

In Poincaré-Birkhoff normal form all of the solutions are rotating waves, so in a rotating frame the problem reduces to a steady-state problem.

Assume $\mathcal{H}$ is fully resonant and is in Poincaré-Birkhoff normal form; thus the eigenvalues are $\pm k\alpha i$. Without loss of generality we can assume a scaling such that $\alpha = 1$. Then

$$\theta v = e^{\theta L} v$$

Let $\mathcal{H}_2$ be the quadratic part of $\mathcal{H}$ and τ be the period perturbation parameter ($T = 2\pi/(1+\tau)$). Then

$$\mathcal{H}_\tau = 2\pi[(1+\tau)\mathcal{H}_2 - \mathcal{H}]$$

and $\mathcal{H}_\tau : \mathbf{R}^{2n} \times \{\tau\} \to \mathbf{R}^n$.

Lemma 10.12 *The periodic trajectories near 0 with period near 2π are the $\mathbf{S}^1$-orbits of critical points of $\mathcal{H}_\tau$.*

Proof. Suppose z is a critical point of $\mathcal{H}_\tau$. then $D\mathcal{H}_\tau(0) = 0$ and so

$$JD\mathcal{H}(z) = (1+\tau)JD\mathcal{H}_2(z) \qquad J = \begin{bmatrix} 0 & -I \\ I & 0 \end{bmatrix}$$

So at the critical point the Hamiltonian vector field points in the same direction as the linear vector field. Now $\mathbf{S}^1$-equivariance implies that the $\mathbf{S}^1$-orbit of the initial point is a periodic solution, indeed a relative equilibrium. $\qquad\square$

10.5 Linear Stability

Classically, stability of a periodic state $v(t)$ is defined in terms of the Floquet operator, the linearization of a Poincaré map. Here the Floquet operator takes the form

$$M_v = \exp\left(\frac{2\pi}{1+\tau}D^2\mathcal{H}_\tau(v(0))\right)$$

For stability (at the linear level) we want the eigenvalues of M_v to be on the unit circle. Because of the symplectic structure, the eigenvalues come in quartets $\mu, \bar{\mu}, \mu^{-1}$ and $\bar{\mu}^{-1}$, as pictured in Figure 10.11. So we have the criterion that v is linearly stable if and only if the linear part

$$v \mapsto JD^2\mathcal{H}_\tau(v)$$

of the $\mathcal{H}_\tau$ flow at $v(0)$, has purely imaginary eigenvalues.

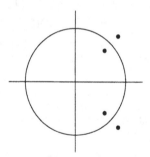

Figure 10.11: A quartet of eigenvalues.

O(3) Symmetry. Consider a liquid drop in space: this system has $\mathbf{O}(3)$ symmetry. For the purposes of this example we consider a single finite-dimensional space of 'vibrational modes' and assume the problem can be reduced onto this space. It is not clear how to achieve this reduction rigorously. In fact, we shall assume that $\mathbf{O}(3)$ acts on its irreducible 5-dimensional representation on spherical harmonics V_2 and $\mathbf{S}^1$ acts on multiplication by $e^{i\theta}$. This yields an action of $\mathbf{O}(3) \times \mathbf{S}^1$ on a 10-dimensional phase space. We can realize the representation in a concrete and computable way. Let V be the space of 3×3 complex, trace zero, symmetric matrices M and let $\mathbf{O}(3)$ act by similarity:

$$\gamma M = \gamma^{-1}M\gamma$$

Let $\mathbf{S}^1$ act by complex multiplication

$$\theta M = e^{i\theta}M$$

Both $-I \in \mathbf{O}(3)$ and $\pi \in \mathbf{S}^1$ act on V as minus the identity. Bearing this in mind, we can replace $\mathbf{O}(3)$ by $\mathbf{SO}(3)$, which we do from now on.

Isotropy Subgroup Σ	Generators in $\mathbf{O}(3) \times \mathbf{S}^1$
$\mathbf{O}(2)$	$(R_\theta, 1), (\kappa, 1)$
$\widetilde{\mathbf{SO}(2)}^1$	$(R_\theta, e^{i\theta})$
$\widetilde{\mathbf{SO}(2)}^2$	$(R_\theta, e^{2i\theta})$
$\tilde{\mathbf{D}}_4$	$(R_{\pi/2}, -1), (\kappa, 1)$
$\tilde{\mathbf{T}}$	$(\rho, 1), (\zeta, e^{2\pi i/3})$

Table 10.1 C-axial subgroups of $\mathbf{O}(3) \times \mathbf{S}^1$.

Group Theory. We define the following elements of $\mathbf{SO}(3) \times \mathbf{S}^1$

$$\kappa = \begin{bmatrix} 1 & 0 & 0 \\ 0 & -1 & 0 \\ 0 & 0 & -1 \end{bmatrix} \quad \rho = \begin{bmatrix} -1 & 0 & 0 \\ 0 & -1 & 0 \\ 0 & 0 & 1 \end{bmatrix} \quad \zeta = \begin{bmatrix} 0 & 0 & 1 \\ 1 & 0 & 0 \\ 0 & 1 & 0 \end{bmatrix}$$

$$R_\theta = \begin{bmatrix} \cos\theta & -\sin\theta & 0 \\ \sin\theta & \cos\theta & 0 \\ 0 & 0 & 1 \end{bmatrix}$$

With this notation, the C-axial subgroups of $\mathbf{SO}(3) \times \mathbf{S}^1$ are given in Table 10.5.

Theorem 10.3 implies that there exist periodic solutions with symmetry Σ. The isotropy lattice for $\mathbf{SO}(3) \times \mathbf{S}^1$ on V is given in Figure 10.12.

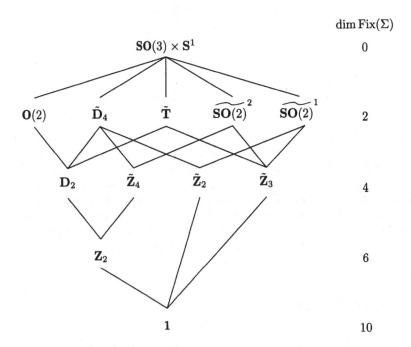

Figure 10.12: Isotropy lattice for $\mathbf{SO}(3) \times \mathbf{S}^1$.

Degree	Invariants
2	$N = \mathrm{tr}(M, M)$
4	$P = (\mathrm{tr}\, M^2)(\mathrm{tr}\, M^2)$
	$Q = \mathrm{tr}(M^2 \bar{M}^2)$
	N^2
6	$R = (\mathrm{tr}\, M^3)(\mathrm{tr}\, M^3)$
	$S = (\mathrm{tr}\, M^2 \bar{M})(\mathrm{tr}\, M \bar{M}^2)$
	N^3
	NP
	NQ

Table 10.2 Invariants for $SO(3)$ up to order 8.

Isotropy	Stability	Floquet Exponent	Multiplicity
$O(2)$	s/u	0	6
		$\pm|u|^2\sqrt{-(24B+4C)(G+H)}$	2
$\tilde{D}_4$	u/s	0	8
		$\pm|u|^2\sqrt{(24B+4C)(G+H)}$	1
$\widetilde{SO(2)}^1$	s	0	4
		$\pm i|u|^2[\tfrac{1}{3}C]$	1
		$\pm i|u|^2[4B+C]$	1
		$\pm i|u|^2[\tfrac{1}{2}C]$	1
$\widetilde{SO(2)}^2$	s	0	4
		$\pm i|u|^2[\tfrac{2}{3}C]$	1
		$\pm i|u|^2[4B+2C]$	1
		$\pm i|u|^2[C]$	1
$\tilde{T}$	s	0	8
		$\pm i|u|^2[4B+\tfrac{2}{3}C]$	1

Table 10.3 Floquet exponents for action of $SO(3)$.

The invariants for this system up to order 8 are listed in Table 10.2. Thus the general $SO(3)$ Hamiltonian (normalized at degree 2) is

$$\mathcal{H} = N + AN^2 + BP + CQ + DN^3 + ENP + FNQ + GR + HS + O(8)$$

We may now compute the Floquet exponents (Montaldi *et al.* [408]) and they are listed in Table 10.3.

At this point it is reasonable to ask why coefficients of sixth order terms appear, and why there is an exchange of stability for $O(2)$ and $\tilde{D}_4$. We can understand these features by considering the submaximal isotropy subgroup D_2 contained in $O(2), \tilde{D}_4$ and $\tilde{T}$. The space $\mathrm{Fix}(D_2)$ is invariant under the flow. On it, we get invariance under $N(D_2)/D_2 \cong D_3$. The important eigenvalues live on this space. For D_3 we know that there exist three maximal isotropy subgroups $\tilde{Z}_3, Z_2$ and $\tilde{Z}_2$ as in the Hénon-Heiles Hamiltonian. Moreover, there is an exchange of stability principle for Z_2 and $\tilde{Z}_2$. So all we need do is match up the subgroups:

The Spring Pendulum. The spring pendulum is a planar coupled oscillator that combines features of the usual rigid-rod pendulum and an oscillating spring. See

$$\tilde{\mathbf{T}} \quad \longleftrightarrow \quad \tilde{\mathbf{Z}}_3 \qquad \text{s}$$

$$
\left.
\begin{array}{ccc}
\mathbf{O}(2) & \longleftrightarrow & \mathbf{Z}_2 \\[2ex]
\tilde{\mathbf{D}}_4 & \longleftrightarrow & \tilde{\mathbf{Z}}_2
\end{array}
\right\} \text{s/u}
\quad
\begin{array}{l}
\text{stability determined} \\
\text{at order 6 in BNF} \\
\text{of } \mathcal{H}
\end{array}
$$

Figure 10.13: Correspondence between isotropy subgroups.

Figure 10.14. In fact, it is a pendulum whose rod can stretch or compress (but only in the direction of the rod — no bending). It possesses both a 'spring mode', in which it behaves like a spring, and a 'pendulum mode', in which it behaves like a pendulum. When these modes interact (which in the Hamiltonian case is usually termed a resonance rather than a mode interaction) it turns out that the nonlinear interactions give rise to either 2, 4 or 6 families of periodic solutions. Moser-Weinstein theory gives only two of these, but the Equivariant Moser-Weinstein Theorem gives them all (Montaldi *et al.* [408]).

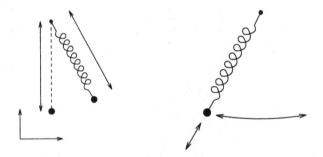

Figure 10.14: The spring pendulum and its two primary modes.

First we abstract the system. There is a spatial $\mathbf{Z}_2$ reflection symmetry κ, a time-reversal symmetry ρ, and a circle group symmetry $\mathbf{S}^1$ from the Birkhoff normal form. Thus the symmetry group Γ acting on the Hamiltonian $\mathcal{H}$ on $\mathbf{C}^2 = \{(z_1, z_2)\}$ is $\Gamma = \mathbf{Z}_2 \times \mathbf{O}(2)$ with actions:

$$
\begin{array}{rcl}
\kappa(z_1, z_2) & = & (z_2, z_1) \\
\rho(z_1, z_2) & = & (-\bar{z}_1, -\bar{z}_2) \\
\theta(z_1, z_1) & = & (e^{i\theta} z_1, e^{i\theta} z_2)
\end{array}
$$

The isotropy lattice and corresponding fixed-point subspaces for Γ are given in Figure 10.15.

The forms of the solutions which correspond with the isotropy subgroups (excluding the trivial solution) are shown in Figure 10.16. The corresponding isotropy subgroups are listed in Table 10.4.

We briefly describe how to find the conditions for these solutions to exist, and their linear stabilities. Using the action defined above, generators for the invariants

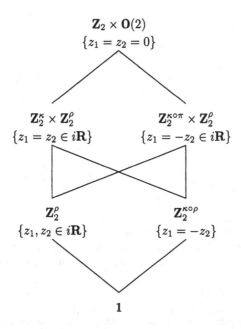

Figure 10.15: Isotropy lattice for $\mathbf{Z}_2 \times \mathbf{O}(2)$.

Isotropy subgroup	solution type
$Z_2^\kappa \times Z_2^\rho$	spring mode
$Z_2^{\kappa\pi} \times Z_2^\rho$	pendulum mode
Z_2^ρ	brake mode
$Z_2^{\kappa\rho}$	symmetric ellipse mode
1	asymmetric ellipse mode

Table 10.4 Isotropy subgroups of $\mathbf{Z}_2 \times \mathbf{O}(2)$ and corresponding solution types.

are:
$$N = |z_1|^2 + |z_2|^2$$
$$P = |z_1|^2 |z_2|^2$$
$$Q = \mathrm{Re}(z_1, \bar{z}_2)$$

Thus the general Hamiltonian to degree four is
$$\mathcal{H} = \mathcal{H}_2 + \mathcal{H}_4,$$
where
$$\mathcal{H}_2 = N$$
$$\mathcal{H}_4 = \alpha_1 N^2 + \alpha_2 P + \beta_1 Q^2 + \beta_2 N Q$$

The Hamiltonian is $\mathbf{U}(2)$-symmetric if $\alpha_2 = \beta_2 = \beta_2 = 0$ (where $\mathbf{U}(2)$ is the unitary group on $\mathbf{C}^2$), $\mathbf{O}(2)$-symmetric if $\beta_1 = \beta_2 = 0$, and it is $\mathbf{D}_4$-symmetric if $\beta_2 = 0$. This suggests that order four is the 'right' place to truncate.

We can determine conditions on the coefficients $\alpha_1, \alpha_2, \beta_1$ and β_2 for the existence and stability of the five solution types shown in Table 10.5.

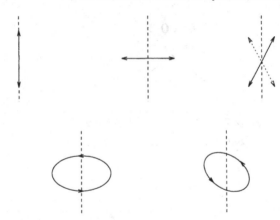

Figure 10.16: Periodic oscillations of the spring pendulum. (Upper left) spring; (upper middle) pendulum; (upper right) brake (pendulum and spring modes in 1:1 resonance); (lower left) symmetric ellipse; (lower right) asymmetric ellipse.

Mode	# of Solutions	Exists	Stable
Spring	1	Yes	$(\beta_1 + \beta_2)(\alpha_2 + \beta_1 + \beta_2) > 0$
Pendulum	1	Yes	$(\beta_1 - \beta_2)(\alpha_2 + \beta_1 - \beta_2) > 0$
Brake	2	$|\beta_2| < |\alpha_2 + \beta_1|$	$\alpha_2(\alpha_2 + \beta_1) > 0$
Symmetric Ellipse	2	$|\beta_2| < |\beta_1|$	$\alpha_2\beta_1 > 0$
Asymmetric Ellipse	None		

Table 10.5 Existence and stability of solutions in spring pendulum.

If we perform Poincaré-Birkhoff normal form reduction to the above form on the general Hamiltonian

$$\mathcal{H}(p_1, p_2, q_1, q_2) = \frac{1}{2}(p_1^2 + p_2^2) + V(q_1, q_2)$$

where V is $\mathbf{Z}_2$-invariant and is of the form

$$V(q_1, q_2) = \frac{1}{2}(q_1^2 + q_2^2) + a_1 q_1^3 + a_2 q_1 q_2^2 + (b_1 q_1^4 + b_2 q_1^2 q_2^2 + b_3 q_2^4) + \cdots$$

then we obtain

$$\alpha_1 = -\frac{5}{4}\left(\frac{3}{4}a_1^2 + \frac{1}{2}a_1 a_2 + \frac{5}{12}a_2^2\right) + \frac{3}{8}(b_1 + b_2 + b_3)$$
$$\alpha_2 = -(a_1 a_2 - 2a_2^2) = b_2$$
$$\beta_1 = -\frac{1}{4}(15a_1^2 - 14a_1 a_2 + 3a_2^2) + \frac{1}{2}(3b_1 - b_2 + 3b_3)$$
$$\beta_2 = -\frac{5}{4}(3a_1^2 - \frac{1}{3}a_2^2) + \frac{3}{2}(b_1 - b_3)$$

We now specialize this Hamiltonian to the spring pendulum using the variables shown in Figure 10.14. The Hamiltonian is

$$\mathcal{H} = \frac{1}{2m}(p_1^2 + p_2^2) + mgq_1 + \sigma(l)$$

# of Solution Families	Region	Spring (1 solution)	Pendulum (1 solution)	Brake (2 solutions)	Symmetric Ellipse (2 solutions)
2	A	stable	stable		
4	B	unstable	stable	stable	
6	C	stable	stable	stable	unstable

Table 10.6

where $l = \sqrt{(l_0 - q_1)^2 + q_2^2}$ and $\sigma(l)$ is the energy in the spring, $\sigma'(1) = 1$ (for equilibrium at the origin). If we let $c = \sigma''(1), d = \sigma'''(1)$ and $f = \sigma^{iv}(1)$ then the Taylor coefficients are:

$$a_1 = -\frac{d}{6} \quad a_2 = 0 \quad b_1 = \frac{f}{24} \quad b_2 = \frac{d}{4} \quad b_3 = 0$$

and the normal form coefficients are:

$$\alpha_1 = \tfrac{1}{192}(3f - 5d^2 + 18d)$$
$$\alpha_2 = -\frac{d}{4}$$
$$\beta_1 = \tfrac{1}{48}(3f - 5d^2 - 6d)$$
$$\beta_2 = \tfrac{1}{48}(3f - 5d^2)$$

The non-degeneracy conditions for this system are

$$\left(\frac{3f}{d} - 5d\right) \neq \infty, 3, 6, 9, 18$$

The Poincaré-Birkhoff normal form analysis of this system is due to Kummer [334].

   It turns out that asymmetric ellipses do not occur as solutions. The existence and stability of the other four types of periodic solution is as shown in Figure 10.17 and Table 10.5.

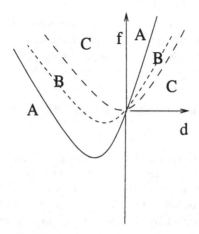

Figure 10.17: Parameter space for the spring pendulum showing regions of existence and stability of various oscillatory modes.

10.6 Molecular Vibrations

We end this chapter with a discussion of the application of symmetric Hamiltonian systems to molecular vibrational spectra, due to Montaldi and Roberts [404]. The traditional modeling assumptions in this area are to treat the molecule as a system of point particles — atomic nuclei and electrons — interacting though classical forces. Such models are hard to analyze: for example, even a molecule as simples as water (H_2O has three nuclei, ten electrons, and thus a 39-dimensional phase space. The Born-Oppenheimer approximation reduces the model to one of the nuclei alone, interacting with each other through a potential energy function that incorporates the effect of the electrons. For water, this reduces the dimension of the phase space to 9.

A further simplification is the 'classical' approximation, which decouples the molecule's vibrational motions from rigid rotations. The mechanics of rigid bodies shows that if the three principal moments of inertia of the body are all distinct, then there are six motions in which the body rotates about a fixed axis (Abraham and Marsden [1]): rotations in each of the two senses (clockwise or counterclockwise) about any of the three principal axes of inertia. Such motions are *relative equilibria* — the time-evolution of the configuration remains within the orbit of the initial configuration under the action of the symmetry group of the system. Here this means that in a suitably corotating frame the configuration is an equilibrium.

Montaldi and Roberts study relative equilibria that bifurcate from a genuine equilibrium as the total angular momentum of the molecule increases from zero. Suppose that S is the group of permutations of identical nuclei (a direct product of symmetric groups, one for each type of nucleus): then the overall symmetry group of the equations that determine the configurations of the molecule is $O(3) \times S$, and the symmetry group of a given configuration is some subgroup Γ. The projection Γ^* of Γ into $O(3)$ is a group of rigid motions in $\mathbf{R}^3$. Each rotation in Γ^* has an axis, which we call an *axis of rotation* of the molecule, and each reflection in Γ^* has an axis — the line perpendicular to its fixed-point space, which is a plane — which we call an *axis of reflection* of the molecule. The main result, proved for the full Born-Oppenheimer model, is:

Theorem 10.13 *Consider a molecule having a nondegenerate equilibrium with symmetry $\Gamma \subseteq O(3) \times S$. For all sufficiently small nonzero $\mu \in \mathbf{R}^3$ there exist at least 6 relative equilibria with angular momentum μ. Indeed, for each axis α of rotation or reflection in Γ^* there exist at least two relative equilibria with angular momentum μ and dynamical axis α, one rotating in each sense.*

The isotropy subgroups of these relative equilibria are maximal isotropy subgroups of Γ.

The Model and Its Symmetries. Before applying Theorem 10.13 to examples, we describe the model equations and list their symmetries. The set-up is similar to that for the many-body model in Section 10.2, but the bodies now lie in $\mathbf{R}^3$ and need not be identical. We follow the notation of Montaldi and Roberts [404].

The molecule is modeled by N atomic nuclei in $\mathbf{R}^3$, which are thought of as point masses. The electrons are absorbed into a potential and their configuration is not included explicitly in the model. Configuration space is $\mathbf{R}^{3N}$, and this is identified with

$$\mathcal{C} = \mathbf{R}^N \otimes \mathbf{R}^3 = L(N, 3)$$

where $L(N, 3)$ is the space of real $3 \times N$ matrices. The ith column of a configuration matrix Q represents the position q_i of the ith atom. The phase space of the system is therefore the cotangent bundle $\mathcal{P} = \mathrm{T}^*\mathcal{C} \cong \mathbf{R}^{6N}$, which can be identified with pairs of matrices (P, Q) in $L(3, N)$. The columns p_i of P are the momenta of the nuclei.

Assuming that nucleus i has mass m_i, the Hamiltonian of the molecule is

$$H(P, Q) = \sum_{i=1}^{N} |p_i|^2 + V(q_1, \ldots, q_N) = \tfrac{1}{2} \operatorname{tr}(PM^{-1}P^T) + V(Q)$$

where

$$M = \operatorname{diag}[M_1, \ldots, m_N]$$

is the mass matrix. For any motion $Q(t)$, the corresponding momentum is given by $P = \dot{Q}M$. If there are no external forces acting on the molecule, then its center of mass moves in an inertial frame, which can be assumed to be fixed and at the origin — that is, total momentum can be set to zero. Configuration space becomes

$$\mathcal{C}_0 = L_0(N, 3) = \{Q \in L(N, 3) : \sum_j q_{ij} = 0\}$$

and phase space becomes

$$\mathcal{P}_0 = \mathrm{T}^* L_0(N, 3) \cong L_0(N, 3) \oplus L_0(N, 3)$$

There are three kinds of symmetry: Euclidean motions, relabeling of particles, and time-reversal. The Euclidean translations have already been removed by fixing the center of mass, so $\mathbf{O}(3)$ remains. View $A \in \mathbf{O}(3)$ as a 3×3 orthogonal matrix: then its action on $\mathcal{C}_0$ is given by left multiplication AQ. Since there are no external forces, the potential V is $\mathbf{O}(3)$-invariant.

Relabeling symmetries are given by a subgroup $\Sigma \subseteq \mathbf{S}_N$. A permutation $\sigma \in \Sigma$ if and only if nuclei i and $\sigma(i)$ are identical for all i. Thus $\sigma \in \Sigma$ if and only if

$$V(q_{\sigma(1)}, \ldots, q_{\sigma(N)}) = V(q_1, \ldots, q_N) \qquad m_{\sigma(i)} = m_i$$

for all i and for all $(q_1, \ldots, q_N) \in \mathcal{C}_0$. We can identify σ with an $N \times N$ permutation matrix, acting on $\mathcal{C}_0$ by right multiplication $Q\sigma^T$. The matrix σ^T commutes with M. Thus overall we obtain an action of $\mathbf{O}(3) \times \Sigma$ on $\mathcal{C}_0$ given by

$$(A, \sigma)Q = AQ\sigma^T$$

and V is invariant under $\mathbf{O}(3) \times \Sigma$. The induced action on P is the same, $(A, \sigma)P = AP\sigma^T$. Therefore

$$H((A, \sigma)(P, Q)) = \tfrac{1}{2} \operatorname{tr}(AP\sigma^T M^{-1} \sigma P^T A^T) + V(AQ\sigma^T) = H(P, Q)$$

since M and σ commute. It follows that H is $\mathbf{O}(3) \times \Sigma$-invariant.

Remark 10.14 The relabeling symmetry group Σ is, in general, different from the obvious 'symmetry group' of the molecule — its Euclidean symmetries in a static configuration. For example, buckminsterfullerene has 60 identical carbon atoms, so $\Sigma = \mathbf{S}_{60}$, but its Euclidean symmetries in a static configuration form the icosahedral group. In fact, the obvious symmetries are those in the isotropy subgroup of the molecule, as we show below. $\diamond$

The Hamiltonian of the molecule is in 'kinetic + potential' form, and therefore is invariant under the time-reversal symmetry

$$\tau(P, Q) = (-P, Q)$$

This generates a group $\mathbf{Z}_2^\tau$, and when this is included the symmetry group of the system becomes $\mathbf{O}(3) \times \Sigma \times \mathbf{Z}_2^\tau$. Note that $\mathbf{Z}_2^\tau$ acts anti-symplectically, not symplectically: it reverses the sign of the symplectic form. Because of the $\mathbf{SO}(3)$ symmetry, angular momentum is conserved.

The Euclidean symmetries of the molecule in a static configuration are a combination of Euclidean rotations and reflections, and relabelings of identical molecules. They therefore correspond to the *pairs* $(A, \sigma) \in \mathbf{O}(3) \times \Sigma$ that map the configuration Q to itself: this is clearly the isotropy group

$$\Sigma_Q = \{(A, \sigma) : AQ\sigma^T = Q\}$$

In normal parlance, this group is identified with its projection into $\mathbf{O}(3)$. However, this identification ignores the labels on nuclei.

Applications to Molecules.

Planar Molecules. Suppose that the molecule lies in a plane, as is always the case, for example, if there are three atoms, Figure 10.18(left). Its isotropy subgroup contains the reflection r_s in that plane, and the configuration has one axis of reflection, orthogonal to the plane. If the atoms are all distinct, this is the only symmetry. Theorem 10.13 implies that there are two families of bifurcating relative equilibria with dynamical axis equal to the reflection axis, plus at least four further families. Montaldi and Roberts [404] show the generically there are precisely four more families overall, whose dynamical axes are close to the other two principal axes of inertia.

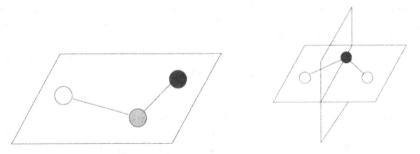

Figure 10.18: (Left) planar molecule; (right) water molecule.

Water. The water molecule H_2O is planar but non-collinear, and has a second reflectional symmetry in a plane orthogonal to the one in which it lies, Figure 10.18 (right). This symmetry interchanges the two hydrogen nuclei and fixed the oxygen nucleus. Thus there are two reflection axes. There is also a rotational symmetry, obtained by composing the two reflections, and the axis of this rotation is orthogonal to the other two axes. Now Theorem 10.13 implies that there are six families of bifurcating relative equilibria. Four of them have dynamical axes equal to the corresponding reflection axes.

Ionized Hydrogen. Ionized hydrogen H_3^+ is a planar triatomic molecule with isotropy subgroup $\mathbf{D}_3$. It has the reflection symmetry r_s, together with three more reflections which flip the equilateral triangle formed by the nuclei about each of its symmetry axes, Figure 10.19(left). Composing any of these with r_s gives a rotational symmetry of order 2. There is also a rotational symmetry of order 3 which cycles the three nuclei. For low angular momentum, there will be at least 14 families of bifurcating relative equilibria. It turns out that generically there are no more than this.

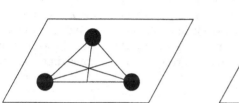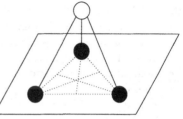

Figure 10.19: (Left) ionized hydrogen molecule; (right) ammonia molecule.

Ammonia. Ammonia NH_3 has $\mathbf{D}_3$ isotropy, but it is not planar. In equilibrium, the three hydrogen nuclei form an equilateral triangle, and the nitrogen nucleus is equidistant from the vertices of this triangle but lies out of the plane of the triangle, Figure 10.19(right). Thus the analysis is similar to that for H_3^+, but reflection in the plane is missing. Theorem 10.13 implies that there are at least eight families of bifurcating relative equilibria. These are *not* the only relative equilibria of ammonia near equilibrium: Montaldi and Roberts [404] Example 2.5 proves that there are at least six more, a total of at least 14. They are related to certain submaximal isotropy subgroups, and are non-geometric in that their precise location depends on the inter-atomic forces.

Ammonia has a second equilibrium state, the reflection of the one described in the plane of the equilateral triangle. These two states are separated by a potential barrier — an external force is required to flip the molecule between them — but this barrier is low, leading to an 'inversion flip' in ammonia that is well known to chemists. Between the two equilibria there is an unstable equilibrium configuration with $\mathbf{D}_3 \times \mathbf{Z}_2$ symmetry. Each of these three equilibria will spawn 14 families of relative equilibria. The local analysis of Theorem 10.13 is therefore *very* local — other nearby equilibria will interact with any particular one, and the analysis is unlikely to extend to high angular momentum.

In other molecules with the symmetry of ammonia, such as CHD_3 (methane with three hydrogens replaced by deuterium) the energy barrier is higher, and the neighborhood in which the analysis holds will be larger, so that the momentum levels to which the bifurcating relative equilibria persist will be higher.

Methane. The methane molecule CH_4 has tetrahedral symmetry Γ^*. The tetrahedron has

- Three axes of 2-fold rotational symmetry, joining the midpoints of opposite edges
- Four axes of 3-fold rotational symmetry, joining a vertex to the midpoint of the opposite face
- Six axes of reflection passing through the carbon nucleus (the center) and containing an edge of the tetrahedron

making a total of 13 axes. Theorem 10.13 implies that there exist at least 26 families of relative equilibria, two for each of the 13 axes. The same goes for any other molecule with tetrahedral symmetry, for example white phosphorus, P_4.

A similar symmetry analysis holds for molecules with cubic (octahedral) symmetry, such as sulphur hexafluoride SF_6, or for icosahedral symmetry, such as C_{60}.

Stability. Which (if any) of these relative equilibria are stable depends on the molecule in question. Montaldi and Roberts describe methods for computing linear stabilities. They also develop methods for finding relative equilibria with submaximal isotropy. For example in ammonia (NH_3), which has $\mathbf{D}_3 \times \mathbf{Z}_2$ symmetry — the $\mathbf{Z}_2$ is generated by reflection in the plane of the molecule and introduces an extra axis — there are 8 relative equilibria given by the above theorem, hence having maximal isotropy subgroups, and a further six with submaximal isotropy. Another molecule with the same symmetry as ammonia is ionized hydrogen, H_3^+, which occurs in interstellar space and the atmospheres of gas giant planets.

Bibliography

[1] R. Abraham and J.E. Marsden. *Foundations of Mechanics*, Benjamin/Cummings, New York 1978.

[2] G. Ahlers and K.M.S. Bajaj. Rayleigh-Bénard convection with rotation at small Prandtl numbers, in *Pattern Formation in Continuous and coupled Systems* (M.Golubitsky, D.Luss, S.H.Strogatz eds.), IMA volumes in Math. and Appns. **115** Springer-Verlag, New York 1999, 1–9.

[3] J.C. Alexander, I. Kan, J.A. Yorke, and Z. You. Riddled Basins, *Internat. J. Bif. Chaos* **2** (1992) 795–813.

[4] E. Allgower, K. Böhmer and M. Golubitsky. *Bifurcation and Symmetry*, ISNM **104**, Birkhäuser, Basal, 1992.

[5] D.C.D. Andereck, S.S. Liu, and H.L. Swinney. Flow regimes in a circular Couette system with independently rotating cylinders. *J. Fluid Mech.* **164** (1986) 155–183.

[6] I.S. Aranson, L. Aranson, L. Kramer, and A. Weber. Stability limits of spirals and traveling waves in nonequilibrium media, *Phys. Rev. A* **46** (1992) R2992–2995.

[7] D. Armbruster and P. Chossat. Remarks on multi-frequency oscillations in (almost) symmetrically coupled oscillators. *Phys. Lett. A* **254** (1999) 269–274.

[8] D. Armbruster and G. Dangelmayr. Coupled stationary bifurcations in nonflux boundary value problems, *Math. Proc. Camb. Phil. Soc.* **101** (1987) 167–192.

[9] D. Armbruster, J. Guckenheimer, and P. Holmes. Heteroclinic cycles and modulated travelling waves in system with O(2) symmetry, *Physica D* **29** (1988) 257–282.

[10] D.G. Aronson, M. Golubitsky, and M. Krupa. Coupled arrays of Josephson junctions and bifurcation of maps with S_N symmetry, *Nonlinearity* **4** (1991) 861–902.

[11] D.K. Arrowsmith and C.M. Place. *An Introduction to Dynamical Systems*, Cambridge U Press, Cambridge 1990.

[12] W. Arthur. *Mechanisms of Morphological Evolution*, Wiley, Chichester 1984.

[13] P. Ashwin. Symmetric chaos in systems of three and four forced oscillators, *Nonlinearity* **3** (1990) 603–617.

[14] P. Ashwin. High corank steady-state mode interactions on a rectangle, in *Bifurcation and Symmetry* (E. Allgower, K. Böhmer, and M. Golubitsky), ISNM **104** Birkhäuser Verlag, Basel, 1992, 23–33.

[15] P. Ashwin. High corank steady-state mode interactions on a rectangle, *Internat. Ser. Numerical Mathematics* **104**, Birkhäuser, Basel 1992, 23–33.

[16] P. Ashwin. Attractors stuck on to invariant subspaces, *Phys. Lett A* **209** (1995) 338–344.

[17] P. Ashwin. Chaotic intermittency of patterns in symmetric systems, in *Pattern Formation in Continuous and coupled Systems* (M.Golubitsky, D.Luss, S.H.Strogatz eds.), IMA volumes in Math. and Appns. **115** Springer-Verlag, New York 1999, 11–24.

[18] P. Ashwin, P. Aston, and M. Nicol. On the unfolding of a blowout bifurcation, *Physica D* **111** (1998) 81–95.

[19] P. Ashwin, J. Buescu, and I. Stewart. Bubbling of attractors and synchronisation of oscillators, *Phys. Lett. A* **193** (1994) 126–139.

[20] P. Ashwin, J. Buescu, and I. Stewart. From attractor to chaotic Saddle: a tale of transverse instability, *Nonlinearity* **9** (1996) 703–737.

[21] P. Ashwin and P. Chossat. Attractors for robust heteroclinic cycles with continua of connections, *J. Nonlin. Sci.* **8** (1998) 103–129.

[22] P. Ashwin, P. Chossat, and I. Stewart. Transitivity of orbits of maps symmetric under compact Lie groups, *Chaos, Solitons and Fractals* **4** (1994) 621–634.

[23] P. Ashwin, G. Dangelmayr, I. Stewart, and M. Wegelin. Oscillator networks with the symmetry of the unit quaternion group, in *Dynamics, Bifurcation, and Symmetry* (P.Chossat ed.), Proceedings, Cargèse 1993, NATO ASI Series C **437** Kluwer, Dordrecht 1994, 35–48.

[24] P. Ashwin, G.P. King, and J.W. Swift. Three identical oscillators with symmetric coupling, *Nonlinearity* **3** (1990) 585–601.

[25] P. Ashwin and I. Melbourne. Symmetry groups of attractors, *Arch. Rational Mech. Anal.* **126** (1994) 59–78.

[26] P. Ashwin and I. Melbourne. Noncompact drift for relative equilibria and relative periodic orbits. *Nonlinearity* **10** (1997) 595–616.

[27] P. Ashwin, I. Melbourne, and M. Nicol. Drift bifurcations of relative equilibria and transitions of spiral waves. *Nonlinearity* **12** (1999) 741–755.

[28] P. Ashwin, I. Melbourne, and M. Nicol. Drift for Euclidean extensions of dynamical systems. In: *Equadiff 99, International Conference on Differential Equations*, (B. Fiedler et al., eds.) World Scientific, Singapore, 2000, 145–150.

[29] P. Ashwin, I. Melbourne, and M. Nicol. Hypermeander of spirals; local bifurcations and statistical properties. *Physica D* **156** (2001) 364–382.

[30] P. Ashwin and J. Montaldi. Group theoretic conditions for existence of robust relative homoclinic trajectories. *Math. Proc. Camb. Phil. Soc.* (2002). To appear.

[31] P. Ashwin, I. Moroz, and M. Roberts. Bifurcations of stationary, standing, and travelling waves in triply diffusive convection, *Physica D* **81** (1995) 374–397.

[32] P. Ashwin and A. Rucklidge. Cycling chaos: its creation, persistence, and loss of stability in a model of nonlinear magnetoconvection, *Physica D* **122** (1998) 134–154.

[33] P. Ashwin and P. Stork. Permissible symmetries of coupled cell networks, *Math. Proc. Camb. Phil. Soc.* **116** (1994) 27–36.

[34] E. Av-Ron, H. Parnas, and L.A. Segel. A basic biophysical model for bursting neurons, *Biol. Cybernetics* **69** (1993) 87–95.

[35] G. Auchmuty. Bifurcation analysis of reaction-diffusion equations V. Rotating waves on a disk. In: *Partial Differential Equations and Dynamical Systems* (W.E. Fitzgibbon III, ed.) Res. Notes Math. **101** Pitman, London 1984, 35–63.

[36] A.K. Bajah and P.R. Sethna. Flow induced bifurcations to three dimensional oscillatory motions in continuous tubes, *SIAM J. Appl. Math.* **44** (1984) 270–286.

[37] T. Bal, F. Nagy, and M. Moulins. Muscarinic modulation of a pattern-generating network: control of neuronal properties, *J. Neurosci.* **14** (1994), 3019–3035.

[38] C.R. Bantock and J.A. Bayley. Visual selection for shell size in *Cepea* (Held), *J. Anim. Ecol.* **42** (1973) 247–261.

[39] C.R. Bantock, J.A. Bayley, and P.H. Harvey. Simultaneous selective predation on two features of a mixed sibling species population, *Evolution* **29** (1975) 636–649.

[40] M. Bär and M. Or-Guil. Alternative scenarios of spiral breakup in a reaction-diffusion model with excitable and oscillatory dynamics, *Phys. Rev. Lett.* **82** (1999) 1160–1163.

[41] E. Barany, M. Dellnitz, and M. Golubitsky. Detecting the symmetry of attractors, *Physica D.* **67** (1993) 66–87.

[42] D. Barkley. A model for fast computer-simulation of waves in excitable media, *Physica D* **49** (1991) 61–70.

[43] D. Barkley. Linear stability analysis of rotating spiral waves in excitable media, *Phys. Rev. Lett.* **68** (1992) 2090–2093.

[44] D. Barkley. Euclidean symmetry and the dynamics of rotating spiral waves, *Phys. Rev. Lett.* **72** (1994) 165–167.

[45] D. Barkley. Spiral meandering, in: [304] (1995) 163–189.

[46] D. Barkley and I.G. Kevrekidis. A dynamical systems approach to spiral wave dynamics. *Chaos* **4** (1994) 453–460.

[47] D. Barkley, M. Kness and L.S. Tuckerman. Spiral-wave dynamics in a simple model of excitable media: The transition from simple to compound rotation. *Phys. Rev. A* **42** (1990) 2489–2492.

[48] R.S.K. Barnes. Life-history strategies in contrasting populations of the coastal gastropod *Hydrobia* III. Lagoonal versus intertidal-marine *H. Neglecta, Vie Milieu* **43** (1993) 73–83.

[49] R.S.K. Barnes. Investment in eggs in lagoonal *Hydrobia ventrosa* and life-history strategies in north-west European *Hydrobia* species, *J. Mar. Biol. Ass.* **74** (1994) 637–650.

[50] R.S.K. Barnes. Breeding, recruitment and survival in a mixed intertidal population of the mudsnails *Hydrobia ulvae* and *H. neglecta*, *J. Mar. Biol. Ass.* **76** (1996) 1003–1012.

[51] L. Bauer, H.B. Keller, and E.L. Reiss. Multiple eigenvalues lead to secondary bifurcation, *SIAM Rev.* **17** (1975) 101–122.

[52] L. Bauer, E.L. Reiss, and H.B. Keller. Axisymmetric buckling of rigidly clamped hemispherical shells, *Int. J. Nonlin. Mech.* **8** (1973) 31–39.

[53] L. Bauer, E.L. Reiss, and H.B. Keller. Axisymmetric buckling of hollow spheres anbd hemispheres, *Commun. Pure Appl. Math.* **23** (1970) 529–568.

[54] A. Bayliss and B.J. Matkowsky. Nonlinear dynamics of cellular flames. *SIAM J. Appl. Math.* **52** (1992) 396–415.

[55] A. Bayliss, B.J. Matkowsky, and H. Riecke. Structure and dynamics of modulated traveling waves in cellular flames. *Physica D* **74** (1994) 1–23.

[56] A. Bayliss, B.J. Matkowsky, and H. Riecke. Symmetries in modulated traveling waves in combustion: jumping ponies on a merry-go-round. *Pattern Formation: Symmetry Methods and Applications* (J. Chadam and W. Langford eds.) Fields Institute Communications **5**, AMS, 1995, 19–43.

[57] R.S.A. Beauchamp and P. Ullyett. Competitive relationships between certain species of fresh-water triclads, *J. Ecol.* **20** (1932) 200–208.

[58] M. Benedicks and L. Carleson. The dynamics of the Hénon map, *Ann. of Math.* **133** (1991) 73–169.

[59] A.N. Beris and B.J. Edwards. *Thermodynamics of Flowing Systems*. Oxford U Press, Oxford, 1994.

[60] A. Bers. Space-time evolution of plasma instabilities — absolute and convective. In: *Handbook of Plasma Physics* (M.N. Rosenbluth and R.Z. Sagdeev, eds.), vol. 1, North-Holland, Amsterdam 19?? 451–517.

[61] R. Bertram. A computational study of the effects of serotonin on the molluscan burster neuron R_{15}, PhD Thesis, Florida State U 1993.

[62] E. Bierstone. *The Structure of Orbit Spaces and the Singularities of Equivariant Mappings*, IMPA, Rio de Janeiro 1980.

[63] V.N. Biktashev, A.V. Holden, and E.V. Nikolaev. Spiral wave meander and symmetry of the plane, *Internat. J. Bif. Chaos* **6** (1996) 2433–2440.

[64] J. Binney and S. Tremaine. *Galactic Dynamics*, Princeton U Press, Princeton NJ 1987.

[65] G.D. Birkhoff. Proof of the ergodic theorem, *Proc. Nat. Acad. Sci.* **17** (1931) 656–660.

[66] W.F. Blair. Character displacement in frogs, *American Zoologist* **14** (1974) 1119–1125.

[67] G. G. Blasdel. Orientation Selectivity, Preference, and Continuity, *Monkey Striate Cortex* **12**, 8 (1992) 3139–3161.

[68] P.T. Boag and P.R. Grant. Heritability of external morphology in Darwin's finches, *Nature* **274** (1978) 793–794.

[69] E. Bodenschatz, J.R. de Bruyn, G. Ahlers, and D.S. Cannell. Transition between patterns in thermal convection, *Phys. Rev. Lett.* **67** (1991) 3078–3081.

[70] S.A. Bories. Natural convection in porous media, *Proc. NATO ASI on Fundamentals of Transport Phenomena in Porous Media* (1985).

[71] I. Bosch Vivancos, P. Chossat, and I. Melbourne. New planforms in systems of partial differential equations with euclidean symmetry. *Arch. Rational Mech. Anal.* **131** (1995) 199–224.

[72] W.H. Bosking, Y. Zhang, B. Schofield, and D. Fitzpatrick. Orientation selectivity and the arrangement of horizontal connections in tree shrew striate cortex, *J. Neurosci.* **17** 6 (1997) 2112–2127.

[73] R. Bowen. Periodic points for Axiom A diffeomorphisms, *Trans. Amer. Math. Soc.* **154** (1971) 377–397.

[74] P.C. Bressloff, J.D. Cowan, M. Golubitsky, and P.J. Thomas. Scalar and pseudoscalar bifurcations: pattern formation in the visual cortex. *Nonlinearity* **14** (2001) 739–775.

[75] P.C. Bressloff, J.D. Cowan, M. Golubitsky, P.J. Thomas, and M.C. Wiener. Geometric visual hallucinations, Euclidean symmetry, and the functional architecture of striate cortex. *Phil. Trans. Royal Soc. B* **356** (2001) 299–330.

[76] P.C. Bressloff, J.D. Cowan, M. Golubitsky, P.J. Thomas, and M.C. Wiener. What geometric visual hallucinations tell us about the visual cortex. *Neural Comput.* To appear.

[77] T. Bröcker and T. tom Dieck. *Representations of Compact Lie Groups*, Springer-Verlag, New York 1985.

[78] H.W. Broer. Formal normal form theorems for vector fields and some consequences for bifurcations in the volume preserving case, in: *Dynamical Systems and Turbulence, Warick 1980* (D.A. Rand and L.-S. Young, eds) Lect. Notes Math. **898**, Springer, New York, 1981, 54–74.

[79] H.W. Broer, G.B. Huitema, and M.B. Sevryuk. *Quasi-Periodic Tori in Families of Dynamical Systems: Order Amidst Chaos.* Lect. Notes Math. **1645** (1996).

[80] H.W. Broer, G.B. Huitema, F. Takens, and B.L.J. Braaksma. *Unfoldings and Bifurcations of Quasi-Periodic Tori, Memoirs Amer. Math. Soc.* **83** (1990)

[81] A. Brown. *The Classification of Bifurcations in Maps with Symmetry*, PhD Thesis, Math. Inst. U Warwick 1992.

[82] J. Buescu and I. Stewart. Sets, lines, and adding machines, in *Dynamics, Bifurcation, and Symmetry* (P.Chossat ed.), Proceedings, Cargèse 1993, NATO ASI Series C **437** Kluwer, Dordrecht 1994, 59–68.

[83] J. Buescu and I. Stewart. Liapunov stability and adding machines, *Ergod. Thy. Dyn. Sys.* **15** (1995) 271–290.

[84] J. Buescu and I. Stewart. Inverse limits and adding machines. In preparation.

[85] P-L. Buono. *A Model of Central Pattern Generators for Quadruped Locomotion*, PhD Thesis, U Houston 1998.

[86] P-L. Buono. Models of central pattern generators for quadruped locomotion: II. Secondary gaits. *J. Math. Biol.* **42** No. 4 (2001) 327–346.

[87] P-L. Buono and M. Golubitsky. Models of central pattern generators for quadruped locomotion: I. primary gaits. *J. Math. Biol.* **42** No. 4 (2001) 291–326.

[88] P.-L. Buono, M. Golubitsky, and A. Palacios. Heteroclinic cycles in systems with D_n symmetry, in: *Bifurcation Theory and its Numerical Analysis* (Z. Chen, S.-N. Chow and K. Li, eds), Springer-Verlag, Singapore 1999, 13–27.

[89] P.-L. Buono, M. Golubitsky, and A. Palacios. Heteroclinic Cycles in Rings of Coupled Cells. *Physica D* **143** (2000) 74–108.

[90] F.H. Busse. *Das Stabilitätsverhalten der Zellarkonvektion bei endlicher Amplitude*, PhD Thesis, U Munich 1962.

[91] F.H. Busse and R.M. Clever. Nonstationary convection in a rotating system, in *Recent Developments in Theoretical and Experimental Fluid Mechanics* (U. Müller, K.G. Roesner, and B. Schmidt eds.), Springer-Verlag, Berlin 1979, 376–385.

[92] F.H. Busse and K.E. Heikes. Convection in a rotating layer; a simple case of turbulence, *Science* **208** (1980) 173–175.

[93] E. Buzano and M. Golubitsky. Bifurcation involving the hexagonal lattice and the planar Benard problem, *Phil. Trans. Roy. Soc. London A* **308** (1983) 617–667.

[94] T.K. Callahan and E. Knobloch. Bifurcations on the fcc lattice, *Phys. Rev. E* **53** No. 4 (1996) 3559–3562.

[95] T.K. Callahan and E. Knobloch. Symmetry-breaking bifurcation on cubic lattices, *Nonlinearity* **10** (1997) 1179–1216.

[96] T.K. Callahan and E. Knobloch. Pattern formation in three-dimensional reaction-diffusion systems. *Physica D* **132** (1999) 339–362.

[97] S. Campbell and P. Holmes. Heteroclinic cycles and modulated travelling waves in a system with D_4 symmetry, *Physica D* **59** (1992) 52–78.

[98] H.L. Carson and A.R. Templeton AR. Genetic revolutions in relation to speciation phenomena: The founding of new populations, *Ann. Rev. Ecol. Systematics* **15** (1984) 97-131.

[99] S.B.S.D. Castro. *Symmetry and Bifurcation of Periodic Solutions in Neumann Boundary Value Problems*, MSc Thesis, U Warwick 1990.

[100] S.B.S.D. Castro. Mode interactions with symmetry, *Dyn. Stab. Sys.* **10** (1995) 13–31.

[101] J. Chadam, M. Golubitsky, W. Langford, and B. Wetton. *Pattern Formation: Symmetry Methods and Applications*, Fields Inst. Comm. 4, Providence 1996.

[102] D.R.J. Chillingworth and M. Golubitsky. Planar pattern formation in bifurcation from a homeotropic nematic liquid crystal. In: *Bifurcations, Symmetry and Patterns* (J. Buescu, S.B.S.D. Castro, A.P.S. Dias and I.S. Labouriau, eds), Birkhäuser. To appear.

[103] P. Chossat, Y. Demay, and G. Iooss. Interactions des modes azimutaux dans le probléme de Couette-Taylor, *Arch Rational Mech. Anal.* **99** (1987) 213–248.

[104] P. Chossat and M. Golubitsky. Symmetry increasing bifurcation of chaotic attractors, *Physica D* **32** (1988) 423–436.

[105] P. Chossat and G. Iooss. Primary and secondary bifurcations in the Couette-Taylor problem. *Japan J. Appl. Math.* **2** No. 1 (1985) 37–68.

[106] P. Chossat and G. Iooss. *The Couette-Taylor Problem.* Springer-Verlag, Appl. Math. Sci. **102**, New York, 1994.

[107] P. Chossat, M. Koenig, and J. Montaldi. Bifurcation générique d'ondes rotatives d'isotropie maximale, *C.R. Acad. Sci. Paris* **320** (1995) 25–30.

[108] P. Chossat, M. Krupa, I. Melbourne, and A. Scheel. Transverse bifurcations of heteroclinic cycles in symmetric systems. *Physica D* **100** (1997) 85–100.

[109] P. Chossat, M. Krupa, I. Melbourne, and A. Scheel. Magnetic dynamos in rotating convection — A dynamical systems approach. *Dynamics of Continuous, Discrete and Impulsive Systems* **5** (1999) 327–340.

[110] P. Chossat and R. Lauterbach. *Methods in Equivariant Bifurcations and Dynamical Systems.* Adv. Ser. Nonlinear Dynamics **15**, World Scientific, Singapore, 2000.

[111] P. Chossat, R. Lauterbach and I. Melbourne. Steady-state bifurcation with O(3)-symmetry, *Arch. Rational Mech. Anal.* **113** (1990) 313–376.

[112] C. Cicogna. Symmetry breakdown from bifurcation, *Lett. Nuovo Cimento* **31** (1981) 600–602.

[113] S. Ciliberto and J. Gollub. Chaotic mode competition in parametrically forced surface waves, *J. Fluid Mech.* **158** (1985) 381–398.

[114] B. Clarke, W. Arthur, D.T. Horsley, and D.T. Parkin. Genetic variation and natural selection in pulmonate molluscs, in *Pulmonates* vol. 2A, *Systematics, Evolution, and Ecology* (V. Fretter, J. Peake eds.), Academic Press, London 1978.

[115] T. Clune and E. Knobloch. Pattern selection in three-dimensional magnetoconvection, *Physica D* **74** (1994) 151–176.

[116] J. Cohen and I. Stewart. *The Collapse of Chaos,* Viking, New York 1994.

[117] J. Cohen and I. Stewart. Polymorphism viewed as phenotypic symmetry-breaking. In: *Nonlinear Phenomena in Physical and Biological Sciences* (S.K. Malik ed.), Indian National Science Academy, New Delhi 2000.

[118] P. Collet and J.-P. Eckmann. *Iterated Maps on the Interval as Dynamical Systems,* Birkhäuser, Basel 1980.

[119] J.J. Collins and I. Stewart. Symmetry-breaking bifurcation: a possible mechanism for 2:1 frequency-locking in animal locomotion, *J. Math. Biol.* **30** (1992) 827–838.

[120] J.J. Collins and I. Stewart. Hexapodal gaits and coupled nonlinear oscillator models, *Biol. Cybern.* **68** (1993) 287–298.

[121] J.J. Collins and I. Stewart. Coupled nonlinear oscillators and the symmetries of animal gaits, *J. Nonlin. Sci.* **3** (1993) 349–392.

[122] J.J. Collins and I. Stewart. A group-theoretic approach to rings of coupled biological oscillators, Biol. Cybern. **71** (1994) 95–103.

[123] K.T. Coughlin and P.S. Marcus. Modulated waves in Taylor-Couette flow. Part 1. Analysis *J. Fluid Mech.* **234** (1992) 1–18. Part 2. Numerical simulation (1992) 19–46.

[124] R. Courant and D. Hilbert. *Methods of Mathematical Physics,* Interscience, New York 1953.

[125] J.D. Crawford. Normal forms for driven surface waves: boundary conditions, symmetry, and genericity, *Physica D* **52** (1991) 429–457.

[126] J.D. Crawford. Surface waves in non-square containers with square symmetry, *Phys. Rev. Lett.* **67** (1991) 441–444.

[127] J.D. Crawford. $D_4 \dotplus T^2$ mode interactions and hidden rotation symmetry, *Nonlinearity* **7** (1994) 697–739.

[128] J.D. Crawford. D_4-symmetric maps with hidden Euclidean symmetry, in *Pattern Formation: Symmetry Methods and Applications*, (J. Chadam *et al.*, eds.) Fields Inst. Comm. 4, Amer. Math. Soc., Providence, 1996.

[129] J.D. Crawford, J.P. Gollub, and D. Lane. Hidden symmetries of parametrically forced waves, *Nonlinearity* **6** (1993) 119–164.

[130] J.D. Crawford, M. Golubitsky, M.G.M. Gomes, E. Knobloch, and I. Stewart. Boundary conditions as symmetry constraints, in *Singularity Theory and Its Applications, Warwick 1989, Part II*. (M. Roberts and I. Stewart, eds.), Lecture Notes in Math. **1463**, Springer-Verlag, Heidelberg 1991, 63–79.

[131] J.D. Crawford and E. Knobloch. Classification and unfolding of degenerate Hopf bifurcations with 0(2) symmetry: no distinguished parameter, *Physica D* **31** (1988) 1–48.

[132] J.D. Crawford and E. Knobloch. Symmetry and symmetry-breaking bifurcations in fluid mechanics. *Ann. Rev. Fluid Mech.* **23** (1991) 341–387.

[133] J.D. Crawford, E. Knobloch, and H. Riecke. Period-doubling mode interactions with circular symmetry, *Physica D* **44** (1990) 340–396.

[134] M.C. Cross and P.C. Hohenberg. Pattern formation outside of equilibrium, *Rev. Mod. Phys.* **65** (1993) 851–1112.

[135] G. Dangelmayr and D. Armbruster. Steady state mode interactions in the presence of 0(2) symmetry and in non-flux boundary conditions, in *Multiparameter Bifurcation Theory* (M. Golubitsky and J. Guckenheimer eds.), Contemporary Math. **56** (1986) 53–68.

[136] G. Dangelmayr, W. Güttinger, and M. Wegelin Hopf bifurcation with $D_3 \times D_3$-symmetry. *Zeitschrift für Angewandte Mathematik und Physik* **44** (1993) 595–638.

[137] G. Dangelmayr, J. Hettel, and E. Knobloch. Parity-breaking bifurcation in inhomogeneous systems, *Nonlinearity* **10** (1997) 1093–1114.

[138] G. Dangelmayr and E. Knobloch. The Takens-Bogdanov bifurcation with 0(2)-symmetry, *Phil. Trans. Royal Soc. London* **322** (1987) 243–279.

[139] G. Dangelmayr and E. Knobloch. Hopf bifurcation with broken circular symmetry. *Nonlinearity* **4** (1991) 399–427.

[140] G. Dangelmayr and M. Wegelin. Hopf bifurcation in anisotropic systems, in *Pattern Formation in Continuous and coupled Systems* (M. Golubitsky, D. Luss, S.H. Strogatz eds.), IMA volumes in Math. and Appns. **115** Springer-Verlag, New York 1999, 33–47.

[141] A. Das and C.D. Gilbert. Topography of contextual modulations mediated by short–range interactions in primary visual cortex, *Nature* **399** (1999) 655–661.

[142] P.G. de Gennes. *The Physics of Liquid Crystals*, Clarendon Press, Oxford, 1974.

[143] M. Dellnitz. *Symmetry Creation by Collisions of Attractors*, Habilitation Thesis, Inst. Angew. Math. U Hamburg 1993.

[144] M. Dellnitz, M. Field, M. Golubitsky, A. Hohmann and J. Ma. Cycling chaos, *Internat. J. Bif. Chaos* **5** (1995) 1243–1247. See also: *IEEE Trans. Circuits Syst.* **42** No. 10 (1995) 821–823.

[145] M. Dellnitz, M. Golubitsky, A. Hohmann and I. Stewart. Spirals in scalar reaction diffusion equations, *Internat. J. Bif. Chaos* **5** (1995) 1487–1501.

[146] M. Dellnitz, M. Golubitsky, and I. Melbourne. Mechanisms of symmetry creation, in *Bifurcation and Symmetry* (E. Allgower, K. Bömer, and M. Golubitsky, eds.), ISNM **104**, Birkhäuser, Basel 1992, 99–109.

[147] M. Dellnitz, M. Golubitsky, and M. Nicol. Symmetry of Attractors and the Karhunen-Loève Decomposition, in *Trends and Perspectives in Applied Mathematics* (L. Sirovich, ed.) *Appl. Math. Sci.* **100**, Springer-Verlag, New York 1994, 73–108.

[148] Y. Demay and G. Iooss. Calcul des solutions bifurquées pour le probléme de Couette-Taylor avec les deux cylindres en rotation, *J. de Mech. Theor. et Appl.* **Numéro Special** (1984) 193–216.

[149] J. Denvir. *Two Topics in Dynamics*, PhD Thesis, Math. Inst. U Warwick 1994.

[150] J.A. DeSimone, D.L. Beil, and L.E. Scriven. Ferroin-collodion membranes: dynamic concentration patterns in planar membranes, *Science* **180** (1973) 946–948.

[151] N.R. Deuel and L.M. Lawrence. Laterality in the gallop gait of horses, *J. Biomechanics* **20** (1987) 645–649.

[152] A.P.S. Dias. *Bifurcations with Wreath Product Symmetry*, PhD Thesis, Math. Inst. U Warwick 1997.

[153] A.P.S. Dias. Hopf bifurcation for wreath products, *Nonlinearity* **11** (1998) 247–264.

[154] A.P.S. Dias and I. Stewart. Hopf bifurcation on a simple cubic lattice. *Dyn. Stab. Sys.* **14** (1999) 3–55.

[155] A.P.S. Dias and I. Stewart. Symmetry-breaking bifurcations of wreath product systems, *J. Nonlin. Sci.* **9** (1999) 671–695.

[156] A.P.S. Dias and I. Stewart. Invariant theory for wreath product groups, *J. Pure Appl. Algebra* bf 150 (2000) 61–84.

[157] A.P.S. Dias and I. Stewart. Hilbert series for equivariant mappings restricted to invariant hyperplanes, *J. Pure Appl. Algebra* 151 (2000) 89–106.

[158] U. Dieckmann and M. Doebeli. On the origin of species by sympatric speciation, *Nature* 400 (1999) 354–457.

[159] B. Dionne. A.P.S. Dias, and I. Stewart. Heteroclinic cycles and wreath product systems, *Dyn. Stab. Sys.* 15 (2000) 353–385.

[160] B. Dionne and M. Golubitsky. Planforms in two and three dimensions, *ZAMP* 43 (1992) 36–62.

[161] B. Dionne, M. Golubitsky, M. Silber, and I. Stewart. Time-periodic spatially-periodic planforms in Euclidean equivariant systems, *Phil. Trans. R. Soc. London A* 352 (1995) 125–168.

[162] B. Dionne, M. Golubitsky, and I. Stewart. Coupled cells with internal symmetry Part 1: wreath products, *Nonlinearity* 9 (1996) 559-574.

[163] B. Dionne, M. Golubitsky, and I. Stewart. Coupled cells with internal symmetry Part 2: direct products, *Nonlinearity* 9 (1996) 575-599.

[164] B. Dionne, M. Silber, and A.C. Skeldon. Stability results for steady, spatially-periodic planforms, *Nonlinearity* 10 (1997) 321–354.

[165] Th. Dobzhansky. *Genetics and the Origin of Species*, Columbia U Press, New York 1937.

[166] M. Doi and F. Edwards. *The Theory of Polymer Dynamics*, Clarendon Press, Oxford, 1986.

[167] R.J. Donnelly. Taylor-Couette flow: the early days, *Physics Today* 44 No. 11 (1991) 32–39.

[168] L. Ehrman. Direct observation of sexual isolation between allopatric and sympatric strains of the different *Drosophila paulistorum* races, *Evolution* 19 (1965) 459–464.

[169] N. Eldredge and S.J. Gould. Punctuated equilibrium: an alternative to phyletic gradualism, in *Models in Palaeobiology* (T.J.M. Schopf ed.), Cooper, San Francisco 1972.

[170] C. Elphick, E. Tirapegui, M.E. Brachet, P. Coulet, and G. Iooss. A simple global characterization for normal forms of singular vector fields, *Physica D* 29 (1987) 95–127.

[171] T. Elmhirst. *Symmetry-breaking Bifurcations of S_N-equivariant Vector Fields and Polymorphism*, MSc Thesis, U Warwick 1998.

[172] T. Elmhirst. *Symmetry and Emergence in Polymorphism and Sympatric Speciation*, PhD Thesis, Mathematics Institute, University of Warwick, 2001.

[173] T. Elmhirst. *Symmetry and Emergence in Polymorphism and Sympatric Speciation*, PhD Thesis, U Warwick. To appear 2001.

[174] M. Eiswirth, M. Bär, and H.H. Rotermund. Spatiotemporal selforganization on isothermal catalysts, *Physica D* 84 (1995) 40–57.

[175] I.R. Epstein and M. Golubitsky. Symmetric patterns in linear arrays of coupled cells, *Chaos* 3(1) (1993) 1–5.

[176] G.B. Ermentrout and J.D. Cowan. A Mathematical Theory of Visual Hallucination Patterns, *Biol. Cybernetics* 34 (1979) 137–150.

[177] T. Erneux and B.J. Matkowsky. Quasi-periodic waves along a pulsating propagating front in a reaction-diffusion system, *SIAM J. Appl. Math.* 44 (1984) 536–544.

[178] U. Eysel. Turning a corner in vision research, *Nature* 399 (1999) 641–644.

[179] M.J. Feigenbaum. Quantitative universality for a class of nonlinear transformations, *J. Stat. Phys.* 19 (1978) 25–52.

[180] T. Fenchel. Factors determining the distribution patterns of mud snails (Hydrobiidae), *Oecologia* 20 (1975) 1–17.

[181] T. Fenchel. Character displacement and coexistence in mud snails, *Oecologia* 20 (1975) 19–32.

[182] B. Fiedler, B. Sandstede, A. Scheel, and C. Wulff. Bifurcation from relative equilibria of noncompact group actions: skew products, meanders and drifts, *Documenta Math.* 1 (1996) 479–505.

[183] B. Fiedler and D. Turaev. Normal forms, resonances, and meandering tip motions near relative equilibria of Euclidean group actions. *Arch. Rational Mech. & Anal.* 145 No. 2 (1998) 129–159.

[184] M.J. Field. Equivariant dynamical systems, *Trans. Amer. Math. Soc.* 229 (1980) 185–205.

[185] M.J. Field. Equivariant bifurcation theory and symmetry breaking, *J. Dyn. Diff. Eqns* 1 (1989) 369–421.

[186] M. Field. Local structure for equivariant dynamics, in *Singularity Theory and Its Applications, Warwick 1989, Part II.* (M. Roberts and I. Stewart, eds.), Lecture Notes in Math. **1463**, Springer-Verlag, Heidelberg 1991, 142–166.

[187] M.J. Field. *Lectures on Bifurcations, Dynamics and Symmetry*, Research Notes in Mathematics **356** Longman, London 1996.

[188] M.J. Field. Heteroclinic cycles in symmetrically coupled systems, in *Pattern Formation in Continuous and coupled Systems* (M.Golubitsky, D.Luss, S.H.Strogatz eds.), IMA volumes in Math. and Appns. **115** Springer-Verlag, New York 1999, 49–64.

[189] M.J. Field and M. Golubitsky. *Symmetry in Chaos: A Search for Pattern in Mathematics, Art, and Nature*, Oxford U Press, Oxford, 1992.

[190] M.J. Field and M. Golubitsky. Symmetries on the edge of chaos, *New Scientist* **1855** (January 9, 1993) 32–35.

[191] M. Field and M. Golubitsky. Symmetric chaos: how and why, *Notices AMS* **42** No. 2 (1995) 240–244.

[192] M.J. Field, M. Golubitsky, and I. Stewart. Bifurcation on hemispheres, *J. Nonlinear Sci.* **1** (1991) 201–223.

[193] M. J. Field, I. Melbourne, and M. Nicol. Symmetric Attractors for Diffeomorphisms and Flows, *Proc. London Math. Soc.* **72** (1996) 657–696.

[194] M.J. Field and R.W. Richardson. Symmetry breaking and the maximal isotropy subgroup conjecture for reflection groups, *Arch. Rational Mech. Anal.* **105** (1989) 61–94.

[195] M.J. Field and R.W. Richardson. Symmetry breaking and branching patterns in equivariant bifurcation theory I, *Arch. Rational Mech. Anal.* **118** (1992) 297–348.

[196] M.J. Field and R.W. Richardson. Symmetry breaking and branching patterns in equivariant bifurcation theory II, *Arch. Rational Mech. Anal.* **120** (1992) 147–190.

[197] M.J. Field and J.W. Swift. Static bifurcation to limit cycles and heteroclinic cycles, *Nonlinearity* **4** (1991) 1001–1043.

[198] M.G. Forest, Q. Wang, and S.E. Bechtel. 1-D models for thin filaments of liquid crystalline polymers: coupling of orientation and flow in the stability of simple solutions. *Physica D* **99** (1997) 527–554.

[199] M.G. Forest, Q. Wang and H. Zhou. Homogeneous pattern selection and director instabilities of nematic liquid crystal polymers induced by elongational flows. *Physics of Fluids* **12**(3) (2000) 490–498.

[200] J.B. Fraleigh. *A First Course in Abstract Algebra* (3rd. ed.), Addison-Wesley, Reading MA 1982.

[201] S. Friedlander and W.L. Siegmann. Internal waves in a contained rotating stratified fluid, *J. Fluid Mech.* **114** (1982) 123–156.

[202] H. Fujii, M. Mimura, and Y. Nishiura. A picture of the global bifurcation diagram in ecological interacting and diffusing systems, *Physica D* **5** (1982) 1–42.

[203] J.E. Furter, A.M. Sitta, and I. Stewart. Singularity theory and equivariant bifurcation problems with parameter symmetry, *Math. Proc. Camb. Phil. Soc.* **120** (1996) 547–578.

[204] J.E. Furter, A.M. Sitta, and I. Stewart. Algebraic path formulation for equivariant bifurcation problems, *Math. Proc. Camb. Phil. Soc.* **124** (1998) 275–304.

[205] F. Galton. *Hereditary Genius*, Macmillan, London 1869.

[206] P.P. Gambaryan. *How Mammals Run: Anatomical Adaptations*, Wiley, New York 1974.

[207] I.I. Gihman and A.V. Skorohod. *Stochastic Differential Equations*, Springer-Verlag, Berlin 1970.

[208] C.D. Gilbert. Horizontal integration and cortical dynamics, *Neuron* **9** (1992) 1–13.

[209] D. Gillis and M. Golubitsky. Patterns in square arrays of coupled cells, *JMAA* **208** (1997) 487–509.

[210] D. Gillis and M. Golubitsky. A formula for symmetry detectives, *Physica D* **107** (1997) 23–29.

[211] B.J. Gluckman, P. Marcq, J. Bridge, and J.P. Gollub. Time-averaging of chaotic spatiotemporal wave patterns, *Phys. Rev. Lett.* **71** (1993) 75–98.

[212] H.F. Goldstein, E. Knobloch, I. Mercader, and M. Net. Convection in a rotating cylinder. Part 1. Linear theory for moderate Prandtl numbers, *J. Fluid Mech.* **248** (1993) 583–604.

[213] M. Golubitsky and V. Guillemin. *Stable Mappings and their Singularities*, Graduate Texts in Mathematics **14**, Springer-Verlag, New York 1973.

[214] M. Golubitsky, E. Knobloch, and I. Stewart. Target patterns and spirals in planar reaction-diffusion systems, *J. Nonlin. Sci.* **10** (2000) 333–354.

[215] M. Golubitsky, M. Krupa and C. Lim. Time-reversibility and particle sedimentation, *SIAM J. Appl. Math.* **51** No. 1 (1991) 49–72.

[216] M. Golubitsky and W.F. Langford. Classification and unfoldings of degenerate Hopf bifurcations, *J. Diff. Eqns.* **41** (1981) 375–415.

[217] M. Golubitsky and W.F. Langford. Pattern formation and bistability in flow between counterrotating cylinders, *Physica D* **32** (1988) 362–392.

[218] M. Golubitsky, V.G. LeBlanc, and I. Melbourne. Meandering of the spiral tip: An alternative approach, *J. Nonlin. Sci.* **7** (1997) 557–586.

[219] M. Golubitsky, V.G. LeBlanc, and I. Melbourne. Hopf bifurcation from rotating waves and patterns in physical space, *J. Nonlin. Sci.* **10** (2000) 69–101.

[220] M. Golubitsky, J. Marsden and D. Schaeffer. Bifurcation problems with hidden symmetries. In: *Partial Differential Equations and Dynamical Systems* (W.E. Fitzgibbon III, ed.) Res. Notes in Math. **101** Pitman Press, Boston, 1984, 181–210.

[221] M. Golubitsky, J.E. Marsden, I. Stewart, and M. Dellnitz. The constrained Liapunov-Schmidt procedure and periodic orbits, *Fields. Inst. Commun.* **4** (1995) 81–127.

[222] M. Golubitsky and M. Roberts. Degenerate Hopf bifurcation with **O**(2) symmetry, *J. Diff. Eqn.* **69** (1987) 216–264.

[223] M. Golubitsky and D.G. Schaeffer. A theory for imperfect bifurcation via singularity theory, *Commun. Pure Appl. Math.* **32** (1979) 21–98.

[224] M. Golubitsky and D.G. Schaeffer. Imperfect bifurcation in the presence of symmetry, *Commun. Math. Phys.* **67** (1979) 205–232.

[225] M. Golubitsky and D.G. Schaeffer. Bifurcations with **O**(3) symmetry including applications to the Bénard problem, *Commun. Pure Appl. Math.* **35** (1982) 81–111.

[226] M. Golubitsky and D.G. Schaeffer. *Singularities and Groups in Bifurcation Theory I*, Applied Mathematical Sciences **51**, Springer-Verlag, New York 1985.

[227] M. Golubitsky and I. Stewart. Hopf bifurcation in the presence of symmetry. *Arch. Rational Mech. & Anal.* **87** No. 2 (1985) 107–165.

[228] M. Golubitsky and I. Stewart. Symmetry and stability in Taylor-Couette flow, *SIAM J. Math. Anal.* **17** (1986) 249–288.

[229] M. Golubitsky and I. Stewart. Hopf bifurcation with dihedral group symmetry: coupled nonlinear oscillators, in *Multiparameter Bifurcation Theory* (M.Golubitsky and J.Guckenheimer eds.), Proceedings of the AMS-IMS-SIAM Joint Summer Research Conference, July 1985, Arcata; *Contemporary Math.* **56** Amer. Math. Soc., Providence RI 1986, 131–173.

[230] M. Golubitsky and I. Stewart. Generic bifurcation of Hamiltonian systems with symmetry, *Physica D* **24** (1987) 391–405. (With an appendix by J.E.Marsden.)

[231] M. Golubitsky and I. Stewart. An algebraic criterion for symmetric Hopf bifurcation, *Proc. Roy. Soc. London A* **440** (1993) 727–732.

[232] M. Golubitsky and I. Stewart. Symmetry and pattern formation in coupled cell networks, In: *Pattern Formation in Continuous and coupled Systems* (M.Golubitsky, D.Luss, S.H.Strogatz eds.), IMA volumes in Math. and Appns. **115** Springer-Verlag, New York 1999, 65-82.

[233] M. Golubitsky and I. Stewart. Patterns of oscillation in coupled cell systems. In: *Geometry, Dynamics, and Mechanics: 60th Birthday Volume for J.E. Marsden* (P. Holmes, P. Newton, and A. Weinstein, eds.) Springer-Verlag. To appear.

[234] M. Golubitsky, I. Stewart, P.-L. Buono, and J.J. Collins. A modular network for legged locomotion, *Physica D* **115** (1998) 56–72.

[235] M. Golubitsky, I. Stewart, P.-L. Buono, and J.J. Collins. Symmetry in locomotor central pattern generators and animal gaits, *Nature* **401** (1999) 693–695.

[236] M. Golubitsky, I. Stewart and B. Dionne. Coupled cells: wreath products and direct products, in *Dynamics, Bifurcation and Symmetry*, NATO ARW Series, (P. Chossat and J.M. Gambaudo eds.), Kluwer, Amsterdam 1994, 127–138.

[237] M. Golubitsky, I. Stewart, and D.G. Schaeffer. *Singularities and Groups in Bifurcation Theory II*, Applied Mathematics Sciences, **69**, Springer Verlag, New York, 1988.

[238] M. Golubitsky, J.W. Swift and E. Knobloch. Symmetries and pattern selection in Rayleigh-Bénard convection, *Physica* **10D** (1984) 249–276.

[239] M.G.M. Gomes. *Symmetries in Bifurcation Theory: the Appropriate Context*, PhD Thesis, U Warwick 1992.

[240] M.G.M. Gomes, I.S. Labouriau, and E.M. Pinho. Spatial hidden symmetries in pattern formation, in *Pattern Formation in Continuous and coupled Systems* (M.Golubitsky, D.Luss, S.H.Strogatz eds.), IMA volumes in Math. and Appns. **115** Springer-Verlag, New York 1999, 83–99.

[241] M.G.M. Gomes and I. Stewart. Steady PDEs on generalized rectangles: a change of genericity in mode interactions, *Nonlinearity* **7** (1994) 253–272.

[242] M.G.M. Gomes and I. Stewart. Hopf bifurcation on generalized rectangles with Neumann boundary conditions, in: *Dynamics, Bifurcations, Symmetry* (P. Chossat ed.), Proceedings of Conference of the European Bifurcation Theory Group, Cargèse 1993, Kluwer, Dordrecht 1994, 139–158.

[243] M.G.M. Gomes and I. Stewart. Symmetry of generic bifurcations in cubic domains, *Internat. J. Bif. Chaos* **7** (1997) 147–171.

[244] M. Gorman. Private communication.

[245] M. Gorman, C.F. Hamill, M. el-Hamdi and K.A. Robbins. Rotating and modulated rotating states of cellular flames. *Combustion Sci. and Technology.* **98** (1994) 25–35.

[246] M. Gorman, M. el-Hamdi, and K.A. Robbins. Experimental observations of ordered states of cellular flames, *Comb. Sci. & Tech.* **98** (1994) 37–45.

[247] M. Gorman, M. el-Hamdi, and K.A. Robbins. Ratcheting motion of concentric rings in cellular flames. *Phys. Rev. Lett.* **76** (1996) 228–231.

[248] M. Gorman and H.L. Swinney. Visual observation of the second characteristic mode in a quasiperiodic flow. *Phys. Rev. Lett.* **43** (1980) 1871–1875.

[249] M. Gorman and H.L. Swinney. Spatial and temporal characteristics of modulated waves in the circular Couette system. *J. Fluid Mech.* **117** (1982) 123–142.

[250] M. Gorman, H.L. Swinney and D. Rand. Doubly periodic circular Couette flow: experiments compared with predictions from dynamics and symmetry. *Phys. Rev. Lett* **46** (1981) 992–995.

[251] P.R. Grant, B.R. Grant, J.N.M. Smith, I.J. Abbott, and L.K. Abbott. Darwin's finches: Population variation and natural selection, *Proc. Nat. Acad. Sci. USA* **73** (1976) 257–261.

[252] C. Grebogi, E. Ott and J. Yorke. Chaotic attractors in crisis, *Phys. Rev. Lett.* **48** (1982) 1507–1510.

[253] P.J. Greenwood, P.H. Harvey, and M. Slatkin. *Evolution, Essays in Honour of John Maynard Smith*, Cambridge U Press, Cambridge 1987.

[254] J. Greenberg. Axi-symmetric, time-periodic solutions to $\lambda - \omega$ systems, *SIAM J. Appl. Math.* **34** (1978) 391–397.

[255] J. Guckenheimer and P. Holmes. *Nonlinear Oscillations, Dynamical Systems, and Bifurcations of Vector Fields*, Appl. Math Sci. **42**, Springer-Verlag, New York 1983.

[256] J. Guckenheimer and P. Holmes. Structurally stable heteroclinic cycles, *Math. Proc. Camb. Phil. Soc.* **103** (1988) 189–192.

[257] J. Guckenheimer and P. Worfolk. Instant chaos, *Nonlinearity* **5** (1992) 1211–1222.

[258] G.H. Gunaratne, M. el-Hamdi, M. Gorman and K.A. Robbins. Asymmetric cells and rotating rings in cellular flames. *Mod. Phys. Lett. B* **10** (1996) 1379–1387.

[259] H. Haaf, R.M. Roberts, and I. Stewart. A Hopf bifurcation with spherical symmetry, *Z. Angew. Math. Phys.* **43** (1992) 793–826.

[260] P. Hadley, M.R. Beasley, and K. Wiesenfeld. Phase locking of Josephson-junction series arrays, *Phys. Rev. B* **38** (1988) 8712–8719.

[261] P. Hagan. Spiral waves in reaction-diffusion equations, *SIAM J. Appl. Math.* **42** (1982) 762–786.

[262] J.K. Hale. Generic bifurcation with applications, in Nonlinear Analysis and Mechanics (R.J. Knopf ed.), Res. Notes in Math. **17**, Pitman, San Francisco 1978, 59–157.

[263] M. Hall. *The Theory of Groups*, Macmillan, New York 1959.

[264] P.W. Hammer, N. Platt, S.M. Hammel, J.F. Heagy, and B.D. Lee. Experimental observation of on-off intermittency, *Phys. Rev. Lett.* **73** (1994) 1095–1098.

[265] Hao Bai-Lin. *Elementary Symbolic Dynamics and Chaos in Dissipative Systems*, World Scientific, Singapore 1989.

[266] Hao Bai-Lin. *Chaos II*, World Scientific, Singapore 1990.

[267] N. Hartmann, M. Bär, I.G. Kevrekidis, K. Krischer, and R. Imbihl. Rotating chemical waves in small circular domains, *Phys. Rev. Lett.* **76** (1996) 1384–1387.

[268] T.J. Healey and H. Kielhöfer. Symmetry and nodal properties in the global bifurcation analysis of quasi-linear elliptic equations, *Arch. Rational Mech. Anal.* **113** (1991) 299–311.

[269] T.J. Healey and H. Kielhöfer. Preservation of nodal structure on global bifurcating solution branches of elliptic equations with symmetry. *J. Diff. Eqns* **106** no. 1 (1993) 70–89.

[270] T.J. Healey and H. Kielhöfer. Hidden symmetry of fully nonlinear boundary conditions in elliptic equations: global bifurcation and nodal structure, *Results in Math.* **21** (1992) 83–92.

[271] M. Hénon and C. Heiles. The applicability of the third integral of the motion, some numerical experiments, *Astron. J.* **69** (1964) 73–79.

[272] M. Higashi, G. Takimoto, and N. Yamamura. Sympatric speciation by sexual selection, *Nature* **402** (1999) 523–526.

[273] M. Hildebrand. Analysis of tetrapodal gaits: general considerations and symmetrical gaits. In: *Neural Control of Locomotion* (R.M. Herman, S. Grillner, P.S.G. Stein and D.G. Stewart, eds). Plenum Press, New York, 1976, 203–236.

[274] M. Hildebrand. The quadrupedal gaits of vertebrates, *Bioscience* **39** No. 11 (1989) 766–775.

[275] A.S. Hill and I. Stewart. Hopf/steady-state mode interactions with O(2) symmetry, *Dyn. Stab. Sys.* **6** (1991) 149–171.

[276] A.S. Hill and I. Stewart. 3-mode interactions with O(2) symmetry and a model for Taylor-Couette flow, *Dyn. Stab. Sys.* **6** (1991) 267–339.

[277] M. Hirsch and S. Smale. *Differential Equations, Dynamical Systems, and Linear Algebra,* Academic Press, New York, 1974.

[278] P. Hirschberg and E. Knobloch. Complex dynamics in the Hopf bifurcation with broken translation symmetry, *Physica D* **90** (1996) 56–78.

[279] P. Hirschberg and E. Knobloch. A robust heteroclinic cycle in an O(2) × Z_2 steady-state mode interaction. *Nonlinearity* **11** (1998) 89–104.

[280] J.G. Hocking and G.S. Young. *Topology,* Addison-Wesley, Reading MA 1961.

[281] M. Holodniak, M. Kubíček, and M. Marek. Disintegration of an invariant torus in a reaction-diffusion system. In: *Numerical Treatment of Differential Equations* (K. Strehmel, ed) B.G. Teubner Verlag, Stuttgart-Leipzig, 1991, 342–350.

[282] E. Hopf. Abzweigung einer periodischen Lösung von einer stationaären Lösung eines Differential-Systems, *Ber. Math.-Phys. Kl. Sächs Acad. Wiss. Leipzig* **94** (1942) 1–22.

[283] F.C. Hoppensteadt and E.H. Izhikevich. *Weakly Connected Neural Networks,* Appl. Math. Sci. **126**, Springer-Verlag, New York 1998.

[284] C. Hou and M. Golubitsky. An example of symmetry breaking to heteroclinic cycles, *J. Diff. Eqn.* **133** No. 1 (1997) 30–48.

[285] Y. Hu, R.E. Ecke, and G. Ahlers. *Phys. Rev. Lett.* **74** (1995) 5040–.

[286] D.H. Hubel and T.N. Wiesel. Sequence regularity and geometry of orientation columns in the monkey striate cortex, *J. Comp. Neurol.* **158** (1974) 267–294.

[287] D.H. Hubel and T.N. Wiesel. Uniformity of monkey striate cortex: a parallel relationship between field size, scatter, and magnification factor, *J. Comp. Neurol.* **158** (1974) 295–306.

[288] D.H. Hubel and T.N. Wiesel. Ordered arrangement of orientation columns in monkeys lacking visual experience, *J. Comp. Neurol.* **158** (1974) 307–318.

[289] P. Huerre and P.A. Monkewitz. Local and global instabilities in spatially developing flows, *Ann. Rev. Fluid Mech.* **22** (1990) 473–537.

[290] R.B. Huey, G.W. Gilchrist, M.L. Carlson, D. Berrigan, and L. Serra. Rapid evolution of a geographic cline in size in an introduced fly, *Science* **287** (2000) 308–310.

[291] R.B. Huey, E.R. Pianka, M.E. Egan, and L.W. Coons. Ecological shifts in sympatry: Kalahari fossorial lizards (*Typhlosaurus*), *Ecology* **55** (1974) 304–316.

[292] R.B. Huey and E.R. Pianka. Ecological character displacement in a lizard, *Am. Zool.* **14** (1974) 1127–1136.

[293] E. Ihrig and M. Golubitsky. Pattern selection with O(3) symmetry, *Physica* **13D** (1984) 1–33.

[294] M. Impey. *Bifurcation in Lapwood Convection,* PhD Thesis, U Bristol, 1988.

[295] M.D. Impey, D.S. Riley, and K.H. Winters. The effect of sidewall imperfections on pattern formation in Lapwood convection, *Nonlinearity* **3** (1990) 197–230.

[296] M.D. Impey, D.S. Riley, and A.A. Wheeler. Bifurcation analysis of cellular interfaces in uni-directional solidification of a dilute binary mixture, *SIAM J. Appl. Math.* **53** (1993) 78–95.

[297] M. Impey, M. Roberts, and I. Stewart. Hidden symmetries and pattern formation in Lapwood convection, *Dyn. Stab. Sys.* **11** (1996) 155–192.

[298] G. Iooss. Secondary bifurcations of the Taylor vortices into wavy inflow and outflow boundaries, *J. Fluid Mech.* **173** (1986) 273–288.

[299] G. Iooss and M. Rossi. Hopf bifurcation in the presence of spherical symmetry: analytical results. *SIAM J. Math. Anal.* **20** (1989) 511–532.

[300] K. Itô. On Stochastic Differential Equations, *Mem. Amer. Math. Soc.* **4** (1961).

[301] W. Jahnke, W. E. Skaggs, and A.T. Winfree. Chemical vortex dynamics in the Belousov-Zhabotinsky reaction and in the two-variable Oregonator model. *J. Phys. Chem.* **93** (1989) 740–749.

[302] W. Jahnke and A.T. Winfree. A survey of spiral-wave behaviors in the Oregonator model, *Internat. J. Bif. Chaos* **1** (1991) 445–466.

[303] S.L. Judd and M. Silber. Simple and superlattice Turing patterns in reaction-diffusion systems: bifurcation, bistability, and parameter collapse, *Physica D* **136** (2000) 45–65.

[304] R. Kapral and K. Showalter. *Chemical Waves and Patterns.* Kluwer Academic Publishers, Amsterdam, 1995.

[305] J. Keener and J. Sneyd. *Mathematical Physiology,* IAM 8, Springer-Verlag, New York 1998.

[306] I.G. Kevrekidis, B. Nicolaenko, and J.C. Scovel. Back in the saddle again: a computer assisted study of the Kuramoto-Sivashinsky equation, *SIAM J. Appl. Math.* **50** (1990) 760–790.

[307] B.L. Keyfitz, M. Golubitsky, M. Gorman and P. Chossat. The use of symmetry and bifurcation techniques in studying flame stability. In: *Reacting Flows: Combustion and Chemical Reactors* (G.S.S. Ludford, ed.). Lectures in Appl. Math. **24**, Part 2, AMS, Providence, 1986, 293–315.

[308] Y. Kifer. General random perturbations of hyperbolic and expanding transformations, *J. d'Anal. Math.* **47** (1986) 111–150.

[309] G.P. King and I. Stewart. Symmetric chaos, in *Nonlinear Equations in the Applied Sciences* (W.F.Ames and C.F.Rogers eds.), Academic Press, New York 1991, 257–315.

[310] G.P. King and I. Stewart. Phase space reconstruction for symmetric dynamical systems, *Physica D* **58** (1992) 216–228. Reprinted in *Interpretation of Time Series from Nonlinear Systems* (P.G. Drazin and G.P. King eds.), North-Holland, Amsterdam 1992, 216–228.

[311] K. Kirchgässner. Exotische Lösungen Bénardschen Problems, *Math. Meth. Appl. Sci.* **1** (1979) 453–467.

[312] A.A. Kirillov. *Elements of the Theory of Representations,* Grundlehren **220**, Springer-Verlag, Berlin 1976.

[313] V. Kirk and M. Silber. A competition between heteroclinic cycles, *Nonlinearity* **7** (1994) 1605–1621.

[314] S.V. Kiyashko, L.N. Korzinov, M.I. Rabinovich, and L.S. Tsimring. Rotating spirals in a Faraday experiment, *Phys. Rev. E* **54** (1996) 5037–5040.

[315] H. Klüver. *Mescal and Mechanisms of Hallucinations,* U of Chicago Press, Chicago 1966.

[316] E. Knobloch and J. Moehlis. Burst mechanisms in hydrodynamics, in: *Nonlinear Instability, Chaos and Turbulence, vol. 2* (L. Debnath and D.N. Riahi, eds), WIT Press, Southampton, 2000, 237–287.

[317] E. Knobloch and M. Proctor. Nonlinear periodic convection in double-diffusive systems. *J. Fluid Mech.* **108** (1981) 291–316.

[318] E. Knobloch and M. Silber. Oscillatory convection in a rotating layer. *Physica D* **63** (1993) 213–232.

[319] L. Kocarev, A. Shang, and L.O. Chua. Transitions in dynamical regimes by driving: a unified method of control and synchronisation of chaos. *Intl. J. Bifurcation and Chaos* **3** (1993) 479–483.

[320] A.S. Kondrashov and F.A. Kondrashov. Interactions among quantitative traits in the course of sympatric spciation, *Nature* **400** (1999) 351–354.

[321] N. Kopell and G.B. Ermentrout. Symmetry and phaselocking in chains of weakly coupled oscillators, *Comm. Pure Appl. Math.* **39** (1986) 623–660.

[322] N. Kopell and G.B. Ermentrout. Coupled oscillators and the design of central pattern generators, *Math. Biosci.* **89** (1988) 14–23.

[323] N. Kopell and G.B. Ermentrout. Phase transitions and other phenomena in chains of oscillators. *SIAM J. Appl. Math.* **50** (1990) 1014–1052.

[324] N. Kopell and L.N. Howard. Target pattern and spiral solutions to reaction-diffusion equations with more than one space dimension, *Advances in Appl. Math.* **2** (1981) 417–449.

[325] N. Kopell and G. LeMasson. Rhythmogenesis, amplitude modulation, and multiplexing in a cortical architecture, *Proc. Nat. Acad. Sci. USA* **91** (1994) 10586–10590.

[326] L. Kramer, F. Hynne, P. Graae Soerenson, and D. Walgraef. The Ginzburg-Landau approach to oscillatory media, *Chaos* **4** (1994) 443–452.

[327] E.R. Kreuger, A. Gross, and R.C. DiPrima. On the relative importance of Taylor-vortex and nonaxisymmetric modes in flow between rotating cylinders, *J. Fluid Mech.* **24** (1966) 521–538.

[328] M. Kroon and I. Stewart. Detecting the symmetry of attractors for six oscillators coupled in a ring, *Internat. J. Bif. Chaos* **5** (1995) 209–229.

[329] M. Krupa. Bifurcations from relative equilibria, *SIAM J. Math. Anal.* **21** (1990) 1453–1486.

[330] M. Krupa. Robust heteroclinic cycles, *J. Nonlin. Sci.* **7** (1997) 129–176.

[331] M. Krupa and I. Melbourne. Asymptotic stability of heteroclinic cycles in systems with symmetry, *Ergodic Thy. Dyn. Sys.* **15** (1995) 121–147.

[332] M. Krupa and I. Melbourne. Nonasymptotically stable attractors in O(2) mode interactions. In: *Normal Forms and Homoclinic Chaos* (W.F. Langford and W. Nagata, eds.) Fields Institute Communications 4, Amer. Math. Soc., Providence, RI, 1995, 219–232.

[333] A. Kudrolli, B. Pier, and J.P. Gollub. Superlattice patterns in surface waves, *Physica D* **123** (1998) 99–111.

[334] M. Kummer. On resonant Hamiltonian systems with finitely many degrees of freedom, in *Local and Global Methods of Nonlinear Dynamics* (A.W. Saenz, W.W. Zachary, and R. Cawley, eds.) Lect. Notes in Physics 252, Springer-Verlag, Berlin.

[335] G. Küppers and D. Lortz. Transition from laminar convection to thermal turbulence in a rotating fluid layer, *J. Fluid Mech.* **35** (1969) 609–620.

[336] D. Lack. *Darwin's Finches: an Essay on the General Biological Theory of Evolution*, Peter Smith, Gloucester MA 1968.

[337] J.S.W. Lamb. Reversing symmetries in dynamical systems, *J. Phys. A* **25** (1992) 925–937.

[338] J.S.W. Lamb. Crystallographic symmetries of stochastic webs, *J. Phys. A* **26** (1993) 2921–2933.

[339] J.S.W. Lamb. *Reversing Symmetries in Dynamical Systems*, PhD Thesis, U Amsterdam 1994.

[340] J.S.W. Lamb. Stochastic webs with fourfold rotation symmetry, in *Hamiltonian Mechanics* (J. Seimenis ed.), NATO ISI ser. B 331, Plenum, New York 1994, 345–352.

[341] J.S.W. Lamb. Resonant driving and k-symmetry, *Phys. Lett. A* **199** (1995) 55–60.

[342] J.S.W. Lamb. Area-preserving dynamics that is not reversible, *Physica A* **228** (1996) 344–365.

[343] J.S.W. Lamb. Local bifurcations in k-symmetric dynamical systems, *Nonlinearity* **9** (1996) 537–557.

[344] J.S.W. Lamb. k-symmetry and return maps of space-time symmetric flows, *Nonlinearity* **11** (1998) 601–629.

[345] J.S.W. Lamb and I. Melbourne. Bifurcation from periodic solutions with spatiotemporal symmetry, in *Pattern Formation in Continuous and coupled Systems* (M. Golubitsky, D. Luss, S.H. Strogatz eds.), IMA volumes in Math. and Appns. 115 Springer-Verlag, New York 1999, 175–191.

[346] J.S.W. Lamb and I. Melbourne. Bifurcation from discrete rotating waves. *Arch. Rational Mech. Anal.* **149** (1999) 229–270.

[347] J.S.W. Lamb and M. Nicol. On symmetric ω-limit sets in reversible flows, in *Nonlinear Dynamical Systems and Chaos* (H.W. Broer, S.A. van Gils, I. Hoveijn, and F. Takens eds.), Progress In: Nonlinear Differential Equations and Their Applications 19, Birkhäuser, Basel 1996, 103–120.

[348] J.S.W. Lamb and M. Nicol. On symmetric attractors in reversible dynamical systems, *Physica D* **112** (1998) 281–297.

[349] J.S.W. Lamb and J.A.G. Roberts. Time-reversal symmetry in dynamical systems: a survey, *Physica D* **112** (1998) 1–39.

[350] J.S.W. Lamb and G.R.W. Quispel. Reversing k-symmetries in dynamical systems, *Physica D* **73** (1994) 277–304.

[351] J.S.W. Lamb and G.R.W. Quispel. Cyclic reversing k-symmetry groups, *Nonlinearity* **8** (1995) 1005–1026.

[352] W.F. Langford, R. Tagg, E. Kostelich, H.L. Swinney, and M. Golubitsky. Primary instability and bicritically in flow between counterrotating cylinders, *Phys. Fluids.* **31** (1988) 776–785.

[353] A. Lasota and J.A. Yorke. On the existence of invariant measures for piecewise monotonic transformations, *Trans. Amer. Math. Soc.* **186** (1973) 481–488.

[354] D.H. Leach and E. Sprigings. Gait fatigue in the racing thoroughbred, *J. Equine Med. Surg.* **3** (1979) 436–443.

[355] F.M. Leslie. Theory of flow phenomena in liquid crystals, in *Advances in Liquid Crystals*, Academic Press, New York, 1979.

[356] D. Lewis. Nonlinear stability of rotating planar liquid drop, *Arch. Rational Mech. Anal.* **106** (1989) 287–333.

[357] D. Lewis. Bifurcation of liquid drops, *Nonlinearity* **6** (1993) 491–522.

[358] D. Lewis and T. Ratiu. Rotating n-gon/kn-gon vortex configurations, *J. Nonlin. Sci.* **6** (1996) 385–414.

[359] D. Lewis and J.C. Simo. Nonlinear stability of rotating pseudo-rigid bodies, *Proc. R. Soc.Lond. A* **427** (1990) 281–319.

[360] D. Lewis, T. Ratiu, J.C. Simo, and J.E. Marsden. The heavy top: a geometric treatment, *Nonlinearity* **5** (1992) 1–48.

[361] G. Li, Q. Ouyang, V. Petrov, and H.L. Swinney. Transition from simple rotating chemical spirals to meandering and traveling spirals, *Phys. Rev. Lett.* **77** (1996) 2105–2108.

[362] J. Liu and G. Ahlers. Spiral-defect chaos in Rayleigh-Bénard convection with small Prandtl numbers, *Phys. Rev. Lett.* **77** No. 15 (1996) 3126–3129.

[363] K.L. Liu, W.S. Lo, and L.-S. Young. Generalized renormalization group equation for period-doubling bifurcations, *Phys. Lett.* **105** A (1984) 103.

[364] E.N. Lorenz. Deterministic non-periodic flow, *J. Atmos. Sci.* **20** (1963) 130–141.

[365] E. Lugosi. Analysis of meandering in Zykov kinetics. *Physica D* **40** (1989) 331–337.

[366] G. Lunter. *Bifurcation in Hamiltonian Systems*, Thesis, Rijksuniversiteit Groningen.

[367] E. Maeda, H.P.C. Robinson, and A. Kawana. The mechanisms of generation and propagation of synchronized bursting in developing networks of cortical neurons, *J. Neurosci.* **15** (1995) 6834–6845.

[368] R.Mañé. *Ergodic Theory and Differentiable Dynamics*, Springer-Verlag, Berlin 1983.

[369] R.A. Mann. Biomechanics. In: *Disorders of the Foot*, (M.H. Jahss ed.) W.B. Saunders and Co., Philadelphia 1982, 37–67.

[370] R.A. Mann, G.T. Moran, and S.E. Dougherty. Comparative electromyography of the lower extremity in jogging, running and sprinting, *Amer. J. Sports Med.* **14** (1986) 501–510.

[371] M. Manoel and I. Stewart. Degenerate bifurcations with $\mathbf{Z}_2 \oplus \mathbf{Z}_2$ -symmetry, *Internat. J. Bif. Chaos* **9** (1999) 1653–1667.

[372] M. Manoel and I. Stewart. The classification of bifurcations with hidden symmetries, *Proc. London Math. Soc.* **80** (2000) 198–234.

[373] P.S. Marcus. Simulation of Taylor-Couette flow. Part 2. Numerical results for wavy-vortex flow with one travelling wave. *J. Fluid Mech.* **146** (1984) 65–113.

[374] J.E. Marsden. *Lectures on Mechanics*, LMS Lect. Notes **174**, Cambridge U Press, Cambridge 1992.

[375] J.E. Marsden and M. McCracken. *The Hopf Bifurcation and Its Applications*, Springer-Verlag, New York 1976.

[376] J.E. Marsden and A. Weinstein. Reduction of symplectic manifolds with symmetry, *Rep. Math.Phys.* **74** (1974) 121–130.

[377] J. Mather. Differential invariants, *Topology* **16** (1977) 145-155.

[378] R.M. May. (ed.) *Theoretical Ecology: Principles and Applications*, W.B. Saunders, Philadelphia 1976.

[379] R.M. May and W. Leonard. Nonlinear aspects of competition between three species, *SIAM J. Appl. Math.* **29** (1975) 243–252.

[380] J. Maynard Smith. Sympatric speciation, *American Naturalist* **100** (1966) 637–650.

[381] E. Mayr. *Animal Species and Evolution*, Belknap Press, Cambridge MA, 1963.

[382] E. Mayr. *Populations, Species, and Evolution*, Harvard U Press, Cambridge MA 1970.

[383] H.P. McKean. *Stochastic Integrals*, Academic Press, New York 1969.

[384] I. Melbourne. Intermittency as a codimension three phenomenon, *Dyn. Diff. Eqn.* 1 (1989) 347–367.

[385] I. Melbourne. An example of a non-asymptotically stable attractor, *Nonlinearity* 4 (1991) 835–844.

[386] I. Melbourne. Instantaneous symmetry and symmetry on average in the Couette-Taylor and Faraday experiments. *Dynamics, Bifurcations, Symmetry* (P. Chossat, ed.) Kluwer, The Netherlands, 1994, 241–257.

[387] I. Melbourne. Derivation of the time-dependent Ginzburg-Landau equation on the line. *J. Nonlin. Sci.* 8 (1998) 1–15.

[388] I. Melbourne. Steady-state bifurcation with Euclidean symmetry. *Trans. Amer. Math. Soc.* 351 (1999) 1575–1603.

[389] I. Melbourne. Hidden symmetries on partially unbounded domains. *Physica D* 143 (2000) 226–234.

[390] I. Melbourne. Ginzburg-Landau theory and symmetry. In: *Nonlinear Instability, Chaos and Turbulence*, Vol 2 , (L. Debnath and D.N. Riahi, eds.) *Advances in Fluid Mechanics* 25, WIT Press, Southampton, 2000.

[391] I. Melbourne. Validity, universality and structure of the Ginzburg-Landau equation. *Equadiff 99, International Conference on Differential Equations* (B. Fiedler *et al.*, eds.) World Scientific, Singapore, 2000, 186–200.

[392] I. Melbourne, P. Chossat, and M. Golubitsky. Heteroclinic cycles involving periodic solutions in mode interactions with $O(2)$ symmetry, *Proc. Roy. Soc. Edinburgh A* 113 (1989) 315–345.

[393] I. Melbourne, M. Dellnitz, and M. Golubitsky. The structure of symmetric attractors, *Arch. Rational Mech. Anal.* 123 (1993) 75–98.

[394] I. Melbourne and M. Nicol. Stable transitivity of noncompact group extensions. University of Houston Research Report UH/MD-266, 2001. Submitted.

[395] I. Melbourne, M.R.E. Proctor, and A.M. Rucklidge. A heteroclinic model of geody-namo reversals and excursions. In: *Dynamo and Dynamics, a Mathematical Challenge* (P. Chossat *et al.*, eds.) Kluwer, Netherlands, 2001, 363–370.

[396] I. Melbourne and I. Stewart. Symmetric ω-limit sets for smooth Γ-equivariant dynamical systems with Γ° abelian, *Nonlinearity* 10 (1997) 1551–1567.

[397] I. Mercader, M. Net, and E. Knobloch. Binary fluid convection in a cylinder, *Phys. Rev. E* 51 (1995) 339–350.

[398] K.R. Meyer. Periodic solutions of the N-body problem, *J. Diff. Eq.* 39 (1981) 2-38.

[399] W. Miller, Jr. *Symmetry Groups and their Applications*, Pure and Appl. Math. 50, Academic Press, New York, 1972.

[400] J. Milnor. On the concept of attractor, *Commun. Math. Phys.* 99 (1985) 177–195.

[401] J. Milnor and W. Thurston. *On Iterated Maps of the Interval*, Lect. Notes in Math. 1342, Springer-Verlag, New York 1988.

[402] J. Moehlis and E. Knobloch. Bursts in oscillatory systems with broken D_4 symmetry, *Physica D* 135 (2000) 263–304.

[403] T. Molien. Über die Invarianten der Linearen Substitutionsgruppe, *Sitzungsber. König. Preuss. Akad. Wiss.* (1897) 1152–1156.

[404] J.A. Montaldi and R.M. Roberts. Relative equilibria of molecules, *J. Nonlin. Sci.* 9 (1999) 53-88.

[405] J.A. Montaldi, R.M. Roberts, and I. Stewart. Nonlinear normal modes of symmetric Hamil-tonian systems, in *Structure Formation in Physics* (G. Dangelmayr and W. Guttinger, eds.), Springer Verlag, New York, 1987, 354–371.

[406] J.A. Montaldi, R.M. Roberts, and I. Stewart. Periodic solutions near equilbria of symmetric Hamiltonian systems, *Phil. Trans. R. Soc. Lond. A* 325 (1988) 237–293.

[407] J.A. Montaldi, R.M. Roberts, and I. Stewart. Existence of nonlinear normal modes of symmetric Hamiltonian systems, *Nonlinearity* 3 (1990) 695–730.

[408] J.A. Montaldi, R.M. Roberts, and I. Stewart. Stability of nonlinear normal modes of symmetric Hamiltonian systems, *Nonlinearity* 3 (1990) 731–772.

[409] J. Moser. Periodic orbits near equilibrium and a theorem by Alan Weinstein, *Commun. Pure Appl. Math.* 29 (1976) 727–747.

[410] S.C. Müller, T. Plesser, and B. Hess. The structure of the core of the spiral wave in the Belousov-Zhabotinskii reaction, *Science* **240** (1985) 661–663.

[411] S.C. Müller, T. Plesser, and B. Hess. Two-dimensional spectrophotometry of spiral wave propagation in the Belousov-Zhabotinskii reaction, *Physica D* **24** (1987) 71–86.

[412] T. Mullin. Cellular mutations in Taylor flow, *J. Fluid Mech.* **121** (1982) 207–218.

[413] J.D. Murray. *Mathematical Biology*, Biomath. **19**, Springer-Verlag, New York 1989.

[414] W. Nagata. Personal communication.

[415] M. Neveling and G. Dangelmayr. Bifurcation analysis of interacting stationary modes in thermohaline convection, *Phys. Rev.* A **38** (1988) 2536–2543.

[416] P.C. Newell. Attraction and adhesion in the slime mold *Dictyostelium*, in *Fungal Differentiation: a Contemporary Synthesis*, (J.E. Smith ed.), Mycology Series **43**, Marcel Dekker, New York 1983 43–71.

[417] M. Nicol, I. Melbourne, and P. Ashwin. Euclidean extensions of dynamical systems. *Nonlinearity* **14** (2001) 275–300.

[418] P.J. Olver. *Applications of Lie Groups to Differential Equations* (2nd ed.), Springer-Verlag, New York 1993.

[419] J.-P. Ortega and T.S. Ratiu. Persistence and smoothness of critical relative elements in Hamiltonian systems with symmetry, *C. R. Acad. Sci. Paris sér. I Math.* **325** (1997) 1107–1111.

[420] E. Ott, J.C. Sommerer, J.C. Alexander, I. Kan, and J.A. Yorke. Scaling behaviour of chaotic systems with riddled basins. *Phys. Rev. Lett.* **71** (1993) 4134–4137.

[421] E. Ott, J.C. Sommerer, J.C. Alexander, I. Kan, and J.A. Yorke. A transition to chaotic attractors with riddled basins. *Physica D* **76** (1994) 384–410.

[422] E. Ott and J.C. Sommerer. Blowout bifurcations: the occurrence of riddled basins and on-off intermittency. *Phys. Lett.* A **188** (1994) 39–47.

[423] D. Otte. Speciation in Hawaiian crickets, in *Speciation and Its Consequences* (D. Otte, J.A. Endler eds.), Sinauer, Sunderland MA 1989, 482–526.

[424] Q. Ouyang and H.L. Swinney. Transition to chemical turbulence, *Chaos* **1** (1991) 411–420.

[425] A. Palacios, G.H. Gunaratne, M. Gorman, and K.A. Robbins. Cellular pattern formation in circular domains, *Chaos* **7** (1997) 463–475.

[426] G.W. Patrick. Relative equilibria of Hamiltonian systems with symmetry: linearization, smoothness, and drift, *J. Nonlin. Sci.* **5** (1995) 373–418.

[427] G.W. Patrick. Dynamics near relative equilibria: nongeneric momenta at a 1:1 group-reduced resonance, *Math. Z.* **232** (1999) 747–788.

[428] T. Peacock, T. Mullin, and D.J. Binks. Bifurcation phenomena in flows of a nematic liquid crystal. *Intern. J. Bifur. & Chaos* **9** No. 2 (1999) 427–441.

[429] K. Pearson. The control of walking. *Sci. Amer.* **235** No. 6, Dec. (1976) 72–86.

[430] L.M. Pecora and T.L. Carroll. Synchronization in chaotic systems, *Phys. Rev. Lett.* **64** (1990) 821–824.

[431] E. Pennisi. Nature steers a predictable course, *Science* **287** (2000) 207–208.

[432] E.R. Pianka. Competition and niche theory, in *Theoretical Ecology: Principles and Applications* (R.M. May ed.), W.B. Saunders, Philadelphia 1976, 114–141.

[433] A.S. Pikovsky and P. Grassberger. Symmetry breaking of coupled chaotic attractors. *J. Phys.* A **24** (1991) 4587–4597.

[434] N. Platt, E.A. Spiegel, and C. Tresser. On-off intermittency; a mechanism for bursting. *Phys. Rev. Lett.* **70** (1993) 279–282.

[435] N. Platt, S.M. Hammel, and J.F. Heagy. Effects of additive noise on on-off intermittency. *Phys. Rev. Lett.* **72** (1994) 3498–3501.

[436] V. Poénaru. *Singularitiés C^∞ en Présence de Symétrie*, Lect. Notes Math. **510**, Springer-Verlag, Berlin 1976.

[437] J. Porter and E. Knobloch. Complex dynamics in the 1:3 spatial resonance, *Physica D* **143** (2000) 138–168.

[438] I. Prigogine and R. Lefever. Stability and self-organization in open systems, in *Membranes, Dissapative Structures, and Evolution* (G. Nicolis and R. Lefever, eds.) Wiley, New York, 1974.

[439] M.R.E. Proctor and M. Silber. Nonlinear competition between small and large hexagonal patterns, *Phys. Rev. Lett.* **81** (1998) 2450–2453.

[440] D. Rand. Dynamics and symmetry. Predictions for modulated waves in rotating fluids, *Arch. Rational Mech. Anal.* **79** (1982) 1–38.

[441] D.A. Rand, H. Wilson, and J. McGlade. Dynamics and evolution: evolutionarily stable attractors, invasion exponents and phenotype dynamics. *Phil. Trans. R. Soc. Lond.* B **343** (1994) 261–283.

[442] R.H. Rand, A.H. Cohen and P.J. Holmes. Systems of coupled oscillators as models of central pattern generators, in *Neural Control of Rhythmic Movements in Vertebrates* (A.H. Cohen, S. Rossignol and S. Grillner, eds.), Wiley, New York (1988) 333–367.

[443] M. Renardy. Bifurcation from rotating waves, *Arch. Rational Mech. & Anal.* **79** (1982) 49–84.

[444] R. Rice and E.E. Hostert. Laboratory experiments on speciation: what have we learned in 40 years? *Evolution* **47** (1993) 1637–1653.

[445] M. Ridley. *The Problems of Evolution*, Oxford U Press, Oxford 1985.

[446] M. Ridley. *Evolution*, Blackwell, Oxford 1996.

[447] H. Riecke, J.D. Crawford and E. Knobloch. Time-modulated oscillatory convection, *Phys. Rev. Lett.* **61** (1988) 1942–1945.

[448] D.S. Riley and K.H. Winters. Modal exchange mechanisms in Lapwood convection, *J. Fluid Mech.* **204** (1989) 325–358.

[449] D.S. Riley and K.H. Winters. A numerical bifurcation study of natural convection in a tilted two-dimensional porous cavity, *J. Fluid Mech.* **215** (1990) 309–329.

[450] D.S. Riley and K.H. Winters. Time-periodic convection in porous media: the evolution of Hopf bifurcations with aspect ratio, *J. Fluid Mech.* **223** (1991) 457–474.

[451] R.M. Roberts and M.E.R. de Sousa Dias. Bifurcations from relative equilibria of Hamiltonian systems, *Nonlinearity* **10** (1997) 1719–1738.

[452] C.D. Rollo. *Phenotypes*, Chapman and Hall, London 1995.

[453] A.D. Rey and T. Tsuji. Recent advances in liquid crystal rheology. *Macromolecular Th. & Simul.* **7** No. 6 (1998) 623–639.

[454] A.M. Rucklidge and M. Silber. Bifurcations of periodic orbits with spatio-temporal symmetries, *Nonlinearity* **11** (1998) 1435–1455.

[455] D. Ruelle. A measure associated with Axiom A attractors. *Amer. J. Math.* **98** (1976) 619–654.

[456] D. Ruelle and F. Takens. On the nature of turbulence, *Comm. Math. Phys.* **20** (1971) 167–192.

[457] H.D. Rundle, L. Nagel, J.W. Boughman, and D. Schluter. Natural selection and parallel speciation in sympatric sticklebacks, *Science* **287** (2000) 306–308.

[458] S.N. Salthe. *Evolutionary Biology*, Holt, Rinehart and Winston, New York 1972.

[459] B. Sandstede, A. Scheel and C. Wulff. Center-manifold reduction for spiral waves, *C. R. Acad. Sci., Serié I* **324** (1997) 153–158.

[460] B. Sandstede, A. Scheel and C. Wulff. Dynamics of spiral waves in unbounded domains using center-manifold reductions, *J. Diff. Eqns.* **141** (1997) 122–149.

[461] B. Sandstede, A. Scheel and C. Wulff. Dynamical behavior of patterns with Euclidean symmetry, in *Pattern Formation in Continuous and coupled Systems* (M.Golubitsky, D.Luss, S.H.Strogatz eds.), IMA volumes in Math. and Appns. **115** Springer-Verlag, New York 1999, 249–264.

[462] D.H. Sattinger. *Group Theoretic Methods in Bifurcation Theory*, Lecture Notes in Math. **762** Springer-Verlag, New York 1979.

[463] D.H. Sattinger and O.L. Weaver. *Lie Groups and Algebras with Applications to Physics, Geometry, and Mechanics*, Springer-Verlag, New York 1986.

[464] D.G. Schaeffer and M. Golubitsky. Boundary conditions and mode jumping in the buckling of a rectangular plate, *Commun. Math. Phys.* **69** (1979) 209–236.

[465] A. Scheel. Subcritical bifurcation to infinitely many rotating waves, *J. Math. Anal. Appl.* **215** (1997) 252–261.

[466] A. Scheel. Bifurcation to spiral waves in reaction-diffusion systems, *SIAM J. Math. Anal.* **29** (1998) 1399–1418.

[467] G. Schwartz. Smooth functions invariant under the action of a compact Lie group, *Topology* **14** (1975) 63–68.

[468] G. Schöner, W.Y. Jiang, and J.A.S. Kelso. A synergetic theory of quadrupedal gaits and gait transitions, *J. Theor. Biol.* **142** (1990) 359–391.

[469] M. Silber and E. Knobloch. Pattern selection in steady binary-fluid convection, *Phys. Rev. A* **38** (1988) 1468–1477.

[470] M. Silber and E. Knobloch. Pattern selection in ferrofluids, *Physica D* **30** (1988) 83–98.

[471] M. Silber and E. Knobloch. Hopf bifurcation on a square lattice, *Nonlinearity* **4** (1991) 1063–1107.

[472] M. Silber and A.C. Skeldon. New stability results for patterns in a model of long-wavelength convection, *Physica D* **122** (1998) 117–133.

[473] M. Silber and A.C. Skeldon. Parametrically excited surface waves: two-frequency forcing, normal form symmetries, and pattern selection *Phys. Rev. E* **59** (1999) 5446–5456.

[474] M. Silber, D.P. Tse, A.M. Rucklidge, and R.B. Hoyle. Spatial period-multiplying instabilities of hexagonal Faraday waves, *Physica D* **146** (2000) 367–387.

[475] M. Silber, C.M. Topaz and A.C. Skeldon. Two-frequency forced Faraday waves: weakly damped modes and pattern selection, *Physica D* **143** (2000) 205–225.

[476] F. Simonelli and J.P. Gollub. Surface wave mode interactions: effects of symmetry and degeneracy, *J. Fluid Mech.* **199** (1989) 471–494.

[477] G.S. Skinner and H.L. Swinney. Periodic to quasiperiodic transition of chemical spiral rotation. *Physica D* **48** (1991) 1–16.

[478] T.J. Sluckin. The liquid crystal phases: physics and technology, *Contemporary Phys.* **41** No. 1 (2000) 37–56.

[479] S. Smale. Differentiable dynamical systems, *Bull. Amer. Math. Soc.* **73** (1967) 747–817.

[480] S. Smale. A mathematical model of two cells via Turing's equation, in *Some Mathematical Questions in Biology V* (J.D. Cowan, ed.), Amer. Math. Soc. Lecture Notes on Mathematics in the Life Sciences **6** (1974) 15–26.

[481] I.N. Sneddon. *Special Functions of Mathematical Physics and Chemistry*, Oliver & Boyd, Edinburgh 1961.

[482] R.P. Stanley. Invariants of finite groups and their applications to combinatorics, *Bull. Amer. Math. Soc.* **1** (1979) 475–511.

[483] O. Steinbock and S.C. Müller. Chemical spiral rotation is controlled by light-induced artificial cores, *Physica A* **188** (1992) 61–67.

[484] I. Stewart. Stability of periodic solutions in symmetric Hopf bifurcation, *Dyn. Stab. Sys.* **2** (1987) 149–166.

[485] I. Stewart. Bifurcations with symmetry, in *New Directions in Dynamical Systems* (T. Bedford and J.W. Swift eds.), London Math. Soc. Lecture Notes **127**, Cambridge U Press 1988, 235–283.

[486] I. Stewart. *Galois Theory* (2nd ed.), Chapman & Hall, London 1989.

[487] I. Stewart. Symmetry and chaotic data, Scientific Correspondence, *Nature* **354** (1991) 113.

[488] I. Stewart. Broken symmetry and the formation of spiral patterns in fluids, in *Spiral Symmetry* (I. Hargittai and C.A. Pickover eds.), World Scientific, Singapore 1992, 187–220.

[489] I. Stewart. Bifurcation theory old and new, in *Dynamics of Numerics and Numerics of Dynamics* (D.S. Broomhead and A. Iserles eds.), IMA Conference Series **34**, Oxford U Press, Oxford 1992, 31–67.

[490] I. Stewart. Symmetry methods in collisionless many-body problems, *J. Nonlin. Sci* **6** (1996) 543–563; reprinted in *Mechanics: from Theory to Computation* (*J. Nonlin. Sci* eds.), Springer-Verlag, New York 2000, 313–333.

[491] I. Stewart. Symmetry-breaking cascades and the dynamics of morphogenesis and behaviour, *Science Progress* **82** (1999) 9–48.

[492] I. Stewart. Designer differential equations for animal locomotion, *Complexity*. To appear.

[493] I. Stewart and A.P.S. Dias. Toric geometry and equivariant bifurcations, *Physica D* **143** (2000) 235–261.

[494] I. Stewart, T. Elmhirst, and J. Cohen. Symmetry-breaking as an origin of species, in *Conference on Bifurcations, Symmetry, Patterns, Porto 2000*, Birkhäuser, Basel 2001.

[495] I. Stewart and M. Golubitsky. *Fearful Symmetry: Is God a Geometer?* Blackwell, Oxford 1992.

[496] G.G. Stokes. *Mathematical and Physical Papers* Vol. 1 (1880) p. 102; Vol. 5 (1905).

[497] B. Sturmfels. *Algorithms in Invariant Theory*, Springer-Verlag, Berlin 1993.

[498] F. Sutton. Onset of convection in a porous channel with net through flow, *Phys. Fluids* **13** (1970) 1931–1934.

[499] R. Tagg and H.L. Swinney. Critical dynamics near the spiral-Taylor vortex codimension-two point, in preparation.

[500] A. Takamatsu, R. Tanaka, H. Yamada, T. Nakgaki, T. Fuji, and I. Endo. Spatio-temporal symmetry in rings of coupled biological oscillators of *Physarum* plasmodial slime mold. *Phys. Rev. Lett* **87** no. 7, 078102.

[501] G.I. Taylor. Stability of a viscous liquid contained between two rotating cylinders. *Phil. Trans. Roy. Soc. London A* **223** (1923) 289–343.

[502] S.M. Tobias and E. Knobloch. On the breakup of spiral waves into chemical turbulence, *Phys. Rev. Lett.* **80** (1998) 4811–4814.

[503] S.M. Tobias, M.R.E. Proctor, and E. Knobloch. Convective and absolute instabilities of fluid flows in finite geometry, *Physica D* **113** (1998) 43–72.

[504] T. Tregenza and R.K. Butlin. Speciation without isolation, *Nature* **400** (1999) 311–312.

[505] W. Tucker. The Lorenz attractor exists. *C.R. Acad. Sci. Paris* **328** (1999) 1197–1202.

[506] A.M. Turing. The Chemical Basis of Morphogenesis, *Phil. Trans. Roy Soc. London B* **237** (1952) 32.

[507] P.B. Umbanhowar, F. Melo, and H.L. Swinney. Periodic, aperiodic, and transient patterns in vibrated granular layers, *Physica A* **249** (1998) 1–9.

[508] A. Vanderbauwhede. *Local Bifurcation and Symmetry*, Habilitation Thesis, Rijksuniversiteit Gent, 1980, (also Res. Notes Math. **75**, Pitman, Boston 1982).

[509] A. Vanderbauwhede. Centre manifolds, normal forms and elementary bifurcations, *Dynamics Reported* **2** 1989.

[510] A. Vanderbauwhede, M. Krupa, and M. Golubitsky. Secondary bifurcations in symmetric systems, *Differential Equations* (C.M. Dafermos, G. Ladas, and G. Papanicolaou, eds.), Marcel Dekker, New York, 1989, Lect. Notes Pure Appl. Math. **118** 709–716.

[511] V.S. Varadarajan. *Lie Groups, Lie Algebras, and Their Representations*, Graduate Texts in Math. **102**, Springer-Verlag, New York 1984.

[512] M. Viana. What's new on Lorenz strange attractors? *Math. Intelligencer* **22** no. 3 (2000) 6–19.

[513] T.L. Vincent and T.L.S. Vincent. Evolution and control system design, *IEEE Control Systems Magazine* (October 2000) 20–35.

[514] J.K. Waage. Reproductive character displacement in *Calopteryx* (Odonata; Calopterygidae), *Evolution* **33** (1979) 104–116.

[515] M. Wasserman and H.R. Koepfer. Character displacement for sexual isolation between *Drosophila mojavensis* and *Drosophila arizonensis*, *Evolution* **19** (1977) 459–464.

[516] M. Wegelin, J. Oppenlander, J. Tomes, W. Güttinger, and G. Dangelmayr. Synchronized patterns in hierarchical networks of neural oscillators, *Physica D* **121** (1998) 213–232.

[517] E.T. Whittaker and G.N. Watson. *A Course of Modern Analysis*. Cambridge U Press, Cambridge, 1965.

[518] H.R. Wilson and J.D. Cowan. Excitatory and inhibitory interactions in localized populations of model neurons, *Biophys. J.* (1972) **12** 1–24.

[519] A.T. Winfree. Scroll-shaped waves of chemical activity in three dimensions, *Science* **181** (1973) 937–939.

[520] A.T. Winfree. Rotating chemical reactions, *Scientific American*, June 1974, 82–95.

[521] A.T. Winfree. *The Geometry of Biological Time*. Biomathematics **8**, Springer, New York, 1980.

[522] A.T. Winfree, ed. Focus Issue: Fibrillation in Normal Ventricular Myocardium, *Chaos* **8** (1998).

[523] F. Wolf and T. Geisel. Spontaneous pinwheel annihilation during visual development, *Nature* **395** (1998) 73–78.

[524] D. Wood. *Coupled Oscillators with Internal Symmetries*, PhD Thesis, U Warwick 1995.

[525] D. Wood. Hopf bifurcations in three coupled oscillators with internal $\mathbf{Z}_2$ symmetries, *Dyn. Stab. Sys.* **13** (1998) 55-93.

[526] D. Wood. A cautionary tale of coupling cells with internal symmetries, *Internat. J. Bif. Chaos* **11** (2001) 123–132.

[527] P. Worfolk. Zeros of equivariant vector fields: algorithms for an invariant aproach, *J. Symb. Comp.* **17** (1994) 487–511.

[528] C. Wulff. *Theory of Meandering and Drifting Spiral Waves in Reaction-Diffusion Systems*, Thesis, Freie Universität Berlin, 1996.

[529] C. Wulff, J.S.W. Lamb, and I. Melbourne. Bifurcation from relative periodic solutions. *Erg. Th. & Dynam. Sys.* **21** (2001) 605-635.

[530] Z. Xia. Arnold diffusion and oscillatory solutions in the planar three-body problem, *J. Diff. Eq.* **110** (1994) 289-321.

[531] M. Yoshimoto, K. Yoshikawa, and Y. Mori. Coupling among three chemical oscialltors: synchronization, phase death, and frustration, em Phys. Rev. E **47** (1993) 864-874.

[532] E.C. Zeeman. Differential equations for the heartbeat and nerve impulse, in Towards a Theoretical Biology vol. 4 (C.H. Waddington ed.) Edinburgh U Press, Edinburgh 1972, 8-67.

[533] E.C. Zeeman. Stability of dynamical systems, *Nonlinearity* **1** (1988) 115-155.

[534] A.M. Zhabotinskii. A study of self-oscillatory chemical reaction III: space behavior, In: *Biochemical Oscillators, Proceedings of the 1968 Prague Symposium* (eds. B. Chance et al.) 1973.

[535] L.-H. Zhang and H.L. Swinney. Nonpropogating oscillatory modes in Couette-Taylor flow. *Phys. Rev. A* **31** (1985) 1006-1009.

[536] V.S. Zykov. Cycloidal circulation of spiral waves in an excitable medium. *Biofizika* **31** (1986) 862-865.

[537] V.S. Zykov. Kinematics of nonstationary circulation of spiral waves in an excitable medium. *Biofizika* **32** (1987) 337-340.

[538] V.S. Zykov and S.C. Müller. Spiral waves on circular and spherical domains of excitable medium, Physica D **97** (1996) 322-332.

Index

C-axial, 88, 91, 286, 290
C-axial subgroups of $\mathbf{S}_N$, 285
Γ-simple, 90, 281
λ-ω system, 100
k-resonance, 199

absolutely irreducible, 15, 45, 115
action, 7
adding machine, 119, 120
adding machine
  binary, 119
  generalized, 121
  multi-base, 120
admissible, 248
all-to-all coupling, 46
allopatric speciation, 2, 20
ammonia, 275, 299
animal locomotion, 59
anomalous vortices, 207
anti-rolls, 140
anti-squares, 140
Archimedean spiral, 103, 105
Arnold diffusion, 283
Arnold's cat, 256
asynchrony, 269
attractor, 116, 247
average symmetry, 267
axial subgroup,
    17, 50, 132, 134, 138, 144, 153
axis of reflection, 296
azimuthal, 166

Bénard experiment, 148
Banach space, 30, 118
Belousov-Zhabotinskii
  reaction, 87, 88, 100, 184
Bessel function, 102–104
bidirectional ring, 69, 70
bifurcation, 5
bifurcation
  blowout, 271
  bubbling, 271
  degenerate steady-state, 97
  fold, 35
  Hopf, 71, 80, 87, 89, 90, 282
  jump, 35
  local, 89

period-doubling, 116, 202
pichfork, 206
pitchfork, 19, 34, 35, 205
saddle-node, 33, 35
secondary, 53
steady-state, 15, 89
symmetry-increasing, 242
transcritical, 35, 46
bifurcation diagram, 5
binary tree, 120
biped, 75
Birkhoff Ergodic Theorem, 255, 260
body mode, 101, 109
Born-Oppenheimer model, 296
bound, 60
boundary conditions
  Dirichlet, 102, 103, 109, 204, 208
  mixed, 101, 102
  Neumann, 101–103, 109, 204, 205
  periodic, 166, 204, 205
  Robin, 101–103, 105, 109
  spiral, 101, 102
Boussinesq equations, 148, 149
Boussinesq fluid, 158
branch, 5
Brouwer fixed-point theorem, 234
Brusselator, 136
bucking bronco, 75
buckling, 204, 216
buckminsterfullerene, 275, 297
bursting, 222, 227, 234

C-joint, 230, 235
Cantor set, 120, 121, 250
Cartan subgroup, 181, 183
catalysis, 100
cell, 84, 111
cell
  projection, 84
  trajectory, 84
center bundle, 175, 176, 194
center manifold reduction, 109, 155
central configuration, 283
central pattern generator, 60, 62, 68, 72
chaos, 283
chaotic interval, 119

circle group, 7, 89, 116
cokernel, 16
collision of attractors, 246
collisionless Boltzmann equation, 284
combustion, 185
commute, 14
compact Lie group, 6
complex Ginzburg-Landau equation, 100
configuration space, 277, 296
connected component of the identity, 181
conserved quantity, 275
constrained
    Liapunov-Schmidt procedure, 282
contralateral, 72
convection, 100, 148
convection cell, 211
corange, 16
correlation matrix, 262
Couette flow, 166
Couette-Taylor system, 166, 179, 184
coupled cell network, 83, 89, 110
coupled cell system, 69, 83, 85, 237, 261
CPG, 60, 62, 68, 72
crisis, 119
critical group orbit, 176
critical wave number, 126
cubic symmetry, 299
cyclic group, 6
cycling chaos, 237
cyclospectral, 283

Darwin's finches, 20, 25
DBC, 204
degeneracy condition, 33
detective, 258, 260
diffeomorphism, 265
dihedral group, 6, 42
discrete rotating wave, 64, 70
dislocation, 101
dispersion curve, 56, 124, 126, 136, 144
doubly periodic, 128
drift, 185, 192
dual lattice, 126, 128, 129
dual lattice vector, 129
dual wave number, 128
dual wave vector, 127
dynamic symmetry, 116, 118

E-joint, 230, 235
eigenfunction, 132
eigenvalue crossing condition, 91
elementary symmetric function, 41, 46
elliptic, 283
energy level, 277, 278
energy surface, 282
ENTWIFE, 211
equilibrium, 2
equivariant, 8, 14, 23, 41, 42, 47
Equivariant Branching Lemma, 17,
    18, 24, 33, 34, 39, 42, 88, 91, 117, 223

Equivariant
    Hopf Theorem, 88, 89, 91, 94, 110
Equivariant Liapunov Center Theorem, 282
Equivariant Moser-Weinstein
    Theorem, 280, 282, 292
ergodic, 255
ergodic sum, 254, 260
ergodic theory, 245
Euclidean group, 7, 88, 123, 125, 126, 193
Euclidean symmetry, 284
Euler method, 52
even, 132
evolution, 53
exchange of stability, 33, 34
excitable medium, 108
exponential map, 181

Faraday experiment, 202, 209
fast foliation, 231
Feigenbaum number, 121
Feigenbaum point, 119
fixed-point subspace, 12
Floquet
    equation, 97
    exponent, 291
    operator, 289
Fokker-Planck equation, 28
fold point, 33
Fourier series, 128, 130
Fourier-Bessel function, 101, 102
fractal, 119
Fredholm operator, 31
fundamental domain, 129

gait, 59
gene-flow, 2
generator, 181
generic, 15, 201
genotype, 20
Gorman-Swinney flow, 189
group orbit, 10, 161
Guckenheimer-Holmes system, 222, 237

Hénon-Heiles Hamiltonian, 276, 279, 291
Hénon-Heiles system, 279
Haar measure, 8
hallucination, 141
Hamilton's Equations, 277
Hamiltonian, 277
Hamiltonian system, 275, 287
hemisphere, 216
heteroclinic cycle, 222, 225, 226
heteroclinic cycle
    group-theoretic signature, 225
    perturbed, 228
hexagons, 148, 149, 153, 158, 159
hidden rotational symmetry, 204
hidden symmetry, 201, 204, 209
Hilbert basis, 41
Hilbert-Weyl Theorem, 41

holohedry, 126, 129, 130
homeotropic, 150
hyperbolic, 283
hypercolumn, 141
hysteresis, 35, 36

icon, 242
icosahedral group, 219, 297
icosahedral symmetry, 299
identical cells, 85
imperfect symmetry, 27
Implicit Function Theorem, 16
inhibitory, 141
inner pattern, 191
instantaneous symmetry, 267
intermittency, 230
interpenetrating spirals, 170
invariant, 23, 41, 46
invariant 2-torus, 97
invariant measure, 254
invariant subspace, 13, 14, 34
inverse limit, 121
ionized hydrogen, 298
ipsilateral, 72
irreducible, 14, 37, 38, 130
iso-orientation patch, 141
isomorphic, 37
isotropy lattice, 10, 11
isotropy subgroup, 8, 9, 11, 34, 50, 53
isotypic component, 37, 50, 127, 144
isotypic decomposition, 34, 38, 43
iterate, 247

Jacobian
    eigenvalues of, 33, 38
    symmetry of, 37
    zero eigenvalue, 34, 39
joint, 228, 230
jump, 73, 75

Küppers-Lortz instability, 221, 222
Karhunen-Loève decomposition, 108
Krupa's Theorem, 177, 180, 182
Kuramoto-Sivashinsky equation, 176

label symmetry, 284
Laplacian, 28
Lapwood convection, 211
lattice, 126, 128, 130
lattice
    hexagonal, 129, 130, 153
    oblique, 129
    rectangular, 129
    rhombic, 129, 130
    square, 129, 130, 152
Liapunov Center Theorem, 276, 277
Liapunov-Schmidt
    reduction, 16, 25, 30, 49, 117, 118
Lie algebra, 180
linear eigenfunction, 109
linear stability, 289

linearized flow, 282
liquid crystal, 149
liquid crystal
    nematic, 150
liquid drop, 289
logistic map, 89, 119, 241
loop space, 89, 92, 112, 116, 118, 121, 281
Lorenz attractor, 256

magnetoconvection, 211
many-body problem, 283
meandering, 192, 196
measure-preserving, 254
methane, 275, 299
mode interaction, 90, 168
mode number, 208
model-independent, 45, 46, 59, 166, 168
modulated wavy vortices, 189
molecular vibration, 296
molecule, 275, 296
momentum, 277
momentum space, 277
monogenic, 181
Morris-Lecar equations, 73
Moser-Weinstein Theorem, 276, 278
moving frame, 183
multifrequency oscillation, 78
multirhythm, 59, 111

Navier-Stokes equations, 139, 148, 166
NBC, 204
nearest neighbor coupling, 69, 71
neuron, 222
nodal circle, 102
nodal line, 102, 103
Noether's Theorem, 275
noise, 53
non-absolutely irreducible, 115
non-resonant, 277, 278
nondegenerate minimum, 277
normal hyperbolicity, 29
normal/tangential decomposition, 174

observable, 102, 190
observation, 254
octahedral group, 219
odd, 132
odd logistic map, 242
on-off intermittency, 271
one-parameter subgroup, 181
orbit space, 275
orthogonal group, 7
outer pattern, 191

pace, 60, 63, 72
partial differential equation, 29, 54
partition, 11
patchwork quilt, 149
pattern, 102, 190
pattern on average, 245, 267
PBC, 204

period, 59
period-q point, 118
period-doubling
    cascade, 89, 118, 119, 121, 241
period-multiplying cascade, 122
periodic, 59, 65, 225
periodic boundary conditions, 98
permutation group, 68
permutation representation, 110
petality, 192
phase space, 277
phase-amplitude equations, 97
phenotype, 10, 20, 48
phosphorus, 299
phyllotaxis, 220
pipe system, 228, 230, 234
pitchfork of revolution, 205
Placeholder for Organism Dynamics, 22
plane wave, 126, 129
plane wave factor, 127, 130
planform, 124, 129, 137, 144, 154
Poénaru's Theorem, 42
POD, 22
Poincaré Recurrence Theorem, 254
Poincaré section, 279
Poincaré-Birkhoff
    normal form, 89, 95, 97, 288
point to point coupling, 71
pointwise symmetry, 243
position coordinate, 277
Prandtl number, 148
preimage set, 250
prenormal form, 211
primary gait, 73
primitive ricocheting jump, 75
probability measure, 243
probability space, 254
procyclic group, 122
profinite group, 121
pseudoscalar, 132, 138, 139
punctuated equilibrium, 23

quadruped, 72
quartet, 283
quasi-static, 35
quasiperiodic, 173, 184
quaternion, 115
quaternionic, 281

ramp, 6, 52
random noise, 28
rank, 173
Rayleigh number, 124, 148
Rayleigh-Bénard convection, 124, 126, 158
reaction matrix, 136
reaction-diffusion equation,
    29, 100, 101, 105, 126, 135, 193
reaction-diffusion equations, 98
rectangles, 158, 159
reduction, 275

reflection, 248
regular, 259, 260
regular element, 181
regular triangles, 149
relabeling symmetry, 297
relative equilibrium,
    164, 171, 173, 175, 296, 298–300
relative periodic orbit, 176
relaxation oscillator, 231
renormalization, 122
representation, 6, 7
reresentation theory, 37
resonance, 82, 197
resonance space, 282
resonant, 278
restricted 3-body problem, 283
retina, 141
reverse cascade, 119
rhombs, 158
ribbons, 168, 225
riddled basin, 271
ring of cells, 68
robust, 230
rodeo, 75
rolls, 140, 148, 152, 153, 158, 159
rotary gallop, 60
rotating wave, 64, 94, 99, 102, 184, 279
rotation group, 7
Runge-Kutta method, 52

saddle-node bifurcation, 35
SBR measure, 256
scalar, 132
Schur's Lemma, 15
Schwartz's Theorem, 41
secondary branch, 33, 40
secondary gait, 73
setwise symmetry, 243
shift-twist representation, 143
Sinai-Bowen-Ruelle measure, 256
singular perturbation theory, 228
slime mold, 79, 100
slow dynamic, 231
slow manifold, 231
spatial symmetry, 62, 65
spatio-temporal chaos, 244, 269
spatio-temporal
    symmetry, 59, 62, 63, 65, 168, 287
special Euclidean group, 195
special orthogonal group, 7
speciation, 20, 48
species, 40
spherical harmonic, 217, 289
spherical pendulum, 279, 283
spiral vortices, 166
spiral wave, 87, 88, 100, 192
spring pendulum, 291
squares, 152, 158
stability, 33, 34, 50, 154
standing wave, 64, 94, 99, 102, 103, 279

steady state, 2
stochastic differential equation, 28
stream function, 139
streamline, 139
stripes, 149
stroboscopic map, 202
strongly admissible, 249
structural stability, 230
subcritical, 34
submaximal, 157
sulphur hexafluoride, 299
supercritical, 34
supports Hopf bifurcation, 112
symmetric chaos, 241
symmetric group, 6, 46
symmetry, 8
symmetry group, 282
symmetry-breaking, 1, 2, 5, 88
symmetry-creation, 263
sympatric speciation, 2, 20, 21
symplectic
    action, 281
    form, 281
    irreducible, 281
    representation, 276
    structure, 281
    vector space, 281
synchrony, 84, 268
synchrony on average, 269

target pattern, 87, 88, 100, 104
Taylor vortices, 166, 179, 207
tetrahedral group, 219
tetrahedral symmetry, 275, 299
thickened attractor, 257
topologically cyclic, 181
torus group, 7
trajectory, 247
transcritical, 5
transversality, 66
transverse gallop, 60
transverse Liapunov exponent, 271
triangles, 158, 159
trivial solution, 4
trot, 60, 63, 72
tube, 228, 230
tubular neighborhood, 173
tunneling, 53
twisted subgroup, 62, 112
twisted vortices, 225
type of representation, 114

unfold, 33
unidirectional coupling, 111
unidirectional ring, 69

van der Pol oscillator, 231
vibrated sand, 100
vibrational spectrum, 275, 296
visual cortex, 141

walk, 60, 62, 72
wall mode, 101, 106, 107, 109
water, 275, 296
wave number, 126, 202, 208
wave vector, 126, 130
wavy inflow, 180
wavy outflow, 179
wavy vortices, 166, 225
Whitney Embedding Theorem, 66
Wiener-Cowan model, 143
Wilson-Cowan equation, 143, 144
winner-take-all, 141, 144

Zhang-Swinney flow, 189